Quantitative Methods in Finance

Quantitative Methods in Finance

Terry J. Watsham and Keith Parramore

SOUTH-WESTERN
CENGAGE Learning

Australia • Brazil • Japan • Korea • Mexico • Singapore • Spain • United Kingdom • United States

SOUTH-WESTERN
CENGAGE Learning

Quantitative Methods in Finance
Terry J Watsham and Keith Parramore

Publisher: Patrick Bond
Marketing Manager: Anne-Marie Scoones
Manufacturing Manager: Helen Mason
Senior Production Controller: Maeve Healy
Typesetter: Gray Publishing, Tunbridge Wells, Kent

For product information and technology assistance, contact **emea.info@cengage.com**.

For permission to use material from this text or product, and for permission queries, email **clsuk.permissions@cengage.com**.

British Library Cataloguing-in-Publication Data
A catalogue record for this book is available from the British Library.

ISBN: 978-1-86152-367-9

Cengage Learning EMEA
Cheriton House, North Way, Andover, Hampshire. SP10 5BE. United Kingdom

Cengage Learning products are represented in Canada by Nelson Education Ltd.

For your lifelong learning solutions, visit
www.cengage.co.uk

Purchase e-books or e-chapters at:
http://estore.bized.co.uk

Printed by Seng Lee Press, Singapore
12 – 11 10 09

Contents

Preface

Over the past 30 years or so the theory and practice of finance has increasingly drawn on the quantitative skills of the mathematician, statistician and econometrician. This has accompanied and also facilitated the increasing use of quantitative analysis of the behaviour of financial markets. Over the last decade this increased quantification of analysis has been enhanced by the development and greater use of desktop computers by academics and financial professionals.

Parallel to, and partly because of, these developments, a whole "new" body of literature has developed regarding risk management and derivatives instruments such as options, futures and swaps. This has been accompanied by the development of several "new" areas of financial practice, the derivatives industry, financial risk management and quantitative investment analysis.

This new literature and the new practices have utilized quantitative skills that were previously only applied in the physical sciences, and at the same time have developed anew or adapted techniques applicable to economics.

These intellectual developments have proceeded in tandem with a considerable increase in the number of people employed in the financial markets and financial services industry needing a thorough grounding in the quantitative techniques underlying their business. Only a minority of these people have a quantitative background. There is therefore a growing need for undergraduates, postgraduates and finance professionals to have the means to develop their personal quantitative skills in order to be fully functional in today's financial markets.

Our objective in writing this book is to facilitate that process by bringing together a number of the more important quantitative techniques as they are applied to finance. To achieve that objective, this book is unashamedly a teaching vehicle. It is written in a style that combines quantitative and financial theory with applications. It is written in a style that will make it applicable to undergraduate and postgraduate students and academic researchers in finance as well as to practitioners in the financial markets.

We are aware that there is a great diversity of mathematical competence both among the students and the practitioners of finance. We have addressed this issue, both in the content of the book and in the style of presentation of that content. We have, of necessity, had to be selective in our choice of topics. We have therefore chosen topics according to their usefulness in developing those quantitative skills most applicable to financial research.

This book has 11 chapters in an order that begins with the basics and proceeds steadily to the more advanced techniques that are applied in the analysis of risk management. The style of the writing will make comfortable those readers that need to understand quantitative techniques currently in use, but have less confidence in their quantitative abilities. Yet the variety of topics covered makes the book a good introduction for those that are already familiar with quantitative techniques, but need to widen their knowledge of those techniques currently used in finance.

The first chapter covers the mathematics of interest rates and asset returns, Chapter 2 explains a number of descriptive statistics that are met in financial practice, while Chapters 3 and 4 introduce calculus and probability respectively. Chapter 5 extends the application of probability to hypothesis testing and confidence intervals, two applications that are very important in risk management. Chapter 6 analyses regression analysis and Chapter 7 introduces some modern techniques for analysing time series data, in particular it covers ARCH, GARCH and cointegration. Chapter 8 introduces a number of "numerical methods" that are increasingly being applied in finance. In particular it covers numerical integration and Monte Carlo simulation. Chapter 9 introduces methods of optimization and their application to portfolio construction. Chapter 10 introduces and applies stochastic calculus which is the basis of many of the best-known option pricing modes including the Black–Scholes model and its variants. Chapter 11 covers multivariate analysis, in particular principal components analysis and factor analysis, the former being particularly applicable to modern risk management.

Writing a book such as this requires much sacrifice, none less than from our wives, Jo and Joan, to whom we extend our thanks.

Thanks are also due to Sarah Henderson and the production team at Cengage Learning for facilitating this project.

Acknowledgements

The *F* distribution tables on pp. 382–385 of the Appendix are reproduced from D. B. Owen, *Handbook of Statistical Tables*, © 1962 Addison-Wesley Publishing Company, Inc. Reprinted by permission of Addison-Wesley Longman Publishing Company, Inc.

The χ-distribution tables on pp. 380–381 of the appendix are reproduced from J. White, A. Yeats and G. Shipworth, *Tables for Statisticians*, published by Stanley Thornes (Publishers) Ltd., Cheltenham, Glos.

Interest rates and asset returns

Introduction

Much of the study of finance is concerned with analysing the benefits of making investments. The objective of holding such investments is to enhance the wealth and or income of the investor. This enhancement is referred to as the return, and when expressed as a percentage of the original value of the investment, it is referred to as the rate of return. In addition to measuring returns, financial analysts are concerned with the uncertainty of achieving those returns. Risk analysis is concerned with such uncertainty. In this chapter we will concentrate on the measurement of returns. Later chapters will cover the various topics of risk.

Investors buy assets such as company shares, bonds or property in the hope of making a rate of return, either by selling that asset at a higher price and or by receiving a dividend, interest payment or rental payment. Lenders lend money in the hope of making a rate of return by receiving an interest payment and finally the full repayment of the loan from the borrower. Thus lenders and investors have similar objectives: they hope to get a rate of return, often referred to as a yield, on the funds they have invested or loaned. Consequently in this chapter the term interest will be synonymous with yield and return, lender will be synonymous with investor and loans will be synonymous with investments.

This chapter will begin with a brief explanation of the economic theory of interest which will be followed by the mathematical analysis of interest rates and asset yields. This mathematical analysis will then be applied to short-term money market instruments such as bank deposits and loans, certificates of deposit, Treasury bills, bankers acceptances and commercial paper. The chapter will also look at the various yield measures used in the bond markets and equity markets. Finally this chapter will analyse the rates of interest on bank and building society mortgages and the value of annuities.

The economic theory of interest

Interest rates are probably the most widely recognized financial variable. Most readers will have borrowed money at some stage in their lives and will have paid interest on that borrowing. Many readers will have deposited money with a bank or other financial institution, and received interest in return. In the process they will have noticed a great variety of interest rates payable on loans and deposits. Not only do these rates differ in amount but also in the method of calculating the amount of interest payable. Some rates are fixed for the life of the financial commitment. Others are reset at pre-agreed intervals to the then prevailing rates. Others, such as domestic mortgages, are reset at the will of the lender. But why is interest payable on loans and deposits anyway?

To answer that question we have to understand that money generally provides utility, or well-being, only indirectly, in that it acts as a medium of exchange. That means that it has to be exchanged for other goods and services which provide utility directly. Thus money itself, i.e. the notes, coins and bank balances, provides little by way of life's needs or luxuries. These are provided by exchanging money for goods and services such as food, clothing and shelter.

Consequently, when a person lends or invests money he or she is giving up the opportunity to convert those funds into goods and services that provide utility directly. As a result that person will be enjoying a lower level of utility than would be possible if he or she had not lent or invested the money but used it to purchase goods or services. That loss of potential utility must be compensated for. This is one of the major functions of interest.

Furthermore, the lender faces considerable uncertainty relating the value of that money when it is repaid, i.e. the future value of that money is uncertain. The quantification of that uncertainty is known as risk. There are several types of risk faced by lenders and these can detract from the utility or well-being that the lender feels. One of the functions of interest is to compensate for the loss of utility due to the existence of risk.

The first risk that we shall consider is the risk of a loss of purchasing power, usually known as inflation risk. If the economy suffers from inflation during the period of the loan or investment, the sum of money involved will buy a smaller quantity of goods and services when repaid at a future date than it could have bought at the time of the investment. Thus the investor will require to be compensated for this loss of purchasing power.

The second risk is that the money will not be returned, either because the project which it financed has not been successful, or because of the dishonesty of the borrower or other unforeseen circumstances. This is the risk of default. Again the investor has to be compensated for bearing this risk.

When the loss of potential utility, the inflation risk and the default risk are combined we have a group of factors that make current possession of money preferable to possession only in the future. People prefer to hold money now rather than promises of money later. They are said to have **positive time preference**.

Interest compensates the lender for not being able to satisfy that preference when funds are loaned or invested. Borrowers are willing to pay for the use of funds because it enables them to enjoy the additional utility of premature consumption which has been made possible by the transfer of funds from the investor.

Although one often glibly talks about the rate of interest, it should be clear from what has just been said that there are in fact many rates of interest. At any one time there will be a variety of interest rates in the financial markets. Thus it is useful to classify those factors that determine interest rates into two groups, those factors that determine the general level of interest rates and those factors that determine the pattern of interest rates.

Factors that influence the level of interest rates are:

- government policy
- the supply of money
- expectations of future inflation.

Factors that influence the pattern of interest rates are:

- the term to maturity of the financial commitment
- default risk
- the liquidity of the financial commitment
- taxation
- other miscellaneous factors specific to the financial commitment, e.g. whether it is secured on assets, or there are financial options embedded in the agreement

To illustrate the influence of the above factors on interest rates consider Table 1.1. It shows the interest rates that related to a number of financial instruments or markets at the end of December 1995.

Table 1.1 Interest rates (%), December 1995

Building society mortgage	7.85
Seven-day notice bank deposit (retail)	4.5
Unsecured bank overdraft (personal)	22.0
Three-month bank certificate of deposit	6.375
Three-month commercial paper	6.45
Three-month UK government Treasury bill	6.32
Six-month interbank deposit	6.34
UK government bond with five years to maturity	7.0
UK government bond with ten years to maturity	7.4
German government bond with ten years to maturity	5.88
UK five-year corporate bond (secured debenture)	8.1
UK five-year unsecured corporate bond	9.2
UK five-year convertible bond	6.5

The time value of money

The time value of money relates to the process of calculating the present value, i.e. the value today of a sum of money promised at some point in the future, or to the calculation of the future value, i.e. the value at some time in the future of a sum of money received or paid today. The process of determining the present value is known as **discounting**, and the process of determining the future value is known as **compounding**.

This present value is calculated by discounting each of the payments due by the interest rate that could be earned if the funds were received today. Common applications of discounting are the valuation of bonds by discounting the expected future interest and redemption payments, and the valuation of equities by applying dividend discount models. Both of these applications will be illustrated later in this chapter.

Financial assets are valued by calculating the present value of the future cash flows expected from the asset. Some financial instruments such as futures contracts and forward contracts are priced off the future value of a sum of money. The **future value of money is calculated by compounding** all the interest payments that the sum of money would earn until the future date.

The mathematical techniques of compounding and discounting will now be analysed in detail. We will begin by looking at the future value of money.

The future value of money

Consider placing a sum of money in an interest-bearing bank deposit today. The value of that deposit will grow over time as interest accrues. Consequently, the expected future value of that bank deposit will be greater than the current value by the amount of the interest earned. For example, the future value of a one-year deposit of 1000 earning 6% per annum to be paid at the maturity of the deposit is 1060, i.e. 1000 plus 60 earned as interest.

This example can be generalized in that the actual amount of interest earned, and thus the **future value** of a given sum of money, will depend on the **rate of interest**, the **time to maturity** of the deposit, and whether the interest rate is quoted as a **simple interest** rate or a **compound interest** rate. If it is compound interest, the **frequency of the compounding** of the interest payments will also be important. These terms are described and illustrated below. Whether simple interest or compound interest is applicable to a particular financial instrument will depend on the terms of that instrument.

Simple interest. If the interest rate is quoted on a simple basis, the interest payable on the deposit or loan is calculated as the annual rate of interest multiplied by the number of years (or fractions thereof) to the maturity of the deposit. To illustrate this, if the funds had been placed on deposit for 6 months, at 6% per annum, the interest earned under simple interest would be

$$(0.06 * 0.5)1000 = 30$$

The future value of the deposit would be

$$1000[1 + (0.06 * 0.5)] = 1030$$

Note how the 6% is handled in these calculations: 6% is 6/100 or 0.06. Thus the interest rate is expressed in decimals, i.e. 6% = 0.06, 10% = 0.10 and so on.

Thus the general formula for the future value of money earning simple interest is given by

$$FV = P[1 + (r * n)] \tag{1.1}$$

where

FV= the future value of money
P = the original principal amount
n = the life of the deposit in years
r = the simple interest rate in decimals, i.e. 10% is inserted as 0.1 and 6% is 0.06.

Generally, if an instrument has a life of more than one year, compound interest is applied. This is discussed next.

Compound interest. Compounding refers to the periodic adding of accrued interest to the principal sum such that when the interest accruals are added to the principal, they form part of an enlarged principal and earn the same rate of interest. For example, consider a sum of 1000 placed on deposit for three years with interest compounded annually at the rate of 6% per annum. At the end of the first year, 60 (i.e. 1000 * 0.06) will be added to the principal giving a value of 1060. This can also be expressed as 1000 * 1.06 = 1060. During that second year 1060 will be earning interest. At the end of the second year 63.6 (i.e. 1060 * 0.06) will be added to the 1060 so that in the third year 1123.6 (i.e. 1000 * 1.06 * 1.06) will be earning interest and so on.

The value of the deposit in three years' time, i.e. the future value of the original principal, is calculated as 1000 * 1.06 * 1.06 * 1.06 = 1191.02. This is an example of what is known as geometric growth and can be written as follows:

$$1000(1.06)^3 = 1191.016$$

The general formula for calculating the future value of a sum of money when interest is compounded annually is

$$FV = P(1 + r)^n \qquad (1.2)$$

where the notation is the same as in equation (1.1).

In many financial transactions the compounding is more frequent than once per year. For example, the interest payments may be added to a deposit or charged to a loan quarterly or monthly. The future value will be greater because interest will be added to the principal after shorter time periods and, therefore, will itself start earning interest earlier.

To account for compounding more frequently than annually, equation (1.2) has to be amended. The annual rate of interest is divided by the number of compounding periods per year, and the power (n) to which (1 + r) is raised is multiplied by the number of compounding periods per year. More formally that is

$$P\left(1 + \frac{r}{m}\right)^{mn} \qquad (1.3)$$

where m is the number of compounding periods per year.

The following example, which assumes quarterly compounding over three years, will illustrate this

$$1000\left(1 + \frac{0.06}{4}\right)^{4 \cdot 3} = 1000(1.015)^{12} = 1195.618$$

Note that the change from yearly to quarterly compounding gives a higher future value, or an additional return – in the case of our example, 4.602 over the three years.

It might be thought that continuing to increase m, i.e. compounding more and more often, would lead to an unlimited increase in the value of

$$P\left(1+\frac{r}{m}\right)^{mn}$$

In fact this is not the case because the expression

$$\left(1+\frac{r}{m}\right)^{mn} \tag{1.4}$$

approaches a limiting value as m increases. Table 1.2 demonstrates this in the case of $r = 1$ and $n = 1$. Note that as $n = 1$, $mn = m$ and therefore n is ignored in this example.

Table 1.2

m	2	3	4	5	10	20	100	1000	10 000
$(1+(1/m))^m$	2.25	2.370	2.441	2.488	2.593	2.653	2.705	2.717	2.718

Thus with $m = 2$ we have $(1 + 0.5)^2 = 2.25$. By increasing the frequency of compounding to 10 we get 2.593, by increasing the frequency to 100 we get 2.704, to 1000 we get 2.717 and so on. The limiting answer is a mathematical constant of great importance, as we shall see later in applications of calculus. It is an irrational number, i.e. one with an infinite decimal expansion. It cannot therefore be written exactly as a decimal, and it has to be given a name – mathematicians call this limiting value the exponential constant and give it the symbol "e". It is, of course, possible to write down decimal approximations to e, such as 2.71828182845904523536287, but even this is not perfectly accurate.

We can generalize the effect of increasing the frequency of compounding, m, by saying that

$$\left(1+\frac{r}{m}\right)^{mn}$$

approaches e^{r*n} as m increases.

At the extreme it can be assumed that the compounding becomes so frequent that the interest payments are continuously added to the principal, and therefore causes the principal to grow exponentially. This is known as **continuous compounding**. The future value of a sum of money where interest is continuously compounded is given as

$$P \cdot e^{r*n} \tag{1.5}$$

where

e = the exponential constant = 2.71828...
n = the number of compounding periods in years or fractions thereof.

Continuous compounding is the assumption made in much of the financial theory and is applied in models that value options. In addition, as will be shown later, converting interest rates with differing frequencies of compounding to continuous compounding equivalents facilitates meaningful comparisons between interest rates.

Converting a discretely compounded yield to a continuously compounded equivalent. To convert discretely compounded yields to continuously compounded yields, note that a sum of money invested at a continuously compounded rate must have the same future value as a similar sum invested at the **equivalent** discretely compounded rate. Thus

$$Pe^{r_{cc}n} = P\left(1 + \frac{r_{dc}}{m}\right)^{mn} \tag{1.6}$$

where

r_{cc} = the continuously compounding rate of interest
r_{dc} = the discretely compounding equivalent rate of interest
n = the number of years to maturity
m = the number of compounding periods per year.

By removing P and n from both sides of this equation, because they have no effect on the equivalence of the annual rates, we have

$$e^{r_{cc}} = \left(1 + \frac{r_{dc}}{m}\right)^{m} \tag{1.7}$$

from this we get

$$r_{cc} = m\ln\left(1 + \frac{r_{dc}}{m}\right) \tag{1.8}$$

where ln is the natural logarithm.

Note that ln is the inverse of the exponential function. For example: e = 2.71828 ... and ln (2.71828 ...) = 1 and e^2 = 7.38905 ..., so ln (7.38905 ...) = 2. Consequently, $e^{0.1}$ = 1.10517 so ln 1.10517 = 0.1

Thus to derive the continuously compounded rate from a discretely compounded rate we simply divide the annual discrete rate by the number of discrete compounding periods per year, add 1, take the natural logarithm and multiply by the number of discrete compounding periods per year. The following example will illustrate this. The continuously compounded equivalent of 6% per annum, compounded quarterly is:

$$r_{cc} = 4\ln\left(1 + \frac{0.06}{4}\right) = 4*0.014889 = 0.059554 \approx 5.955\%$$

Notice that the continuously compounded rate is below the discretely compounded equivalent because with continuous compounding, interest is compounded more frequently, and so the interest itself earns more interest.

Converting continuously compounded returns to discretely compounded returns. We can derive the equivalent discretely compounded return from a known continuously compounded return as follows:

$$r_{dc} = m(e^{rcc/m} - 1) \qquad (1.9)$$

Thus we raise the exponential function e to the power equal to the known continuously compounded yield divided by the number of discrete compounding periods per year, and subtract 1. This result is then multiplied by the number of discrete compounding periods per year to arrive at the discretely compounded equivalent.

To illustrate this, assume a continuously compounded rate of 12.5% per annum. What is the equivalent rate with compounding only four times per year?

$$r_{dc} = m(e^{0.125/m} - 1) = 4(e^{0.125/4} - 1) = 12.7$$

Notice that the equivalent discrete rate has to be higher than the continuously compounded equivalent because with a discrete rate, interest is added less frequently and will thus earn less interest on interest.

The present value of money

We have already noted that the value today of a sum of money promised at some time in the future will be less than the future value of the promised sum. Even in a riskless world that would be the case, because the funds could be invested to earn the risk-free rate of interest and, therefore, the future value of that sum invested today would be greater than the same sum promised in the future. In the risky world that we live in there are a number of uncertainties regarding the value of the sum of money promised in the future such as inflation and the risk of the promisor defaulting on his or her promised payment.

In order to be able to compare the value of different cash flows due at different dates in the future it is necessary to discount the future cash flows to their present value. The **present value** is that value which, if invested at the currently available rate of interest until the date of the promised payment, would result in a future value equal to the sum promised.

Discrete discounting. In many financial transactions, even short term ones, compound interest rates are used for discounting. Thus, in order to discount a future cash flow to its present value, the cash flow must be divided by (1 + the discount rate, expressed in decimals) raised to the power equal to the number of years until the cash flow is due to be received. This may be stated formally as

$$PV = \frac{CF_n}{(1+r)^n} \qquad (1.10)$$

where

PV = present value of the future cash flow
CF_n = amount of future cash flow due n years from now
r = the discount rate
n = the number of years until cash flow is due.

A numerical example will illustrate equation (1.10). Assume that 1000 will be received in five years' time and the market currently discounts such payments at 10% per annum. The present value would be

$$PV = \frac{1000}{1.10^5} = 620.92$$

If the frequency of discounting is more than once per year, r is divided by the number of periods per year and n is multiplied by the number of periods per year. Using similar notation to that in the section on compounding, the formula is

$$PV = \frac{CF_n}{\left(1 + \dfrac{r}{m}\right)^{mn}} \qquad (1.11)$$

Now consider the earlier numerical example but changing the frequency of discounting to four times per year

$$PV = \frac{1000}{\left(1 + \dfrac{0.1}{4}\right)^{20}} = 610.27$$

Thus the present value of a given cash flow gets smaller the more frequently is the discounting by a given annual rate of interest.

Continuous discounting. As with compounding, the period between receipt of cash flows can be shortened until in the extreme the discounting is continuous.
  The formula for continuous discounting is

$$PV = CF_n \, e^{-r * n} \qquad (1.12)$$

Thus for continuous discounting e is raised to the minus power of $r * n$. For example, the present value of 1000 due in five years' time continuously discounted at 10.0% per annum

$$1000e^{(-0.1 * 5)} = 606.53$$

Spot rates, forward rates and quality spreads

So far we have explained how compounding and discounting work. We now go on to discuss how the actual interest rates that we use in these processes are determined. We will begin with two basic, but very important, types of interest rate: spot rates and forward rates.

Spot rates

The **spot rate of interest** is that rate of interest that the financial markets use to discount a single future cash flow to its **present value**. Take, for example, a cash flow due in one year's time. The one-year spot rate will be that rate that discounts the cash flow to its present market value. If the cash flow were due in say five years' time, the annual rate that discounts it to the present market value would be the five-year spot rate and so on.

Assuming **continuous discounting**, the one- and five-year spot rates will be r_1 and r_5 in each of the following equations

$$PV_1 = CF_1 \, e^{-r_1}$$

$$PV_5 = CF_5 \, e^{-r_5 * 5}$$

and the n period continuously compounded spot rate is calculated as

$$r_n = \frac{\ln\left(\dfrac{CF_n}{PV}\right)}{n} \tag{1.13}$$

However, the practice in most financial markets is to use **discrete discounting**, therefore the n-year spot rate would be the value of r_n that solves the following equation:

$$PV = \frac{CF_n}{(1 + r_n)^n} \tag{1.14}$$

The clearest examples of spot rates are to be found in the yields of zero coupon bonds. These bonds pay no periodic interest during their life and are redeemed at face value on maturity. Consequently they are issued at a discount to face value, and trade at a discount all their life because the current price at any point in time is the present value of the redemption payment.

The interest rate that the market uses to discount that single redemption payment to the current price is the spot rate of the appropriate term to maturity. However, the spot rate itself is not observed in the market place. We only observe the price of the zero coupon bond. The spot rate is derived from the current price of a zero coupon bond by applying the following equation:

$$r_n = \left(\frac{CF_n}{PV}\right)^{\frac{1}{n}} - 1 \tag{1.15}$$

where

r_n = the n period spot rate
CF_n = the cash flow due at time n; this is usually 100, but may not be
PV = the present value or current price of an n-year zero coupon bond.

Applying this equation to a zero coupon bond with 30 years to maturity, redeemable at 100 and currently selling for £16.97, the 30-year spot rate would be calculated as

$$r_n = \left(\frac{100}{16.97}\right)^{\frac{1}{30}} - 1 = 6.09\%$$

The **forward rate of interest** is that rate which will apply in the future on a commitment made today to lend money in the future. Thus the rate of interest on a three-month loan beginning three months from now is the three-month forward rate for three months hence. If the loan were a one-year loan to begin 12 months from now, the rate would be the one-year forward rate for 12 months hence. In the money markets these rates are known as forward forwards and quoted as say 3/6 or 12/24, meaning the forward transaction starts in 3 or 12 months and ends in 6 or 24 months. They are also referred to as 3s against 6s or 12s against 24s.

Forward rates are not observed in the market place. However, with regard to the money markets they can be derived by the judicious use of short-term loans and deposits. For bond markets including Treasury bills, the forward rates can be implied by the relationship between successive spot rates.

The actual calculation of forward rates depends on whether the market practice is to use simple or compound interest. If compound interest is used, then the calculation depends on whether continuous or discrete compounding is assumed. Simple interest rates are used in the money markets for instruments with under one year to maturity. In the bond markets discrete compounding is used but in the markets for bond options continuous compounding is used.

Forward rates using simple interest

A forward rate in the money markets is the rate on a loan or deposit beginning some time in the future. A **forward deposit** rate can be "created" by borrowing for a short period and placing that money on deposit for a long period. A forward loan can be created by borrowing for a long period and depositing for a shorter period.

For example, if your bank commits itself to a rate on a three-month deposit beginning in three months' time, your bank is offering you a forward deposit. The bank has to hedge that forward commitment by borrowing (i.e. taking a deposit) for three months and lending those same funds for six months. The bank's books will be in balance over the first three months because the lending is funded by the borrowing. However, that borrowing (the deposit) has to be repaid in three months whereas the lending will not be repaid for six months. Thus the bank will use the deposit that you have promised them to balance their books for the second three-month period.

The break-even interest rate that the bank can offer you for the forward deposit will be the difference between the interest that the bank earns on the six-month lending and the interest costs it had to pay out on the first three-month deposit.

Consider now the situation if you asked your bank for a commitment on the rate of interest on a three-month loan to start in three months' time. The bank would hedge that position by borrowing (taking deposits) for six months and lending those funds for three months. In three months' time that short loan would be repaid and the bank would lend the funds to you. When you repay your loan at the end of the period, i.e. six months from now, the bank will use the funds to repay its six-month borrowing. The break-even rate at which the bank could lend to you is the difference between the interest cost of the six-month funds to the bank and the interest income on the loan for the first three-month period.

An intuitive illustration will help understanding. Assume that the interest rate on three-month interbank money is 6.0–6.1% per annum (p.a.) and the interest rate on six-month interbank money is 6.15–6.25% p.a. Thus the bank takes in (bids for) three-month money at 6.0% and lends out (offers) at 6.1%, and it takes six-month money at 6.15% and lends it at 6.25%. The bank offers a forward deposit and must calculate the appropriate rate. The cost of the hedge will be as follows: if the bank lends 100 for six months at 6.25% p.a. it will earn 3.125. It borrows funds for three months at 6% p.a. paying 1.5. Thus the amount that can be paid on a deposit over the second three-month period is 1.625, which when annualized by multiplying by four in this example is 6.5% p.a.

In actual fact this example is slightly inaccurate because:

(a) This procedure annualizes the forward rate, yet it is by no means certain that the transaction could be rolled over at the same rate for the required number of periods per year.

(b) For a forward loan the bank borrows for a long period and lends for a short period. The interest on the lending can be reinvested for the second time period, thus reducing the break-even interest rate that must be charged on the loan for the second period and thus the break-even forward rate. However, that adjustment due to reinvestment is not known at the outset. This uncertainty can be mitigated by evaluating the forward rate on a loan as the net interest earnings expressed as an annualized percentage of principal and interest costs.

(c) For a forward deposit the bank hedges by borrowing short and lending long. The interest cost on the first period borrowing (taking a deposit) has to be financed during the second period thus reducing the break-even amount that can be paid on the deposit for the second time period, and thus reducing the break-even forward rate. The forward rate on a forward deposit can be evaluated as the net interest cost expressed as an annualized percentage of principal + interest earnings.

To illustrate this consider deriving the forward rate from a three-month forward loan starting in three months' time in a money market that assumes a 360-day year. The cost of six-month money to the bank is 6.25% p.a., the rate on three-month reinvestment is 6% p.a. The net cost on £100 is

$$6.25 * \frac{180}{360} - 6.0 * \frac{90}{360} = 3.125 - 1.50 = 1.625$$

The sum of principal plus earnings is 100 + 1.50 = 101.50, thus the forward rate is

$$\frac{1.625}{101.5} * \frac{360}{90} = 0.064039 \approx 6.4\%$$

A formula for deriving the forward rate in the money markets that takes account of these uncertainties is as follows:

$$FR = \frac{1}{(n-m)} * \frac{(n * r_n) - (m * r_m)}{1 + \frac{(m * r_m)}{360}} \tag{1.16}$$

where

FR = the forward rate
n = the maturity of the longer dated instrument in days
m = the maturity of the shorter dated instrument in days
r_n = the rate of interest on the longer dated instrument
r_m = the rate of interest on the shorter dated instrument.

To illustrate this using the above example:

$$FR = \frac{1}{90} * \frac{(180 * 0.0625) - (90 * 0.06)}{1 + \frac{(90 * 0.06)}{360}} = 0.064039 \approx 6.4\%$$

Day count conventions are particularly important in the calculation of forward rates of interest. The above example is applicable to most non-sterling money markets. If it were to be applied to the sterling interbank market a 365-day year would have to be assumed. Day count conventions are discussed in detail on page 15.

Forward rates using continuous compounding

Using continuous compounding, the forward rate is calculated as follows:

$$FR_{m,n} = \frac{nr_n - mr_m}{n - m} \qquad (1.17)$$

This calculation can be illustrated with the following example which assumes that the 180-day rate is 11% p.a. and the 90-day rate is 10% p.a. The forward rate on a 90-day instrument beginning in 90 days time is calculated as

$$FR_{90,180} = \frac{(180*11) - (90*10)}{180 - 90} = 12\%$$

Forward rates using discrete compounding

Assuming discrete compounding and using successive spot rates to derive the implied forward rates is simply a matter of applying the following equation, where rn and rm are the appropriate consecutive spot rates, and n and m are now being measured in years

$$FR_{m,n} = \sqrt[n-m]{\frac{(1+r_n)^n}{(1+r_m)^m}} - 1 \qquad (1.18)$$

A numerical example will be helpful here. Assume that the three-year spot rate is 12.75% p.a. and the two-year spot rate is 11.63% p.a. The forward rate will be

$$FR_{2,3} = \sqrt[3-2]{\frac{(1.1275)^3}{(1.1163)^2}} - 1 = 15.04$$

In this example $n - m$ equals one, because the difference in time periods was a whole year. However, had we used rates for periods with say six months or three months between then $n - m$ would be 0.5 or 0.25, respectively.

Clearly, as we live in an uncertain world the forward rates implied by these calculations may not be, indeed probably will not be, the short-term rates that actually apply in the future time period. What these forward rates actually represent is the market's current expectations of what the rates will be in the future.

To explain this consider a market in one-year and two-year zero coupon bonds. An investor wishing to invest for two years has a choice between investing for one year and rolling over his or her investment for the second year. Alternatively, the investor will invest in the two-year instrument initially.

The investor's choice will be determined by the forward rate implicit in the two-year instrument compared with his or her expectations of what one year yields will be in one year's time. If the investor expects yields to be higher in one year's time than the forward rate currently implied in the two-year spot rate the investor will buy the one-year bond and roll over one year hence. However, if rates are expected to be lower in the future than currently implied by the forward rate, the investor will buy the two-year bond.

In markets where investors can trade between instruments according to their expectations, the forward rate will reflect the market's current consensus expectation of the future spot rate.

Spot rates and forward rates compared

Table 1.3 below shows the prices of zero coupon instruments, the resulting spot rates and the implicit one-year forward rates calculated assuming both continuous and discrete compounding (CC and DC, respectively).

Table 1.3. Spot and forward rates of interest

Maturity years	Price of zero	Spot rate (CC)	Spot rate (DC)	Forward rate (CC)	Forward rate (DC)
1	90.48	10.0	10.52	N/A	N/A
2	80.25	11.0	11.63	12.0	12.75
3	69.76	12.0	12.75	14.0	15.04
4	59.45	13.0	13.88	16.0	17.34
5	49.66	14.0	15.03	18.0	19.71

Alternatively, to avoid rounding errors, the last column may be computed directly from the bond price column rather than from the computed spot rates. Thus, for instance

$$19.71 = 100 \left(\frac{59.45}{49.66} - 1 \right)$$

The spot rates are discount rates; they are also the annual rate of return from investing in a particular zero coupon bond at the price stated. For example, the return from investing in a one-year zero is

$$(100 - 90.48)/90.48 = 10.52\%$$

The spot rates and the forward rates are clearly related. Indeed it can be seen that the spot rate is the geometric mean of the current short-term spot rate and all the short-term forward rates relevant to a given term to maturity. For example, the five-year spot rate of 15.03% (discrete compounding) given above is the geometric mean of the current one-year spot rate and all four one-year forward rates given in Table 1.3 above. This is shown in the following equation:

$$\sqrt[5]{(1.1052)(1.1275)(1.1504)(1.1734)(1.1971)} - 1 = 0.1503$$

or 15.03%.

Spreads reflecting quality differentials

Remembering that interest rates compensate for the disutility resulting from loss of purchasing power and the various risks associated with lending, it is reasonable to assume that the greater the disutility, the greater interest rate required in compensation. Indeed we observe that lenders require higher interest rates the higher their expectation of inflation or default risk, and the lower their perception of the quality of the investment, for example the liquidity of the market.

The expectation of inflation will affect all interest rates, but the additional compensations for risk and other qualitative differences will be specific to each investment. With regard to the term structure of interest rates, the spot rates are derived from default risk-free bonds. Such bonds are those issued by a government in the domestic currency of that government. Such bonds are assumed to be free of default risk because the government has the power to tax and issue currency in order to meet its commitments.

The bonds and other securities issued by corporate and other non-governmental bodies do bear default risk. As such the rate of interest required by investors is greater than that required on otherwise identical default-free bonds. This additional return is known as a quality spread. That part of the total quality spread due to default risk is known as a credit spread. However, the quality of an investment is influenced by more than default risk. Liquidity of the particular market and tax liabilities attached to the instrument are two examples. Thus the quality spread is often constituted of more than the default risk spread. This is reflected schematically in Fig. 1.1.

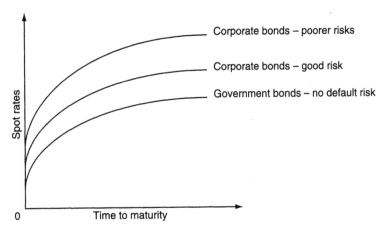

Fig. 1.1 Spot rate curves.

Practical applications of interest rates in financial markets

Valuing financial instruments

Earlier in this chapter it was noted that financial instruments are valued as the sum of the present values of all the expected future cash flows from the particular instrument. In addition futures and forward contracts are priced as the future value of a current cash flow. In this section we will illustrate the use of interest rates in discounting and compounding by valuing bank certificates of deposit, Treasury bills, zero coupon bonds, coupon paying bonds and company shares. We will also price forward and futures contracts. However, first we must consider some market conventions regarding the way in which interest accrues on the major types of debt instruments available in financial markets.

Day count conventions and simple interest

One complication that enters into simple interest rate calculation is the day count convention that applies to a particular financial market. Interest accrues on a daily basis on all financial instruments. However, some markets calculate interest on the basis that the year has 360 days rather than 365 or 366 in leap years. In addition some markets assume that all months have 30 days irrespective of the actual number of days. In the UK sterling money markets, a 365-day year is assumed in the calculation of interest. In most other centres and for non-sterling instruments in London a 360-day year is used.

Thus in those markets that assume a 360-day year, the actual interest paid on, say, a 45-day deposit is calculated with the holding period calculated as 45/360ths of a year. This is referred to as **actual over 360**. In those markets where interest payments are calculated on a 365-day year, interest on a 45-day deposit is calculated at 45/365ths of a year. This method is referred to as **actual over 365**.

To illustrate the impact of these conventions, consider the eurocurrency market where the year is assumed to have 360 days but interest accrues on an actual day count basis, i.e. actual/360. Thus a deposit placed for two 31-day months would accrue interest on a daily basis for 62 days, but the return would be

$$(62/360) * \text{the quoted rate of interest}$$

For example, if the quoted rate of a 62-day eurocurrency deposit is 6% p.a., the interest earnings on a 10 000 000 deposit would be

$$10\,000\,000 * \frac{62}{360} * 0.06 = 10\,000\,000 * 0.0103333 = 103\,333.33$$

Thus the interest earned would be 103 333. Compare that with the sterling domestic interbank market which calculates interest over a 365-day year. The interest would be

$$10\,000\,000 * \frac{62}{365} * 0.06 = 10\,000\,000 * 0.010191781 = 101\,917.81$$

Thus a 6% deposit in the eurocurrency market is not the same as a 6% deposit in some domestic interbank markets such as the sterling market, where interest accrues on an actual/365-day basis. To make the annual rates comparable it is necessary to multiply the eurocurrency rate by 365/360, or multiply the domestic interbank rate by 360/365. Thus the 6% p.a. deposit in the euromarkets will earn 6.083%, i.e. [(365/360) * 6], in a full year, whereas the domestic interbank deposit will only earn the stated 6%.

Given the differing day count conventions between markets and between instruments, we must distinguish between the **nominal** rates of interest and the **effective** rate of interest. The nominal rate is that published or otherwise specified. The effective rate is the amount of interest actually received by the investor, expressed as a percentage of the amount invested.

Clearly, it is crucially important to determine the interest rate convention applicable to the particular security and market! Day count conventions also apply to the calculation of accrued interest in bonds. This will be discussed later in this chapter.

As the convention in the money markets is to use simple interest when calculating both future and present values, as well as the fact that the day count conventions must also be applied, equation (1.14) must be adapted for use in the money markets to

$$PV = \frac{CF_n}{\left[1 + \left(\frac{D}{B} * r_n\right)\right]} \tag{1.19}$$

where

D = the number of days the instrument is being discounted
B = 360 or 365 according to the day count convention of the market in question
PV = present value
CF = the future cash flow – it is also the terminal or future value
r_n = the annual, n period, simple interest rate used to discount.

We are now in a position to consider the valuation of the financial instruments listed earlier.

Bank certificates of deposit

Certificates of deposit (CDs) are issued by banks in return for funds placed with them. Such deposits differ from the traditional bank deposit only in that a certificate evidencing the deposit is issued which can be traded in a secondary market. The CDs are issued at par and interest is added on at maturity, when the deposit is repaid.

Assume that two CDs, one a domestic sterling CD, the other a euro-sterling CD, each had an initial life of one year, an initial yield of 6% and have been trading for 91 days. A £1 000 000 CD will have a redemption value of principal plus 365 days interest. However, the domestic CD will accrue interest at the rate of actual/365, whereas the euro CD will accrue interest at the rate actual/360. Thus the future value of each will be as follows:

$$FV_{CD} = 1\,000\,000 * \left[1 + \left(0.06 * \frac{365}{365}\right)\right] = 1\,060\,000$$

$$FV_{euroCD} = 1\,000\,000 * \left[1 + \left(0.06 * \frac{365}{360}\right)\right] = 1\,060\,833.33$$

Note that the effective yield of the euro CD will be 6 * (365/360) = 6.0833.

The CDs are traded in the secondary market, and are valued as the present value of the expected future cash flows. The expected future cash flow will, in the absence of default, be the future value as calculated above.

The discount rate to be used in calculating the present value of that future cash flow must be that required by the market for similar risks and similar maturities. Assume that the yields on CDs of similar risk and with a maturity of 274 (365 – 91) days in both the euro-sterling CD markets have risen to 6.1%.

The current price of the two CDs will be the terminal values as calculated above discounted by 6.1%, although the effective yield of the euro CD will be 6.1847% (6.1 * 365/360)

$$P_{CD} = \frac{1\,060\,000}{\left[1 + \left(0.061 * \frac{274}{365}\right)\right]} = 1\,013\,586$$

$$P_{euroCD} = \frac{1\,060\,833.33}{\left[1 + \left(0.061 * \frac{274}{360}\right)\right]} = 1\,013\,766$$

Treasury bills

The Treasury bill (T-bill) is a generic term for short-term government paper, typically with an initial maturity of 90 days, but also issued with maturities as short as one month and as long as one year. The markets tend to be highly liquid, with new issues being made by the government at frequent intervals – often weekly.

The bills are issued at a discount and redeemed at par, therefore the discount represents the absolute return to the investor. In the cash T-bill market the trading quotations are in the form of a discount rate and not a security price. Thus a typical bid offer quote may be 7.0–6.9, meaning that the trader will bid (buy) at a discount yield of 7% p.a. and will offer (sell) at 6.9% p.a. The discount yield is calculated from a known discount as

$$\text{Discount yield} = \left(\frac{\text{discount}}{\text{face value}}\right) * \left(\frac{365}{n}\right) \tag{1.20}$$

assuming an actual over 365-day count convention.

In the calculation of the discount yield it is important to take account of the actual life of the bill and the appropriate day count convention as explained above.

The price paid for the T-bill is the **invoice price**. This is the discounted price that corresponds to a particular discount yield quotation. This is calculated as:

$$INVP = PAR\left[1 - r\left(\frac{n}{365}\right)\right] \tag{1.21}$$

where

$INVP$ = invoice price
PAR = par value
r = discount yield quoted in decimals
n = number of days to maturity of bill.

To illustrate the calculation of the invoice price of T-bills, consider the following example relating to a US T-bill where an actual over 360-day count convention applies. The bill has 44 days to maturity and is trading on a discount yield of 8.25%. The invoice price is

$$INVP = 1\,000\,000\left[1 - 0.0825\left(\frac{44}{360}\right)\right] = 989\,916.67$$

Zero coupon bonds and coupon paying bonds

A zero coupon bond is a bond which makes no interest payments (known as coupons) during its life, hence the term zero coupon. It is issued and traded at a discount because the only expected cash flow is the redemption payment. The return is generated by the difference in capital value at the time of issue or purchase and the value at redemption or sale. Thus the current value or price of such a bond is the present value of the single redemption payment. The formula for calculating the price is

$$P_z = \frac{CF_T}{(1 + r_T)^T} \tag{1.22}$$

where

P_z = the current price of the zero coupon bond
CF_T = the cash flow (single redemption payment) due at time T
r_T = the spot rate for discounting payments due at time T.

For example, assume that the bond has five years to redemption and will be redeemed at 100, and that the spot rate used by the market to discount a single payment due five years hence, the five-year spot rate, is currently 8% p.a. The present value of the zero coupon bond is thus

$$P_z = \frac{100}{(1.08)^5} = 68.06$$

Coupon paying bonds that have paid their penultimate coupon are also effectively zero coupon bonds, because the final coupon and redemption payment are a joint single payment due in the future. For example, a 10% semi-annual coupon bond due for redemption at par in four months time will have a final cash flow of 105, which will be discounted at the four-month spot rate.

The valuation of a zero coupon bond, i.e. the discounting of a single future payment by the appropriate spot rate, is the foundation stone for the valuation of all bonds, as will be illustrated in the next section.

Valuation of a coupon paying bond

The majority of bonds promise regular payments of interest known as coupons. In the major government and corporate bond markets, coupons are paid half yearly; however, in the eurobond markets, coupons are paid annually.

Thus a coupon paying bond entitles the holder to a series of periodic interest payments and a redemption repayment of the principal. Each of the promised cash flows is a single cash flow due at a given time in the future. As such each cash flow is not conceptually different from the single cash flow due under the zero coupon bond. It will have a present value derived by discounting cash flow at the appropriate spot rate. Thus the valuation of a coupon paying bond can be achieved by treating the bond as a portfolio of zero coupon bonds. Each coupon and the redemption payment is individually discounted at its appropriate spot rate, which need not be the same for all payments. These present values are added together to derive the current value of the bond. This is formally stated as

$$P_B = \sum_{1}^{T} \frac{CF_T}{(1+r_T)^T} \tag{1.23}$$

To illustrate this consider a five-year bond paying an annual coupon of 10 and a redemption payment of 100. The appropriate spot rates are 10, 11, 12, 13 and 14% p.a. for payments due in years 1–5, respectively.

The discounted value of the cash flows is

$$P_B = \frac{10}{1.1} + \frac{10}{(1.11)^2} + \frac{10}{(1.12)^3} + \frac{10}{(1.13)^4} + \frac{110}{(1.14)^5} = 87.59$$

This example implicitly assumes that a coupon has just been paid and that the next coupon is one full year away. In reality bonds have to be valued between coupon dates. This is straightforward. Assume that the bond pays annual coupons; if the next coupon is due in three months' time, that cash flow should be discounted by the three-month spot rate and the coupon after that, which will be due in 15 months' time, would be discounted using the 15-month spot rate and so on.

To illustrate this, consider the valuation of a bond due in 4.25 years, paying 8% annually. The last coupon was paid nine months previously, so the next coupon is three months or 0.25 years away. The three-month spot rate, and those at annual intervals thereafter, are 9, 10.5, 11.1, 12.2 and 13.5%. The bond is valued as

$$P = \frac{8}{(1.09)^{0.25}} + \frac{8}{(1.105)^{1.25}} + \frac{8}{(1.111)^{2.25}} + \frac{8}{(1.122)^{3.25}} + \frac{108}{(1.135)^{4.25}} = 89.76$$

Note: this price is the "dirty" price of the bond. Markets quote "clean" prices which are dirty prices minus interest that has accrued since the last coupon payment.

Equities

One technique for valuing equities is to calculate the present value of all the expected future dividends, the so-called dividend discount models – see Williams (1938), Gordon (1962) and Fuller and Hsia (1984).

Although, equity values are generally considered to be a function of expected future earnings, **dividend discount models** treat dividends as a proxy for earnings and thus account for future earnings implicitly. This is acceptable given that a firm may either pay out what is earned or may reinvest those earnings within the firm. If the earnings are reinvested and an increasing dividend policy is assumed, future dividends will be greater than current dividends. Thus dividends will grow as long as some profits are ploughed back into the business.

The theoretical price of an equity can be stated formally as

$$P_0 = \frac{D_1}{1+k} + \frac{D_1(1+g_2)}{(1+k)^2} + \frac{D_1(1+g_2)(1+g_3)}{(1+k)^3} + \ldots + \frac{D_1(1+g_2) * \ldots * (1+g_n)}{(1+k)^n} \qquad (1.24)$$

where

  P_0 = the theoretical price of the equity
  D_i = the dividend due at the end of period i
  g_i = the growth rate of dividends or earnings in period i
  k = the risk adjusted rate of return required by the market, i.e. the rate used by the market to discount the future cash flows.

Thus the current price is the present value of all future dividends. The future dividend values are derived by taking the dividend payable at the end of the current time period and compounding that dividend by the successive expected future growth rates. For example, the dividend paid at the end of the third time period will be the dividend expected to be paid at the end of the first period, compounded for one period at the growth rate for that period, g_2, and then compounded for the third period at the expected growth rate for that period, g_3.

As earnings can be reinvested in the business or paid out as dividends, $D = E * (1 - b)$. Equation (1.24) can therefore be restated in terms of earnings in the following way:

$$P_0 = \frac{E_1(1-b)}{1+k} + \frac{E_1(1-b)(1+g_2)}{(1+k)^2} + \frac{E_1(1-b)(1+g_2)(1+g_3)}{(1+k)^3} + \ldots + \frac{E_1(1-b)(1+g_2) * \ldots * (1+g_n)}{(1+k)^n}$$

$$(1.25)$$

where

E_i = earnings in period i
b = the retention ratio, i.e. the proportion of earnings not paid out as dividends.

This equation assumes that the retention ratio, b, is constant, but it is an easy matter to change it in each time period.

Where does the growth of earnings and therefore dividends come from? The answer is the amount of new investment and the rate of return on that investment. In a world that assumes no external financing, g is the product of the retention ratio, b, and the return on equity, r. Thus

$$g = br$$

The growth in earnings, and therefore dividends, depends on the availability of funds to invest and the availability of investment opportunities. If external funding is used, growth would be enhanced by the returns of the externally funded projects. Thus $g = r(b + f)$, where f represents the external funding as a proportion of earnings.

As equities are perpetuities, the models set out in equations (1.24) and (1.25) are somewhat intractable because they require forecasts of the variables out to infinity. A number of approaches have been developed to overcome this problem but discussion of them is beyond the scope of this book. Interested readers should consult Watsham (1993).

The pricing of forward and futures contracts

Forward and futures contracts are both contracts to buy or sell a given quantity of a certain asset at a given date in the future but at a price agreed at the time the contract is initiated. We have already noted that an agreement regarding the interest rate to be applied on a loan or deposit starting at a future date is a forward interest rate. An agreement to buy or sell foreign currency at some date in the future but at a rate agreed today is a forward exchange contract. Similar agreements to trade bonds and shares are also available.

The difference between forward contracts and futures contracts is simply that futures are traded on organized futures exchanges whereas forwards are traded in an informal market, for example between a bank and its customers. There is very little price difference between a future and a similar forward contract on the same type of asset, therefore we will use the term futures contract to apply to forward contracts as well.

At this point it is appropriate to explain the concept of arbitrage. Arbitrage is the process of simultaneously buying and selling the same asset which is priced differently in separate markets, and thus make a risk-free profit. For example, consider a share in company "A" that is traded both in London and New York. In London the price is £2.0, in New York it is $3.0 and the exchange rate is $1.40 to the pound. In such a situation the same share is trading at a different value in the two markets. If we assume no transaction costs, an arbitrage profit is possible by selling one share in New York for $3.0, buying £2.0 for $2.80, buying one share in London and keeping $0.20 profit. If such an arbitrage opportunity existed many arbitragers would be trading until the price of the shares in the two markets were equal, once allowance had been made for the exchange rate.

Now to consider the pricing of futures. Recall that earlier we calculated the future value of a sum of money as the principal compounded by the interest payable at the current rate. Well, the price of a future on any financial instrument is the current price of that asset compounded at the net yield of holding that asset. The term net yield, also known as the net cost of carry, is the cost of borrowing funds to buy the underlying asset minus any income that will be received while holding that asset.

Assuming the process of arbitrage between the future and the underlying asset, i.e. the currency, the share or bond, etc., is possible, the fair price of the futures contract is where there is no profit and no loss from arbitrage. If the future is overvalued arbitragers will short it (i.e. sell what they do not own), buy the underlying asset with borrowed funds and deliver the underlying into the futures contract at delivery. If on the other hand the future is cheap, arbitragers will buy it and sell the underlying asset against it.

As an example take the situation where the spot price of an asset is 100. This asset pays no income and the only carrying cost is interest, currently at 10% p.a. The fair price for delivery one year hence would be the cost of buying that asset, i.e. the current spot price, plus the cost of carrying the asset for one year. In the above example, the spot price is 100 and the carrying costs are 10% of 100, i.e. 10. Therefore, the fair futures price is 110. If there are no hindrances to arbitrage, the activities of arbitragers should keep the future at this fair price. If, for example, the future was trading above its fair price, say at 115, arbitragers would short the future, agreeing to deliver the underlying asset and receive a price of 115 at delivery. Simultaneously they would buy the asset for 100 with the aid of a bank loan paying 10 in interest. The asset would be held for one year and delivered into the futures contract, thus guaranteeing a risk-free profit, after one year, of 5. This arbitrage is known as **cash and carry arbitrage.**

If the future were cheap to its fair price, say 108, The arbitragers would engage in **reverse cash and carry**. They would buy the future, selling the underlying asset for 100 and depositing the sale proceeds to earn 10 over the year. At delivery, the arbitrager would take delivery of the underlying asset, paying 108, but after netting out interest earned the net cost is 98 compared with the current cost in the market of 100. A risk-free profit of 2 is made.

The principles of the above example can be generalized to pricing forwards or futures and if simple interest is assumed can be formally stated as

$$F = P[1 + (C * (T - t))] \tag{1.26}$$

If compound interest is assumed, equation (1.26) would become

$$F = P(1 + C)^{T-t} \tag{1.27}$$

where

    C = the cost of carry expressed as a rate and quoted in decimals
    $T - t$ = the time to maturity of the future or forward expressed in years.

At this stage it will be instructive to apply equation (1.26) to the valuation of a three-month forward or future.

Assume that the spot price of the underlying asset is 94.6. The interest rate on a three-month loan is 10% p.a., the custody costs are 0.02% p.a. charged quarterly, and the asset will earn income in a steady stream at the rate of 5% p.a.

From the above data the net carry costs are 5.02% p.a., i.e. (10 + 0.02 − 5), thus the futures price should be:

$$F = 94.6[1 + (0.0502 * 0.25)] = 94.6(1.01255) = 95.79$$

Now consider equations (1.26) and (1.27) applied to futures contracts on foreign currencies, bonds and stock exchange indices. In each case P is the spot price of the underlying asset, i.e. the current exchange rate, the bond price and the current index level. C, the net cost of carry, will differ in its make up for each of the types of asset considered. In the case of currency futures, the cost of borrowing funds to buy the foreign currency will be the domestic rate of interest, but that foreign currency can be invested at the foreign rate of interest until it is delivered into the futures contract. Custody costs for currency are non-existent. Thus C is the domestic rate of interest minus the foreign interest rate.

In the case of stock exchange indices, buying the shares in the index will entitle the holder to any dividends payable on the shares during the life of the future. Thus the net cost of carry, C, is the domestic rate of interest minus the dividends due, expressed as a percentage of the value of the index.

Similarly with bonds, they pay a rate of interest and thus C will be the domestic rate of interest minus the current or running yield on the bonds.

It is the **possibility of cash and carry arbitrage** that maintains the fair price relationship between the forward or future and the underlying security. This relationship is fundamental to the pricing of forwards and futures. There is a great number of different futures contracts traded in a number of the world's financial markets. For a more detailed analysis, interested readers are referred to Watsham (1992).

Returns from holding securities

We now turn to a discussion of the various measures of return that are used by investors. Returns can be historical returns, i.e. actually achieved, or they can be expected returns, i.e. hoped to be achieved in the future. It is expected returns that are important in the investment process. Historical returns are important only in as much as they are a guide to the future.

The holding period return

The **holding period return** (HPR) is the total return from an investment over a given holding period. It encompasses all income and capital gains or losses over the holding period. The expected HPR is that which the investor hopes to achieve. The historical HPR is what he or she actually achieved. The method of calculating each is similar.

To calculate the HPR consider that

$$\text{HPR} = \frac{INC + V_T}{V_t} - 1 \tag{1.28}$$

where

INC = income received during holding period
V_T = value of investment at end of holding period

V_t = value of investment at beginning of holding period.

For example, if the initial investment cost $10, one dollar was received in dividends or interest and the value of the investment is $11 at the end of the holding period, the HPR is as follows:

$$\text{HPR} = (1 + r) - 1 = \frac{1 + 11}{10} - 1 = 1.2 - 1 = 0.2 = 20\%$$

The return of 20% relates to the complete holding period, not to any annual sub-periods. In other words it is not an annual return, except when the holding period is one year.

Now let us assume that the holding period has been a number years. Take the example of £1000 invested in a growth fund assuming all dividends had been reinvested. If today were the end of the holding period and the value of the fund were £1800 the total HPR would be

$$1 + r = \frac{1800}{1000} = 1.8$$

$$1.8 - 1 = 0.8 = 80\%$$

Assume that the 80% return was achieved over a holding period of five years. This total return over five years is made up of annual returns in the following functional form:

$$1 + \text{HPR} = (1 + r_1)(1 + r_2)(1 + r_3)(1 + r_4)(1 + r_5)$$

The HPR can be converted to an annual return of r by solving as follows:

$$r = \sqrt[T]{(1 + r_1)(1 + r_2)(1 + r_3) \dots (1 + r_T)} - 1 \tag{1.29}$$

where T is the number of annual time periods. In such circumstances the return would be an annual compound return. If the time period represented by T were reduced to say months then the results would be a monthly compound return.

Returning to the example above the annual compound rate would be

$$r = \sqrt[5]{1.8} - 1 = 0.1247 = 12.47\%$$

This is in fact the geometric average rate of return over the five-year period.

Yields on bank deposits

Eurocurrency deposits, domestic interbank deposits and certificates of deposit are traded on a yield basis. However, as the amount to be deposited is paid in full at the outset and the principal is repaid at maturity with **interest added on**, the yield on this type of instrument in known as an **add-on yield.** Moreover, the rates are quoted on a simple interest basis.

To illustrate this, assume that the 90-day sterling domestic interbank deposit rate is quoted at 10% p.a. The holding period return over the 90 days of the deposit's life is

$$\text{Add-on yield} = \left(0.1 * \frac{90}{365}\right) = 0.02466 = 2.466\% \qquad (1.30)$$

Yields on discount instruments

It has already been noted that **Treasury bills and commercial paper and bankers acceptances** are quoted in the market by yield and not by price. The quoted yield is known as a **discount yield,** which is the discount expressed as a percentage of **par or face value.** For example, assume that a Treasury bill with 90 days to maturity is priced at 97.5, i.e. at a discount of 2.5 from par, the discount yield is calculated as follows:

$$\text{Discount yield} = \left(\frac{\text{discount}}{\text{face value}}\right) * \frac{365}{n} = \frac{2.5}{100} * \frac{365}{90} = 10.1389 \qquad (1.31)$$

where n is the number of days to maturity.

Equation (1.31) has shown how the bill is valued from the discount yield.

As the market practice is to quote these yields on a simple interest basis and not an annualized compound basis, the 2.5/100 in the above example is multiplied by 365/90 rather than $[1 + (2.5/100)]$ raised to the power of 365/90.

It must be noted that the discount yield is not the yield that the investor enjoys. That yield is calculated as the discount divided by the **purchase price** of the bill, all multiplied by 365/n or 360/n, according to market convention. To determine the actual yield on Treasury bills and to make comparison with that on bank deposits it is necessary to convert the Treasury bill discount yield to an add-on yield.

For example, a **discount yield** can be converted to an **add-on yield** equivalent as follows:

$$\text{Add-on yield} = \left(\frac{\text{discount}}{\text{purchase price}}\right) * \left(\frac{365}{n}\right) \qquad (1.32)$$

A numerical example will illustrate this. If the discount is 2.5 the add-on yield will be

$$\frac{2.5}{97.5} * \frac{365}{90} = 10.3989$$

Dividend yield on equities and current or running yield on bonds

The dividend yield on equities measures the dividend as a percentage of the current share price. The dividend used may be the historical, i.e. dividends paid over the last year, or it may be the expected dividend. It is calculated simply as

$$\text{Dividend yield} = \frac{D}{P} \qquad (1.33)$$

where D is the historical or expected value of dividends according to choice. Usually the dividends are included gross of tax in order to avoid complication due to the different tax liabilities of different types of investors.

The current or running yield on bonds measures the expected annual coupon income as a percentage of the clean price, i.e. excluding accrued interest, of the bond and is calculated as follows:

$$Y_c = \frac{\text{Coupon}}{\text{Clean price}} \tag{1.34}$$

where

Y_c = current or running yield.

Neither of these concepts of yield take account of reinvestment of income nor the profit or loss on the bond at maturity (bond selling at discount or premium to purchase price) nor profit or loss on the share when sold.

Note that the practice, particularly in the UK, of using the clean price of the bond overestimates the actual running yield because when an investor buys a bond he or she has to pay the dirty price of the bond, i.e. the clean or market quoted price plus any accrued interest due. Accrued interest is the daily accrual of the coupon rate for the number of days since the last coupon was paid.

Yield to maturity or gross redemption yield

The yield to maturity is widely used by market practitioners as a measure of return and as a measure of relative value. It is the **internal rate of return (IRR)** that will discount the expected future cash flows of a bond to the current price of that bond. The **yield to maturity** is derived by solving the following equation for r through an iterative process

$$P_B = \sum_{i=1}^{n} \frac{CF_i}{(1+r)^i} \tag{1.35}$$

where

CF_i = the cash flow due at the end of time period i (coupon and redemption payment)
r = the yield to maturity.

Take as an example a five-year bond with a 10% coupon paid annually and a market price of 87.59 and a nominal value of £100:

$$\frac{10}{1.1358} + \frac{10}{(1.1358)^2} + \frac{10}{(1.1358)^3} + \frac{10}{(1.1358)^4} + \frac{110}{(1.1358)^5} = 87.59$$

The yield to maturity is calculated to be 13.58%. This figure is derived by an iterative process. To be able to calculate the IRR we need to know the current price as well as the pattern of cash flows. An explanation of the iterative processes used is given in Chapter 8.

Most bonds, including the vast majority of government bonds, pay coupons half yearly, in which case the CFs will refer to the semi-annual coupon paid in each period, and the is will be in half-yearly periods. Consequently, the yield to maturity as calculated by equation (1.35) would be a half-yearly yield. To annualize this yield, the market convention, particularly in the United States and the UK, is simply to double the half-yearly yield to maturity to arrive at the **bond equivalent yield**. This method actually understates the effective yield. For example, assume a bond is priced at par (i.e. 100% of its face value) and paying a coupon of 10% p.a. in semi-annual coupons of 5. The half-yearly yield to maturity would be 5%. Annualized by doubling it would be 10%. However, if we assume that the first coupon in each year could be reinvested for the second half of each year at 10% p.a. compounded half yearly the **effective** yield to maturity would be 10.25% p.a.

The above example assumed that a coupon had just been paid and the next coupon was a whole coupon period in the future. In practice it may be necessary to calculate the yield to maturity between coupon dates. This is achieved by solving the following equation:

$$P_B = \sum_{i=1}^{n} \frac{CF_i}{(1+r)^v (1+r)^i} \tag{1.36}$$

where

v = the proportion of the period between coupons represented by the number of days until the next coupon.

There is an exception to the need to solve equations (1.35) and (1.36) iteratively and that is when only one single payment is due as for example in the case of a zero coupon bond. Then the yield is calculated as

$$Y = \left(\frac{CF_n}{P_z}\right)^{1/n} - 1 \tag{1.37}$$

where

CF_n = the cash flow at redemption
P_z = the current price of the bond.

To illustrate this consider a zero coupon bond redeemable at par in two years' time and currently selling for 80. The yield to maturity is given as

$$Y = \left(\frac{100}{80}\right)^{1/2} - 1 = 11.8\%$$

This will be recognized as similar to equation (1.15). This is because the yield to maturity of a zero coupon bond is in fact a spot rate.

It was noted above that the yield to maturity is often used to compare the promised yields on alternative bonds. However, it has some serious drawbacks in this respect. In particular, the actual return to the investor will only equal the yield to maturity if:

(a) the bond is held until maturity and
(b) all the coupon receipts are reinvested at a rate equal to the yield to maturity until the bond is redeemed.

Clearly, not all investors will expect to hold the bond until maturity and certainly it is unlikely that all the coupons can be reinvested at that same rate throughout the life of the bond.

Moreover, the yield to maturity tells us nothing about the relative value of a bond. To calculate it we need to know the market price of the bond, the contracted cash flows and the timing of those cash flows. Thus we need to know the price of the bond before we can compute the yield to maturity, therefore it cannot help us value the bond.

It should also be noted that the yield to maturity is coupon specific. To illustrate this, consider the yield to maturity on a five-year bond with a 6% annual coupon. The discounted value of the cash flows using the same spot rates used earlier to value the 10% coupon bond is

$$P_B = \frac{6}{1.1} + \frac{6}{(1.11)^2} + \frac{6}{(1.12)^3} + \frac{6}{(1.13)^4} + \frac{106}{(1.14)^5} = 73.33$$

Discounting the cash flows to that same price gives a yield to maturity of 13.72% p.a.

$$\frac{6}{1.1372} + \frac{6}{(1.1372)^2} + \frac{6}{(1.1372)^3} + \frac{6}{(1.1372)^4} + \frac{106}{(1.1372)^5} = 73.33$$

Compare that with the yield to maturity on a five-year bond paying annual coupons of 10 calculated on page 19 above. The present value of the bond paying coupons of 10 is 87.59, using the same spot rates as above, and its yield to maturity is 13.58%, whereas on the bond paying coupons of 6 it is 13.72. Yet both streams of coupons are payable at the same future points in time.

Consequently, not only does using the yield to maturity mean that the same-sized cash flows due at different future points in time are discounted at the same rate, but different-sized cash flows due at the same point in time in the future are discounted at different rates. Both these points are contrary to financial theory and practice, the first because we never observe flat term structures of interest rates, the second because if two cash flows of the same quality were priced differently in the same market, traders would engage in arbitrage by buying the cheap cash flow and selling the expensive one, making a risk-free arbitrage profit. Consequently, the yield to maturity cannot meaningfully be used to make comparisons between bonds.

Realized compound yield

Given the conceptual weaknesses of the yield to maturity as a summary measure of the bond an alternative measure is required which will take account of the investor's desired holding period and the investor's assumptions of the rates at which the coupons can be reinvested.

When an investor wishes to hold an asset and fully compound the income to a future date, the **realized compound yield** is the appropriate yield measure for comparing between alternative interest bearing assets.

Unlike the previous measures of yield, the **realized compound yield (RCY)** includes the influence of "interest on interest" in the terminal value of the investment. However, because of that, the actual yield calculated is crucially dependent on the assumed reinvestment rate. Although the future reinvestment rates will be uncertain, they must be explicitly stated to facilitate calculation of this measure of yield. Consequently, providing expected reinvestment rates are consistent between instruments, meaningful comparisons can be made.

When the desired holding period is shorter than the life of the bond an assumption regarding the level of bond yields at the time of sale will also be required. If the bond has a life shorter that the desired holding period, an assumption about the reinvestment of the redemption proceeds must be made.

The **realized compound yield** is the compound rate of growth in total value during the holding period expressed as an annualized rate of interest. Recalling the calculation of the **annualized holding period return** it will be seen that it is the same as the realized compound yield. In fact both measures can be derived in their *ex ante* form in which case the result would be more accurately described as the expected realized compound yield, or expected holding period return.

The steps required to calculate the **realized compound yield** are:

1. Calculate the total coupon payments and the interest upon reinvestment. This latter may be derived using the forward yield curve to represent consensus expectations of future interest rates.
2. Assume the price of the bond at the end of the holding period. This may or may not be equal to the redemption price, depending on whether the holding period is shorter than the life of the bond.
3. Derive the terminal value of the investment by summing (1) and (2) above and solve as follows:

$$\text{RCY} = \left(\frac{TV}{P}\right)^{1/n} - 1 \qquad (1.38)$$

where

TV = terminal value of investment
P = the original purchase price of the bond
n = the number of coupon paying periods. If the coupons are paid annually the result will be an annual RCY. However, if coupons are paid half yearly, the result will be a semi-annual RCY. In the latter case the result must be doubled to arrive at an annual bond equivalent RCY.

To illustrate the calculation of the RCY, consider a five-year bond paying an 8% coupon annually with the first coupon due in exactly one year, and trading at 80.46. In addition assume that the forward rates are as given in Table 1.3 above.

The periodic coupons are assumed to be reinvested at the appropriate forward rates as follows:

```
8 * 1.1275 * 1.1504 * 1.1734 * 1.1971 = 14.5758
8 * 1.1504 * 1.1734 * 1.1971         = 12.9275
8 * 1.1734 * 1.1971                  = 11.2374
8 * 1.1971                           = 9.5768
Final payment and coupon            = 108.0
```

Total   156.3175

The RCY is calculated as

$$\text{RCY} = \left(\frac{156.3175}{80.45}\right)^{\frac{1}{5}} - 1 = 0.142050 \approx 14.21\%$$

Annualized yields: the usefulness of continuous compounding

Both the annualizing of periodic simple interest yields by simple multiplication and the method of compound annualizing implicitly **assume that it is possible to reinvest the deposit and interest at the current rate periodically over the next year!** For example, a six-month deposit earning 6% p.a. earns 30 per 1000 of principal. That same principal will only earn 60 over the whole year if it is invested at 6% p.a. for the second six-monthly period. Thus annualizing the 30 to be a return of 6% is only correct if:

1. the interest is not compounded
2. the deposit is again placed for six months at 6% p.a.

This clearly may not be the case. For example, the original principal and interest may be reinvested for the second six months at 6% p.a. in which case the total interest earned over the year will be 60.90 or 6.09% p.a. Alternatively, the funds, with or without interest, may be reinvested for a longer or shorter period, and or higher or lower rates, in which case the interest earned over the year will again be different. Consequently, the yields on instruments of different maturities are not really comparable because they assume a different number of implicit reinvestment periods at the current yield.

Therefore it is useful to transform yields to a common form for comparison. This is best achieved by converting market yields to **continuously compounded add-on yields** on a 365.25-day year.

To achieve this comparison, firstly, the yields on instruments traded on a 360-day calendar basis must be converted to a 365.25-day year basis. Thus the actual interest expected to be earned during the holding period must be calculated in the following manner:

$$\text{Interest} = R * (n/360) * \text{principal}$$

where R is the quoted interest rate on the instrument and n is the actual number of days from settlement until maturity.

To illustrate this, assume a deposit of \$1 million, at 10% p.a., and placed for 90 days. The expected interest receivable will be

$$(0.1)(90/360) * \$1\,000\,000 = \$25\,000.00$$

The continuously compounded add on yield is then calculated as follows:

$$R = \frac{365.25}{n} * \ln\left[\frac{\text{principal} + \text{interest}}{\text{principal}}\right]$$

$$= \frac{365.25}{90} * \ln\left[\frac{1\,025\,000}{1\,000\,000}\right] = 10.0211\%$$

(1.39)

Note that this is above the 10% rate of simple interest despite being a continuously compounded rate, because the simple interest was based on a 360-day year.

Converting yields to the continuous compounding basis allows consistent comparison between instruments traded according to different yield conventions and different maturities. However, it does assume that the current interest rate is constant over the compounding period.

The APR (annual percentage rate). An alternative method of standardizing the effective rate of interest is the annual percentage rate. The APR is defined to be the equivalent annual percentage rate of interest if periodic compounding occurred only once annually. Thus a credit card company charging 2% per month on outstanding balances would be using a nominal annual rate of 24% but would be compounding at monthly intervals. The equivalent APR would be computed as $(1.02)^{12} - 1 \approx 0.268$, and the company would quote an APR of 26.8%.

Continuously compounded asset returns. The assumption of continuously compounded asset returns has been incorporated into many financial models including those used for pricing options. We have already seen how to convert simple and discretely compounded rates of interest to their continuously compounded equivalents. However, it is often necessary to calculate the continuously compounded return from asset prices.

To do this, one simply takes the natural logarithm of the price relative. For example, if the price of a share today is 10 and the price one week earlier was 8, the continuously compounded weekly return would be

$$R_{cc} = \ln \frac{P_1}{P_0} = \ln \frac{10}{8} = \ln 1.25 = 0.22314 \simeq 22.3\%$$

where

P_1 is the current price
P_0 is the price at the end of the previous time period.

This return relates to the period reflected in the data, in this example one week. However, if daily or monthly data were used, the return would be the continuously compounded daily or monthly return and so on.

When using prices to derive asset returns it is assumed that no interest, dividend or other payment has been received. If such a payment has been received during the period bounded by the price observations it must be added to P_1.

The nature of the term structure of interest rates

Many readers will be familiar with the concept of the **yield curve**: it is a plot of yields-to-maturity against the term to maturity. As coupon paying bonds are the most numerous long-term debt instrument, yield curves usually reflect the yield to maturity of such bonds. Yield curves are frequently published in the financial press and elsewhere, and sometimes are depicted as representing the term structure of interest rates; however, the **term structure of interest rates** actually relates the **spot rates** to the term to maturity.

As we noted earlier, the spot rate can be found in the market place as the yield to maturity of a zero coupon bond. Unfortunately, the complete term structure is not observable because of the insufficient issue of zero coupon bonds. It is, therefore, necessary to derive spot rates from other instruments, for example from the add-on yields of short-term bank deposits, the yields on Treasury bills and the prices and expected cash flows of default-free coupon paying bonds. This is demonstrated below.

Consider the information in Table 1.4 relating to a one-month interbank deposit, a three-month Treasury bill and a number of bonds with maturities from 0.5 to 3 years. The six-month and one-year bonds are zero coupon bonds; the bonds with maturities of 1.5–3 years are coupon paying bonds.

From the information in the first three columns of Table 1.4, it is possible to derive the spot yield curve.

The first thing to do is to convert the add-on yield of the bank deposit to an appropriate spot rate. We must therefore account for the differing day count conventions in the money markets while using equation (1.30). Then we must adjust the monthly add-on return to an annual spot rate for the period in question.

To understand this consider a 100 deposit earning 7.25% p.a. for one month. Assuming a 31-day month, and an actual over a 365-day count convention, the add-on yield would be

$$7.25 * \frac{31}{365} = 0.6157$$

Table 1.4

Maturity (years)	Coupon	Price	Yield to maturity	Spot rate	Fwd rate 0.5 years
1 month					
0.0833	7.25	100	–	7.644	
0.25	zero	98.1	7.98	7.975	
0.5	zero	96.18	8.10	8.101	
1.0	zero	92.21	8.45	8.448	8.796
1.5	8.5	99.18	9.30	9.347	11.166
2.0	9.0	99.37	9.57	9.624	10.461
2.5	11.0	103.16	9.78	9.853	10.771
3.0	9.5	99.11	10.09	10.187	11.878

Thus the terminal value of the deposit will be 100.6157. However, to earn that terminal value after one month, we would have to invest 100 today. Thus the equivalent annual spot rate can be derived using equation (1.15) which is repeated below

$$SR_n = \left(\frac{CF_n}{PV}\right)^{\frac{1}{n}} - 1 \qquad (1.40)$$

Thus the one-month spot rate is

$$SR_{0.0833} = \left(\frac{100.6157}{100}\right)^{\frac{1}{0.08333}} - 1 = 0.07644 = 7.644\%$$

Note that the annual equivalent spot rate is greater than the quoted one-month rate. This is because the spot rate assumes no monthly compounding.

As the next three securities pay no coupons, their yields to maturity are also the appropriate spot rates for those maturities, so no further calculation is required.

However, the instruments with maturities of 1.5–3 years are coupon paying bonds, and to derive the appropriate spot rate we have to do some manipulation.

The spot rate for a payment due in 1.5 year's time can be calculated as follows. Recall that the price of the bond is the present value of all the expected future cash flows, with each cash flow discounted at its appropriate spot rate. Thus the price of the 1.5-year bond is derived by discounting the first half year coupon by the six-month spot rate, discounting the one-year coupon by the one year spot rate, and by discounting the redemption payment and the final coupon by the 1.5-year spot rate

$$\frac{4.25}{(1.081)^{0.5}} + \frac{4.25}{(1.0845)} + \frac{104.25}{(1 + SR_{1.5})^{1.5}} = 99.18$$

Our current problem is that we do not know the 1.5-year spot rate. However, we can derive it by subtracting the present values of all the coupons, except the final payment, from the price of the bond, thus

$$99.18 - \left(\frac{4.25}{(1.081)^{0.5}} + \frac{4.25}{(1.0845)}\right) = 91.17$$

The resulting sum of 91.17 is the present value of the coupon and redemption payment, 104.25, discounted at the spot rate for payments due in 1.5 years from now. Therefore, as $91.17 = 104.25/(1 + R)^{1.5}$, it follows that $(1 + R)^{1.5} = 104.25/91.17 = 1.1435$. From this it follows that $(1.1435)^{1/1.5} - 1 = 0.0935$. This is a 1.5-year spot rate. We can check that this is correct by discounting 104.25 by $(1.0935)^{1.5}$ to get 91.17.

With this information we can repeat the process to determine the two-year spot rate. We use the three known spot rates to discount the first three of the four coupons due on the two-year bond and deduct the present value of these coupons from the current dirty price of the bond. The result is the present value of the final coupon plus redemption payment discounted at the two-year spot rate. The two-year spot rate can then be derived as above.

This procedure, sometime referred to as the "bootstrap" method, can be followed until all the required spot rates are calculated. However, some caveats are in order. Firstly, only coupon paying bonds without any embellishments such as embedded options can be used, otherwise the market prices will represent something different to the straight cash flow due under the bond indenture. In addition where capital gains and coupon income are taxed at different rates, an adjustment is necessary to take account of the tax benefits of low coupon bonds. This problem may be avoided by using only recently issued bonds, but these may not cover all sections of the maturity spectrum.

This **bootstrap** method is rather cumbersome when a large number of coupon bonds with similar maturities but differing characteristics are trading. Consequently a number of complex econometric techniques have been applied to modelling the term structure with differing degrees of success. Once the term structure is being modelled, the model output is only an estimate of the true term structure. Consequently, the resulting spot rates are only estimates of the true spot rates. As a consequence, the bond prices resulting from the use of the estimated spot rates are themselves only estimates of the true bond prices.

Mortgages and annuities

Mortgages. It is a feature of housing mortgages that when a given interest rate is set, the periodic payments are fixed equal amounts throughout the remaining life of the mortgage. If interest rates are reset periodically as they are with floating rate mortgages, the periodic payments will change only when the interest rates change, but will still be equal in each period until the next interest rate reset. The periodic (typically monthly) payment consists of both principal and interest. In the early days of the mortgage when the principal is at its greatest, the monthly payment consists mostly of interest with only a small proportion devoted to repaying principal. As the principal is reduced the proportion of the monthly payment required to cover the interest will fall and therefore the proportion of the periodic payment allocated to principal increases. Each time the interest rate is changed, it is necessary to recalculate the periodic payment which covers the interest and repay the principal over the remaining life of the mortgage.

To illustrate how that is done, consider first of all calculating the equal annual payments that would be required to pay off a mortgage loan of £100 000 with an interest rate of 10% p.a. (annual compounding) over a 20-year period.

Starting with a debt of £100 000 we can see that the first year's interest can be applied to this by multiplying by the compounding factor of 1.10, after which the balance can be reduced by the annual repayment. If we let that repayment be £X we can see that the debt at the start of year 2 is $(100\ 000 * 1.10) - X$.

Applying that same reasoning again to the current debt, $(100\,000 \times 1.10) - X$, we get

$$((100\,000 * 1.10) - X) * 1.10 - X = 100\,000 * 1.10^2 - 1.10X - X$$

This is the debt at the start of year three.

If we repeat this 20 times we will produce an expression for the debt at the start of year 21, which we require to be zero

$$100\,000 \times 1.10^{20} - 1.10^{19}X - 1.10^{18}X - 1.10^{17}X - \dots - 1.10^2X - 1.10X - X = 0$$

$$100\,000 \times 1.10^{20} - X(1.10^{19} + 1.10^{18} + \dots + 1.10^2 + 1.10 + 1) = 0$$

Now, reversing the order of the expression in brackets, $1 + 1.10 + 1.10^2 + 1.10^3 + \dots + 1.10^{18} + 1.10^{19}$ is an example of a geometric progression with a first term of 1 and a common ratio of 1.10. The formula for the sum to n terms of a geometric progression where the common ratio is greater than 1 is given by

$$\frac{1(1.10^n - 1)}{1.10 - 1} \tag{1.41}$$

This can be generalized to

$$a * \frac{[(1 + r)^n - 1]}{1 + r - 1} \tag{1.42}$$

where a is the "first term" and $1 + r$ is the common ratio, which is greater than one since r is positive.

Applying this to our example gives

$$1 * \frac{[1.10^{20} - 1]}{1.10 - 1} \approx 57.275$$

So our equation to calculate the regular mortgage payment, X, reduces to

$$100\,000 * 1.10^{20} - 57.275X = 0$$

and this gives $X = 11\,745.96$.

This generalizes to

$$X = \frac{P(1 + r)^n r}{((1 + r)^n - 1)} \quad \text{i.e.} \quad \frac{100\,000(1.10)^{20}0.10}{(1.10)^{20} - 1} = 11\,745.96 \tag{1.43}$$

Of course, most of us pay our mortgages in monthly instalments rather than by annual payments. To compute the monthly payment for the above mortgage we can use exactly the same approach with 240 monthly steps, provided that we convert the annual interest rate to an equivalent monthly rate. If we let the equivalent monthly compounding factor be C, then C is defined by $C^{12} = 1.10$, so that $C = \sqrt[12]{1.10} \approx 1.007974$.

Our equation for X will then be

$$100\,000 * C^{240} - \frac{C^{240} - 1}{C - 1} X = 0$$

giving

$$X = \frac{100\,000 * 1.007974^{240} * (1.007974 - 1)}{1.007974^{240} - 1} = 936.64$$

(Note that this amounts to some £500 less per annum compared to paying annually, representing saved interest charges.)

This generalizes to the formula for a monthly mortgage repayment as

$$X = \frac{P\left(\sqrt[12]{1+r}\right)^{12n}\left(\sqrt[12]{1+r} - 1\right)}{\left(\sqrt[12]{1+r}\right)^{12n} - 1} \qquad (1.44)$$

where P in the amount borrowed, r is the annual rate of interest and n is the number of years.

Annuities. In the example of a mortgage, the client receives a sum of money at the beginning of the mortgage period, and repays that sum with a series of small cash outflows during the period. However, with an annuity the cash flows are reversed. The client makes an initial payment and in return receives a series of small cash inflows during the period of the annuity. Thus an annuity is a reverse mortgage (or a mortgage is a reverse annuity).

Not surprisingly, the mathematics of annuities is much the same as that of mortgages. Consider, for example, an annuity paying £1200 p.a. over a 20-year period, with the rate of return being set at 10%. What would be the cost of such an annuity?

The current value of the first payment, which we will take as occurring one year in the future, is 1200/1.10, since this is the sum of money which will grow to £1200 in one year at 10%. Similarly the current value of the second payment is $1200/1.10^2$. Thus the sum of the present values of all 20 payments is

$$P = \frac{1200}{1.10} + \frac{1200}{1.10^2} + \frac{1200}{1.10^3} + \dots + \frac{1200}{1.10^{20}}$$

This is an example of a geometric progression, but this time the first term (a in the formula) is 1200/1.10 and the common ratio is 1/1.10. The common ratio is the amount by which the previous expression has to be multiplied by to arrive at the current expression, for example

$$\frac{1200}{1.10^2} * \frac{1}{1.10} = \frac{1200}{1.10^3}$$

The formula for a geometric progression where the common ratio is less than one, i.e. $1/(1 + r)$ or 1/1.10 in our example is

$$a\frac{\left(1 - \frac{1}{(1+r)^n}\right)}{1 - \frac{1}{(1+r)}} \qquad (1.45)$$

where $a = 1200/(1 + r)$. We can multiply the numerator and the denominator by $1 + r$ to get

$$1200\left(\frac{1-\dfrac{1}{(1+r)^n}}{r}\right) \qquad (1.46)$$

Thus the value of an annuity paying 20 annual payments of 1200 with a discount rate of 10%, will be

$$1200\frac{1-\dfrac{1}{1.10^{20}}}{0.10} = 1200\frac{1-0.1448644}{0.10} = 10\ 216.28$$

Since the common ratio is less than one, this opens the intriguing possibility of allowing n, the number of payments, to tend to infinity ($\to \infty$), or at least for it to become very large. This is because, as $n \to \infty$, $1/(1+r)^n \to 0$.

Thus we have the concept of a perpetual annuity, whose value is given by

$$\frac{a}{1-\left(\dfrac{1}{1+r}\right)}$$

With $a = 1200/1.10$ and the common ratio $= 1/1.10$, this gives the value 12 000. In practice it would be quite normal to compute the value of a lifetime annuity, the value of which (depending on the age of the client) may not be far distant from the value of the perpetual annuity.

Monthly payments. As in the case of a mortgage we can compute the cost of providing for 240 monthly payments each of £100. Again this will require us to use a compounding factor given by

$$C = \sqrt[12]{1.10} = 1.007974$$

The expression will be

$$\frac{100}{1.007974}\frac{\left(1-\dfrac{1}{1.007974^{240}}\right)}{\left(1-\dfrac{1}{1.007974}\right)} = 99.2063\frac{1-0.148644}{1-0.99209} = 10\ 676.47$$

giving 10 676.47. That this is more expensive reflects the fact that the cash is effectively being received earlier.

Exercises

1. What factors contribute to the determination of:
   (a) the general level of interest rates
   (b) the term structure of interest rates
   (c) the quality structure of interest rates?

2. Calculate the interest payable on a one-year loan of 10 000 if the interest rate is 6% p.a., and:
   (a) the interest regime is simple interest
   (b) compounding is half yearly
   (c) compounding is quarterly
   (d) compounding is monthly
   (e) compounding is daily
   (f) compounding is continuous?

3. What is the future value of an instrument that pays no income, is currently priced at 25 000 and delivery is due in 210 days? The current 210-day interest rate is 7.5% p.a. Assume a 365-day market.

4. Convert 10% p.a. continuously compounded to the equivalent half yearly compounded rate and the equivalent quarterly compounded rate.

5. From the following zero coupon bond prices determine the periodic **continuous compounded** and **discrete compounded** spot rates and implied three-month forward rates.

   Zero coupon details:

Maturity (years)	Price
0.25	98.4
0.5	96.5
0.75	94.5
1.0	92.4
1.25	90.5
1.5	88.5

6. From the three-month maturity and six-month maturity bond prices in question 5, calculate the simple interest rates that give the equivalent yields and then calculate the three-month forward rate implied by those simple interest rates.
   Assume an actual over 365-day market with each quarter having 91 days.

7. Calculate the effective rate of interest on a 5% p.a. certificate of deposit trading in an actual over 360-day money market.

8. Calculate the current value of a DM5 million 4.5% p.a. CD, initially issued with a life of 270 days and now having a residual maturity of 160 days. The 160-day spot rate is currently 4.3% p.a.

9. A security has a spot price of 102.5, the annual rate of interest is 6% p.a., income is flowing at a steady rate of 4% p.a. and custody costs are three basis points. (A basis point is 100th of 1%.)

Calculate the price of a six-month forward contract assuming:
(a) simple interest
(b) compound interest.

10. Critically evaluate the use of the gross redemption yield as a measure of relative value of bonds. Illustrate your answer with numerical examples where appropriate.

11. Develop a numerical example to illustrate the application of the realized compound yield.

12. Using the data from question 5 as appropriate, use the bootstrap method to calculate the two-year spot rate from a 10% coupon bond paying coupons semi-annually (i.e. 5 per six months) and currently priced at 103.0 and with two years to maturity.

13. Calculate the monthly payment on a $150 000 mortgage with 20 years to final payment and currently attracting an interest rate of 4.5% p.a.

14. Twenty-year annuities assume interest rates of 6% p.a. What will be the price of an annuity paying 1500 per month.

15. What would be the price of the annuity in question 14, if it were to be a perpetuity?

Answers to selected questions

2. Simple interest   600
   Compounding:

	2	609
	4	613.6355
	12	616.7781
	365	618.3129
	continuous	618.3655

3. 26078.77

4. 10.2542
   10.1260

5.

Spot rates		Forward rates	
DC	CC	DC	CC
6.6644	6.4518		
7.3854	7.1254	8.1113	7.7991
7.8345	7.5427	8.7382	8.3773
8.2251	7.9043	9.4055	8.9891
8.3131	7.9856	8.6658	8.3109
8.4854	8.1445	9.3506	8.9389

6. 6.50; 7.25; 7.88

7. 5.06944

8. 5071821.8

9. 103.5404; 103.5351

12. 8.5382

13. 941.6839

14. 212076.5

15. 308163.2086

References

Anderson, N., Breedon, F., Deacon, M., Derry, A. and Murphy, G. (1996) *Estimating and Interpreting the Yield Curve*. John Wiley, Chichester (particularly chapters 1 and 7–9).

Fabbozzi F. (1993) *Fixed Income Mathematics*. Probus Publishing.

Fuller, R. J. and Hsia, C. C. (1984) A simplified common stock valuation model. *Financial Analysts Journal*, September/October, pp. 49–55.

Gordon, M. J. (1962) *The Investment Financing and Valuation of the Corporation*. Richard D. Irwin, Homewood, Ill.

Watsham, T. J. (1992) *Options and Futures in International Portfolio Management*. Chapman & Hall, London.

Watsham, T. J. (1993) *International Portfolio Management: A Modern Approach*. Longman, London.

Williams, J. B. (1938) *The Thory of Investment Value*. Harvard University Press, Cambridge, MA.

2

Presentation of data and descriptive statistics

Introduction

The term "descriptive statistics" is a generic term applied to a group of summary measures which each summarize, in one number, a certain quality of a group of data. For example, in the analysis of financial markets it is frequently necessary to calculate some form of average of the individual returns of a group of assets or an average price of those assets over a given period. In addition we frequently need to measure how spread out or dispersed those returns or prices are. We may also need to know whether the returns or prices are symmetrically distributed around the mean value, whether the returns are skewed or whether the distribution is more or less peaked than expected.

Statisticians refer to the averages as **measures of location** or **measures of central tendency**. The statistics that describe or measure how spread out the data are around the mean are known as **measures of dispersion**. Statistics that describe the symmetry or otherwise of the data are known as **measures of skewness** and those that describe the peakedness of the data are known as measures of **kurtosis**.

Types of data

Data come in various forms and sometimes the form of that data dictates, or at least influences, the choice of method to be used in the statistical analysis. Therefore, before we begin to investigate the various techniques of statistical analysis, it is appropriate to consider different forms of data.

Continuous and discrete data

Data may be classified as **continuous** or **discrete**. Continuous data are those data that can take on any value within a continuum, i.e. the data are **measured** on a continuous scale and the value of that measurement is limited only by the degree of precision. A typical example is the percentage rate of return on an investment. It may be 10%, or 10.1% or indeed 10.0975%. Data relating to time, distance and speed are other examples of continuous data.

Discrete data are data which results from a process of counting. For example, financial transactions are discrete in that half or quarter transactions have no rational meaning. Data relating to asset prices may be discrete due to rules set by individual markets regarding price quotations and minimum price movements. For example, in the UK, government bonds are quoted in thirty-seconds (1/32) of a point or pound. Consequently price data are discrete, changing only by thirty-seconds or multiples thereof. Markets for bonds, equities, futures and options are other examples where minimum price change rules cause the data to be discrete.

We will see later that the distinction between continuous or discrete data sometimes has an influence on the way that we calculate the descriptive statistics.

Cross-sectional and time-series data

Cross-sectional data represent the situation of a group of variables at any one point in time. For example, data relating to, say, the price of each of the 100 constituents of the FTSE 100 equity index at a particular time will be cross-sectional data. Lists of share prices, interest rates or exchange rates published in the business pages of newspapers are also examples of cross-sectional data because the data relate to the prices or rates of a number of variables (shares or currencies, etc.) at a particular point in time.

Time-series data reflect changes over time of one particular variable. For example, data giving the price of a share, a currency exchange rate, or the level of an index each day (or week or month) for two years, would be a daily (or weekly or monthly) time series.

Grouped and ungrouped data

When only a small number of data are being processed, those data can be left in a raw or **ungrouped** state yet the reader can still get the feel of the data. For example, the following data which represents the monthly observations of the FTSE 100 index over a 12-month period are just about manageable:

2407.5, 2289.2, 2160.1, 2311.1, 2422.7, 2345.8,
2238.4, 2221.6, 2117.9, 2391.4, 2372.0, 2339.0

However, now consider the data in Table 2.1 which give the end of month levels, and the monthly returns of the FTSE 100 index from September 1989 to December 1993. As

Table 2.1

	FTSE 100	Returns
September 1989	2407.5	
	2289.2	−5.0386
	2160.1	−5.8048
	2311.1	6.7569
January 1990	2422.7	4.7159
	2345.8	−3.2256
	2238.4	−4.6865
	2221.6	−0.7534
	2117.9	−4.7803
	2371.4	11.3055
	2372	0.0253
	2339	−1.4010
	2166.6	−7.6564
	2030.8	−6.4729
	2028	−0.1380
	2162.7	6.4307
January 1991	2143.5	−0.8917
	2165.7	1.0304
	2386.9	9.7252
	2456.5	2.8742
	2508.4	2.0908
	2515.8	0.2946
	2443.6	−2.9118
	2591.7	5.8842
	2679.6	3.3353
	2645.6	−1.2770
	2549.5	−3.7001
	2414.9	−5.4239
January 1992	2493.1	3.1869
	2560.2	2.6558
	2554.3	−0.2307
	2408.6	−5.8733
	2659.8	9.9205
	2697.6	1.4112
	2493.9	−7.8515
	2420.2	−2.9998
	2298.4	−5.1637
	2572.3	11.2587
	2687.8	4.3923
	2792	3.8035
January 1993	2846.5	1.9332
	2851.6	0.1790
	2882.6	1.0812
	2878.4	−0.1458
	2813.1	−2.2948
	2849.2	1.2751
	2888.8	1.3803
	2941.7	1.8146
	3085	4.7564
	3039.3	−1.4924
	3164.4	4.0336
December 1993	3233.2	2.1509

the Stock Exchange calculates the index to only one decimal place, the levels data are discrete. There are 52 observations of the levels data and 51 observations of the returns data.

It is clear that when the data set is as large as it is here it needs to be summarized in a **frequency table** before the analyst can meaningfully comprehend the characteristics of the data. The resulting table gives the data in groups or class intervals. The example of **discrete grouped data** in Table 2.2 represents the FTSE 100 index levels.

The example of **continuous grouped data** given in Table 2.3 relates to the returns data in Table 2.1.

Table 2.2

Level of FTSE 100	Number of observations	Cumulative frequency
Up to 2000	0	0
2000.1–2100	2	2
2100.1–2200	6	8
2200.1–2300	4	12
2300.1–2400	6	18
2400.1–2500	9	27
2500.1–2600	7	34
2600.1–2700	5	39
2700.1–2800	1	40
2800.1–2900	7	47
2900.1–3000	1	48
3000.1–3100	2	50
3100.1–3200	1	51
3200.1–3300	1	52
	52	

Table 2.3

Monthly return	Number of observations	Cumulative frequency
Up to −8%	0	0
More than −8% and up to −7%	2	2
More than −7% and up to −6%	1	3
More than −6% and up to −5%	5	8
More than −5% and up to −4%	2	10
More than −4% and up to −3%	2	12
More than −3% and up to −2%	3	15
More than −2% and up to −1%	3	18
More than −1% and up to 0%	5	23
More than 0% and up to +1%	3	26
More than +1% and up to +2%	7	33
More than +2% and up to +3%	4	37
More than +3% and up to +4%	3	40
More than +4% and up to +5%	4	44
More than +5% and up to +6%	1	45
More than +6% and up to +7%	2	47
More than +7% and up to +8%	0	47
More than +8% and up to +9%	0	47
More than +9% and up to +10%	2	49
More than +10% and up to +11%	0	49
More than +11% and up to +12%	2	51
	51	

When grouping data, attention must be paid to the class intervals. Firstly, they should not overlap. Secondly, they should be of equal size unless there is a specific need to highlight data within a specific "sub-group", or if data are so limited within a group that they can safely be amalgamated with a previous or subsequent group without loss of meaning. Thirdly, the class interval should not be so large as to obscure interesting variation within the group. Lastly, the number of class intervals should be a compromise between the detail which the data are expected to convey and the ability of the analyst to comprehend the detail.

Presentation of data

This section will look at the following ways of presenting data:

- frequency distributions
- relative frequency distributions
- cumulative frequency distributions or ogives
- histograms.

Frequency distributions

In the previous section raw data were summarized in a frequency table. In this section we will construct frequency distributions to display the "frequency" of the various observations graphically.

The raw data for the returns series were transformed into a frequency table, which has class intervals with a width of 1% ranging from −8% to +12%. These data will be used to plot **frequency distributions, cumulative frequency distributions, relative frequency distributions and histograms**. The data for the frequency distributions are given in Table 2.4.

Table 2.4

Percent returns	frequency	Relative frequency	Cumulative frequency
−8	0	0	0
−7	2	0.039215	2
−6	1	0.019607	3
−5	5	0.098039	8
−4	2	0.039215	10
−3	2	0.039215	12
−2	3	0.058823	15
−1	3	0.058823	18
0	5	0.098039	23
1	3	0.058823	26
2	7	0.137254	33
3	4	0.078431	37
4	3	0.058823	40
5	4	0.078431	44
6	1	0.019607	45
7	2	0.039215	47
8	0	0	47
9	0	0	47
10	2	0.039215	49
11	0	0	49
12	2	0.039215	51

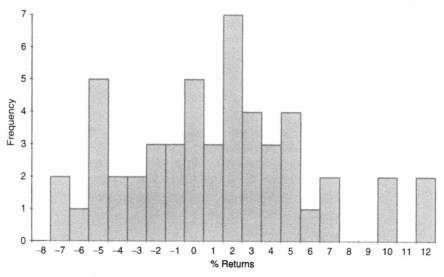

Figure 2.1 Frequency distribution of monthly FTSE 100 returns for September 1989 to December 1993.

The frequency distribution

To plot the frequency distribution, the frequency of observation is set on the vertical axis and the class interval (the returns) is set on the horizontal axis. Each of the frequencies is then plotted and the frequency chart is produced as in Fig. 2.1.

The height of each bar represents the frequency of observation within each class interval.

The relative frequency distribution

To derive the relative frequency of a given group of data the number in that group must be divided by the total number of data points. The relative frequency distributions can then be plotted in a similar manner to the frequency distribution, producing a chart as shown in Fig. 2.2.

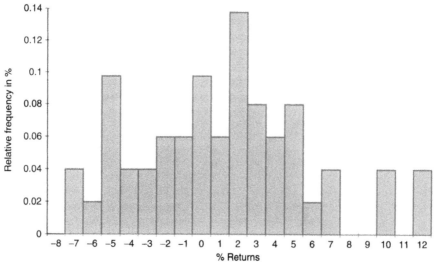

Figure 2.2 Relative frequency distribution of monthly FTSE 100 returns for September 1989 to December 1993.

The height of each bar represents the relative distribution of the particular change in yields.

Cumulative frequency distribution or ogive

To construct cumulative frequency distributions the cumulative frequency is plotted on the vertical axis and the class intervals are plotted along the horizontal axis.

Ogives are constructed by plotting the data in ascending order, and in the case of grouped data, plotting against the upper value of the class interval. This is illustrated in Fig. 2.3.

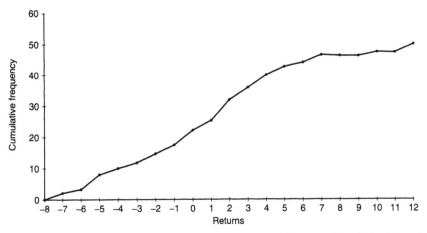

Figure 2.3 Cumulative frequency distribution of monthly FTSE 100 returns for September 1989 to December 1993.

The histogram

The next step in our thinking is to present the relative distribution, not in terms of the height of the bar but in terms of the area represented by each bar. This is known as a **histogram**. To illustrate a histogram consider the following grouped data

Returns	Frequency	Relative frequency
up to −0.04	3	0.015
more than −0.04 and up to −0.02	29	0.144
more than −0.02 and up to −0.01	191	0.946
more than −0.01 and up to 0.00	714	3.538
more than 0.00 and up to 0.01	816	4.004
more than 0.01 and up to 0.02	235	1.165
more than 0.02 and up to 0.04	29	0.144
more than 0.04 and up to 0.05	1	0.005
Total	2018	

The objective is to construct the histogram so that the area of each "column" represents the relative frequency of the distribution and the combined area represents 100% of the distribution. The horizontal axis represents changes in yield of 0.01%. However, two class intervals, "−0.04 and up to −0.02" and "+0.02 and up to +0.04" both represent 0.02%.

In addition we have the problem that the lower class interval is open ended. We will use our prior knowledge and assume that the lower class interval limit is −0.05. In order to ensure that the area of the columns above each class interval represents the relative frequency we must divide the relative frequency by the number of "units" in each class interval.

To illustrate this the 29 observations of returns "more than 0.02 and up to 0.04" represents a class interval of 0.02% whereas the others represent 0.01%. Consequently the height of the "bar" above this class interval is calculated as $(1.44/0.02) = 0.72$. This compares with the height of the "bar" for 235 observations for the preceding class interval. The class interval represents 0.01, and the height is thus $(0.1165/0.01) = 11.65$.

We can summarize this procedure by saying that when constructing a histogram a standard unit of class interval must be determined. Next the height of each "bar" is determined by dividing the relative frequency by the number of standard units in the particular class interval. Thus the height of "bars" relating to class intervals that are less than standard will be higher and those relating to class intervals that are wider will be lower. This is illustrated in Fig. 2.4 below.

A **polygon** is a curve that connects the mid-points of the top of each column. Such a curve is superimposed on Fig. 2.4. If we extend our thinking and consider the class intervals (the ranges of the yield changes) on the horizontal axis to be infinitesimally small, the polygon becomes a smooth curve, the area under the curve representing 100% of the occurrences of the yield changes observed. In Chapter 4 we will see such curves again – we will refer to them as probability densities.

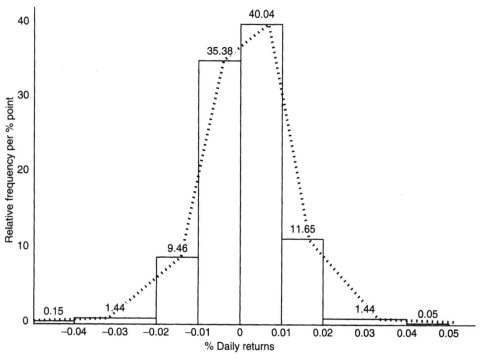

Figure 2.4

Descriptive statistics

Introduction

To provide summary measures of, say, asset returns, asset prices or numbers of financial transactions we use **measures of location or central tendency, measures of dispersion, measures of skewness and measures of kurtosis.** The averages give us a summary measure of the central or most common value found within the data. The measures of dispersion indicate how spread out the data are around that average. Measures of skewness indicate to what extent the frequency distribution has a tail to the left, negative skewness, or to the right, positive skewness. Measures of kurtosis indicate the extent to which the frequency distribution is peaked (leptokurtic) or flat (platykurtic).

At this point it is worth explaining the idea of **moments.** The kth moment about any origin A is given as

$$\frac{\sum_{i=1}^{n}(X_i - A)^k}{n} \tag{2.1}$$

If $A = 0$ and $k = 1$ we have what will be shown later to be the arithmetic mean. So the mean is sometimes said to be the first moment about zero. If A is itself the arithmetic mean and $k = 2$ we have the second moment about the mean, known as the variance, which is a measure of dispersion. If A is the mean and $k = 3$ we have the third moment about the mean which is a measure of skewness, and if $k = 4$ we have the fourth moment about the mean, which measures peakedness.

Measures of location or central tendency

You will be familiar with the concept of an average, whether it be the "average" size of a credit card balance, the "average" number of transactions per account, or indeed the "average" score per over in a cricket match.

In fact there are several measures of "the average" which are of particular interest to us in finance. These are:

- the mode
- the median
- the arithmetic mean
- the geometric mean.

Examples of each will be given below, both for grouped and ungrouped data.

The mode

The mode is simply the most frequently observed value of the variable being studied. To illustrate this, consider the following data which indicates the price of a particular penny share each day over a 15-day period:

10, 12,   9,   8, 10, 15, 14,   12
11, 10, 12, 12, 10, 12, 11.

The mode, being the most frequent observation, is 12.

The median

The median is the value of the observation which is halfway along the array of an ordered list of data, i.e. the middle observation.

The median from ungrouped data. Thus to derive the median from ungrouped data we must first place the data in ascending order. This is shown with the data used for the mode

8, 9, 10, 10, 10, 10, 11, 11, 12, 12, 12, 12, 12, 14, 15.

As there are 15 observations the median value will be the value of the eighth observation, i.e. 11.

If there were an even number of observations, there would be no "middle" observation, thus the middle two would be summed and divided by 2. The result might be a number not actually in the array of data.

The median from grouped discrete data. When grouped data are used we do not know the value of the individual observations. We therefore have to "estimate" the descriptive statistics. To do this we have to make assumptions about how the observations are distributed within each class interval. It is the general practice to assume that the data are evenly distributed throughout each class.

To estimate the median from grouped discrete data it is first necessary to construct a cumulative frequency distribution. This is achieved simply by adding the next frequency to the running total. To illustrate this consider again the data in Table 2.2 which is repeated here as Table 2.5.

To calculate the median number of observations we must first find the median observation by estimating the

$$\left(\frac{n+1}{2}\right)\text{th observation} \tag{2.2}$$

Table 2.5

Level of FTSE 100	Number of observations	Cumulative frequency
Up to 2000	0	0
2000.1–2100	2	2
2100.1–2200	6	8
2200.1–2300	4	12
2300.1–2400	6	18
2400.1–2500	9	27
2500.1–2600	7	34
2600.1–2700	5	39
2700.1–2800	1	40
2800.1–2900	7	47
2900.1–3000	1	48
3000.1–3100	2	50
3100.1–3200	1	51
3200.1–3300	1	52
	52	

Using the above data this is $(52 + 1)/2 = 26.50$. The "26.5th" observation is in the group of frequencies relating to an index level of 2400.1–2500. Thus the median is in the group 2400.1–2500. As it is 26.5th out of a cumulative sum of 27, the median will be very close to 2500. We could interpolate as follows:

$$\left[2400.1 + \left(\frac{8.5}{9} \times 100 \right) \right] \approx 2494.5$$

The median from grouped continuous data. To illustrate this consider the data in Table 2.3 and repeated here in Table 2.6.

Table 2.6

Monthly return	Number of observations	Cumulative frequency
Up to −8%	0	0
More than −8% and up to −7%	2	2
More than −7% and up to −6%	1	3
More than −6% and up to −5%	5	8
More than −5% and up to −4%	2	10
More than −4% and up to −3%	2	12
More than −3% and up to −2%	3	15
More than −2% and up to −1%	3	18
More than −1% and up to 0%	5	23
More than 0% and up to +1%	3	26
More than +1% and up to +2%	7	33
More than +2% and up to +3%	4	37
More than +3% and up to +4%	3	40
More than +4% and up to +5%	4	44
More than +5% and up to +6%	1	45
More than +6% and up to +7%	2	47
More than +7% and up to +8%	0	47
More than 8% and up to +9%	0	47
More than +9% and up to +10%	2	49
More than +10% and up to +11%	0	49
More than +11% and up to +12%	2	51

To estimate the median from this data we apply the formula

$$\text{Median} = L + i \left(\frac{\frac{n+1}{2} - F}{f} \right) \tag{2.3}$$

where

L = the lower bound of the median group
i = the width of the median group
F = the cumulative frequency up to the median group
f = the frequency in the median group.

The median observation is $(n + 1)/2 = 52/2 = 26$. Thus it is in the group of observations between 0 and +1%. The width of this group (i) is 1. The cumulative frequency up to this group (F) is 23 and the frequency of the median group (f) is 3.

The median is therefore calculated as

$$\text{Median} = 0 + 1\left(\frac{26 - 23}{3}\right) = 0 + [1(1)] = 1.0\%$$

The arithmetic mean

The arithmetic mean is the most widely used measure of location and is what most people would consider to be the **average**.

The arithmetic mean from ungrouped data. The arithmetic mean is calculated by summing all the individual observations and dividing by the number of those observations. For example, assume that we wish to calculate the arithmetic mean price of a particular asset over a given five-day period. We observe the following prices during the period in question

$$225, \quad 225, \quad 240, \quad 215, \quad 230$$

The arithmetic mean is calculated by summing the five observations and dividing by the number of observations. More formally the arithmetic mean is calculated as

$$\bar{X} = \frac{\sum_{i=1}^{n} X_i}{n} \tag{2.4}$$

This says that $\bar{X}$ (the mean of the Xs) is the sum of the X_is (the individual observations of X, i.e. the security prices in this example) divided by the number of observations. The Σ, the Greek capital letter sigma, is the **summation operator** and simply says that all the X_is are to be added together. As in our example there are five X_is the summation operator would be set out as

$$\sum_{i=1}^{5}$$

indicating that the first five X_is should be added together.

We can now calculate the arithmetic mean of the five security prices as

$$\frac{225 + 225 + 240 + 215 + 230}{5} = 227$$

Thus the arithmetic mean of the prices is 227.

Now let us apply this to the FTSE returns data in Table 2.7. Data relating to monthly observations of the level of the index is given in the left-hand column. The data in the right-hand column relates to the monthly continuously compounded return.

We begin by summing all 12 returns observations, deriving a total of −10.53. Next we divide this sum by 12. Readers may satisfy themselves that the arithmetic mean of this data series is −0.8775. Thus the arithmetic mean of the continuously compounded monthly return to this index over the period in question was −0.8775%.

The arithmetic mean from grouped data. Often the upper and lower limits of the extreme class intervals are left open. It will therefore be necessary to assume a limit. This assumption must be based upon the characteristics of the data being analysed. For

Table 2.7

FTSE 100	Continuously compounded returns
2407.5	
2289.2	−5.04
2160.1	−5.80
2311.1	6.76
2422.7	4.72
2345.8	−3.23
2238.4	−4.69
2221.6	−0.75
2117.9	−4.78
2371.4	11.31
2372	0.03
2339	−1.40
2166.6	−7.66
Σ	−10.53
Mean	−0.8775

example, the data in Table 2.8 relates to people employed by a financial services company. The lower class interval is labelled "Under 26 years" and the last class interval is labelled "55 years and over". If it was known that the company in question had a policy of not employing people under 17 or over 62 years of age, the appropriate limits to these particular class intervals could be determined from knowledge of that policy.

Table 2.8

Age	Mid-point (X)	f	fX
Under 26 years	21.5	20	430
26 to under 40 years	33.0	35	1155
40 to under 55 years	47.5	42	1995
55 years and over	58.5	15	877.5
		112	4457.5

The arithmetic mean of grouped data is found by applying the following formula

$$\overline{X} = \frac{\sum_{i=1}^{m} fX_i}{n} \tag{2.5}$$

where

X_i = the mid-point of class interval i
f = the frequency of each class interval
m = the number of class intervals
n = the total number of observations or data points.

Applying the above data the average age of the employees as calculated by the arithmetic mean is

$$\overline{X} = \frac{4457.5}{112} = 39.8 \text{ years}$$

The geometric mean

An alternative measure of the mean which is particularly applicable when the average rate of growth is to be measured is the **geometric mean.** To illustrate this, first assume that a stock exchange index grew at the following annual rates over five years: $+10\%$, $+20\%$, $+15\%$, -30%, $+20\%$. The arithmetic mean rate would be $+35/5 = 7$. However, 100 invested in year one would grow to the following values in each year: 110, 132, 151.80, 106.26, 127.51. Thus the actual growth over the whole of the five-year period is only 27.5%. Divide this figure by five and we get 5.5% per year; but is this the correct answer?

Actually no! What we really want is a single measure of a growth rate which if repeated n times will transform the opening value into the terminal value. The correct measure of periodic growth rates is derived by using the geometric mean. To measure the **annual growth rate** over n years, the appropriate equation is as follows:

$$\overline{X}_g = \sqrt[n]{X_1 * X_2 * X_3 \ldots * X_n} - 1 \tag{2.6}$$

where the X_is are $(1 + r)$ and r, the growth rate, is expressed in decimals, i.e. $10\% = 0.1$.

Using the above data, the geometric mean rate of growth for the five years in question will be

$$\overline{X}_g = \sqrt[5]{(1.1) * (1.2) * (1.15) * (0.7) * (1.2)} - 1$$

$$= 1.0498 - 1 = 4.98\%$$

Thus the arithmetic mean overstates the annual average growth rate in this example. Using the geometric average will give the correct "average" growth rate.

Which average to use?

The choice of the appropriate average depends in part on the nature of the data and in part on how the measure will be used.

The arithmetic mean is particularly susceptible to extreme value in one direction, what we call skewed data. Extremely high values increase the mean above what is actually representative of the point of central tendency of the data. Extremely low data have the opposite effect. Sometimes the **truncated mean** is calculated to take account of the influence of extreme data points. This simply requires the removal of the top 5% and bottom 5% of observations, before the arithmetic mean is calculated.

The median and the mode are not affected by extreme observations but are not as useful in further mathematical or statistical analysis.

The geometric mean is the most appropriate measure when an "average" rate of change over a number of time periods is being calculated.

Measures of dispersion

As well as knowing the point of central tendency, we often need to know how the data are dispersed around that central point. The following are the measures of dispersion that we shall study:

- the semi-interquartile range or quartile deviation
- the variance
- the standard deviation.
- the coefficient of variation

The semi-interquartile range or quartile deviation

Just as the observation half-way through the data is known as the median, so the observations 25% and 75% through the data are referred to as the lower and upper quartiles (Q1 and Q3), respectively. The range between these quartiles (Q3 – Q1) is known as the semi-interquartile range or quartile deviation and is a widely used indicator of the dispersion of the data around the median.

Calculation of quartiles from ungrouped data. The location of Q1 and Q3 are found by applying the following

$$Q1 = \frac{n+1}{4}$$

(2.7)

$$Q3 = \frac{3(n+1)}{4}$$

Q1 and Q3 will be those observations that are nearest to $(n + 1)/4$ or $3(n + 1)/4$, respectively.

Using the same ungrouped data relating to penny share prices that we used for the median, which are repeated here

8, 9, 10, 10, 10, 10, 11, 11, 12, 12, 12, 12, 12, 14, 15.

Position of Q1 = $(n + 1)/4$ = 3.75, i.e. 4
Position of Q3 = $3(n + 1)/4$ = 11.25, i.e. 11

Therefore the value of Q1 is 10, and the value of Q3 is 12.

Calculation of quartiles from grouped data. To estimate Q1 and Q3 from grouped data we use a methodology analogous to that used for calculating the median from grouped data, using the data in Table 2.9.
To estimate Q1 and Q3 the from this data we apply the following formulae

$$Q1 = L + i \left(\frac{\frac{(n+1)}{4} - F}{f} \right)$$

(2.8)

$$Q3 = L + i \left(\frac{\frac{3(n+1)}{4} - F}{f} \right)$$

where

L = the lower bound of the quartile group
i = the width of the quartile group
F = the cumulative frequency up to the quartile group
f = the frequency in the quartile group.

The Q1 account is $(n + 1)/4 = 52/4 = 13$. Thus it is in the group between –3 and –2. The width of this group (i) is 1. The cumulative frequency up to this group (F) is 12 and the frequency of the Q1 group (f) is 3.

The Q1 account is therefore estimated as

$$Q1 = -3 + 1\left(\frac{13-12}{3}\right) = -3 + 1(0.333) = -2.667\%$$

The position of the Q3 account is $3(n + 1)/4 = 39, L = +3, F = 37, f = 3$. The value of Q3 is calculated as

$$Q3 = 3 + 1\left(\frac{39-37}{3}\right) = 3 + 1(0.666) = 3.666\%$$

The quartile range is Q3 – Q1 = 3.666 – (–2.667) = 6.333%. The quartile deviation is (Q3 – Q1)/2 = 6.333/2 = 3.1666.

Percentiles

Just as we divided the data points into quarters, so we can divide them into percentiles. The first percentile is that observation that is 1% through the data. Q1 and Q3 coincide with the 25th and 75th percentiles respectively and the median is the 50th percentile.

Table 2.9

Monthly return	Number of observations	Cumulative frequency
Up to –8%	0	0
More than –8% and up to –7%	2	2
More than –7% and up to –6%	1	3
More than –6% and up to –5%	5	8
More than –5% and up to –4%	2	10
More than –4% and up to –3%	2	12
More than –3% and up to –2%	3	15
More than –2% and up to –1%	3	18
More than –1% and up to 0%	5	23
More than 0% and up to +1%	3	26
More than +1% and up to +2%	7	33
More than +2% and up to +3%	4	37
More than +3% and up to +4%	3	40
More than +4% and up to +5%	4	44
More than +5% and up to +6%	1	45
More than +6% and up to +7%	2	47
More than +7% and up to +8%	0	47
More than +8% and up to +9%	0	47
More than +9% and up to +10%	2	49
More than +10% and up to +11%	0	49
More than +11% and up to +12%	2	51

To illustrate the estimation of percentiles from grouped data we will calculate the 90th percentile from the earlier data relating to FTSE 100 returns.

The position of the 90th percentile is $0.9 * 52 = 46.8$. The value of the 90th percentile is

$$90\text{th percentile} = 6 + \left[1 \times \left(\frac{46.8 - 45}{2} \right) \right] = 6 + 0.9 = 6.9\%$$

The variance and standard deviation

If the arithmetic mean is the chosen measure of location, the variance and the standard deviation are the appropriate measures of dispersion. The variance has wide applications in finance as a measure of risk and uncertainty, and has the attractive property of being additive. These points are elaborated on in Chapter 4. The standard deviation has a similar use and is used as the measure of volatility in pricing options, a topic covered in Chapters 8 and 10. However, because the standard deviation is, as you will see later, the square root of the variance, it is not additive.

The variance from ungrouped data. If the individual observations were closely clustered around the mean value the differences between each individual observation, X_i, and the mean, $\bar{X}$, will be small. If the individual observations are widely dispersed the differences between each individual X_i and $\bar{X}$ will be large.

It may be thought that by summing the $(X_i - \bar{X})$s we get a measure of dispersion. Unfortunately that is not the case because $\Sigma(X_i - \bar{X})$ always equals zero. To overcome this problem it is necessary to square the individual $(X_i - \bar{X})$s, and sum the squared figure. Dividing the result by $n - 1$, i.e. the number of observations less one, gives the variance. This can be stated more formally as

$$\sigma^2 = \frac{\Sigma(X_i - \bar{X})^2}{n-1} \tag{2.9}$$

Note that if the variance is derived from the whole population of data the divisor in equation (2.9) would be n. However, when the data we have only represent a **sample** of the population data, as it usually does in empirical research in finance, we must divide by $n - 1$. This is because we must divide by the number of what statisticians call **degrees of freedom** in order to avoid any bias in the result. Degrees of freedom are the number of observations minus the number of parameters estimated from the data. In the case of the variance, the mean is estimated from the same data. Thus when using that estimate in computing the standard deviation, the degrees of freedom are $n - 1$. The effect of not adjusting n by the degrees of freedom is greatest with small samples. It should be intuitive that with large amounts of data, the effect of the adjustment will be minimal.

To illustrate the calculation of the variance from ungrouped data we refer to Table 2.10 below, which shows the first 12 months of the FTSE 100 returns given earlier in Table 2.7. Firstly the deviations from the mean $(X_i - \bar{X})$ are calculated and placed in the third column. Each deviation is then squared and placed in the fourth column. This column is summed to give 361.9. This figure is divided by 11, i.e. $n - 1$, giving 32.90. This figure represents squared percentages, not an intuitively appealing outcome.

Table 2.10

FTSE 100	Returns X	$X_i - \bar{X}$	$(X - \bar{X})^2$
2407.5			
2289.2	−5.04	−4.16	17.31
2160.1	−5.80	−4.93	24.27
2311.1	6.76	7.64	58.3
2422.7	4.72	5.59	31.3
2345.8	−3.23	−2.35	5.51
2238.4	−4.69	−3.81	14.5
2221.6	−0.75	0.13	0.02
2117.9	−4.78	−3.9	15.22
2371.4	11.31	12.18	148.45
2372.0	0.03	0.90	0.82
2339.0	−1.40	−0.52	0.27
2166.6	−7.66	−6.78	45.94
Σ	−10.54	0.00	361.90
Mean	−0.88	Variance	32.90

The variance from grouped data. When using grouped data the formula for the variance is

$$\frac{\Sigma f(X_i - \bar{X})^2}{n-1} \tag{2.10}$$

where f is the frequency in each class interval and X_i is the mid-point of each class interval but $\bar{X}$ is the mean of all the Xs.

To illustrate this calculation consider the grouped data given in Table 2.11 which, you will recall from the earlier example of the mean from grouped data, relates to the recruitment policy of a financial institution.

Table 2.11

Age (X)	Mid-point (X)	f	$X - \bar{X}$	$(X - \bar{X})^2$	$f(X - \bar{X})^2$
Under 26 years	21.5	20	−18.30	334.89	6697.8
26 to under 40 years	33.0	35	−6.80	46.24	1618.4
40 to under 55 years	47.5	42	7.70	59.29	2490.18
55 and over	58.5	15	18.70	349.69	5245.35
Total		112			16051.73

Recalling that $\bar{X} = 39.8$ years, the variance is calculated as

$$\sigma^2 = \frac{16051.73}{112-1} = 144.61$$

The 144.61 is in squared years, which is rather cumbersome to consider.

The standard deviation

The variance is in squared units of the underlying data which makes interpretation rather counterintuitive. This problem is overcome by taking the square root of the variance, which results in the **standard deviation,** designated σ.

The standard deviation from ungrouped data is formally calculated as

$$\sigma = \sqrt{\frac{\sum(X_i - \overline{X})^2}{n-1}} \qquad (2.11)$$

Referring to the calculation of the variance from ungrouped data of the FTSE 100 index returns in Table 2.10, the square root of 32.9 is 5.74.

The standard deviation from grouped data. The formula for calculating the standard deviation from grouped data is

$$\sigma = \sqrt{\frac{\sum f(X_i - \overline{X})^2}{n-1}} \qquad (2.12)$$

Again it is clear that this is just the square root of the variance, this time from grouped data. Remember also that in this example X_i is the mid-point of each class interval but $\overline{X}$ is the arithmetic mean of all the observations. The standard deviation of the grouped data relating to age of recruitment is 12.02 years.

The negative semi-variance and negative semi-deviation

The **negative semi-variance** is similar to the variance but only the negative deviations from the mean are included. The measure of negative semi-variance is sometimes suggested as an appropriate measure of risk in financial theory when the returns are not symmetrical. It is calculated as follows:

$$\overline{\sigma}^2 = \frac{\sum(R_i - \overline{R})^2}{n-1} \qquad (2.13)$$

where

    R_i = periodic portfolio returns
    n = the total number of negative returns
    $\overline{R}$ = the mean return for the period including positive and negative returns
    $\overline{\sigma}^2$ = the semi-variance.

However, note that only terms where $R_i < \overline{R}$ are included. Thus only the negative deviations from the mean are included in the calculation. Using the same data as for the earlier example of the variance from ungrouped data, the negative semi-variance is 20.50.

The square root of the negative semi-variance is the negative semi-deviation depicted as $\overline{\sigma}$; the negative semi-deviation from the same data is 4.53.

Which measure of dispersion?

The choice of measure of dispersion is dictated by the measure of central tendency that is used. If the median is used to measure "the average" so the quartile-deviation will be used as a measure of dispersion. If the arithmetic mean is used, so the variance, the standard deviation or the negative semi-variance will be used.

The coefficient of variation

The standard deviation is expressed in the underlying units of measurement. Thus when comparing the degree of dispersion between variables account must be taken of the differences in the magnitude of the variables. To do this we must calculate the coefficient of variation. This is calculated as

$$CV = \frac{\sigma}{\overline{X}}$$
(2.14)

An example of the application of the coefficient of variation is in comparing the dispersion of changes in the levels of the FTSE 100 index in the UK with those of the S&P composite index in the USA. The former index is at the time of writing about 3700, whereas the latter is around 650. Comparing the dispersion of changes in levels using the standard deviation would be misleading, thus the coefficient of variation should be used. However, note that if we were measuring the dispersion of returns between the two indices, the standard deviation would be appropriate. The reason is that returns calculations already take into account the levels of the underlying data.

Skewness

It is important to consider if there is any bias in the dispersion of the data. This bias is indicated by the skewness of the data. Figure 2.5 below shows two typical forms of skewness and a symmetrical distribution, together with the relative positions of the mean median and mode.

In the case of positive skewness the distribution has a long tail to the right. The mean return is greater than the median return, which in turn is greater than the mode. The mean is greater than the median or mode because it is pulled up by the few very high returns observations.

Negative skewness results in a longer tail to the left, with the mean being below the median and mode. Most observations in the distribution are above the mean, but the mean is pulled down by the few very small observations.

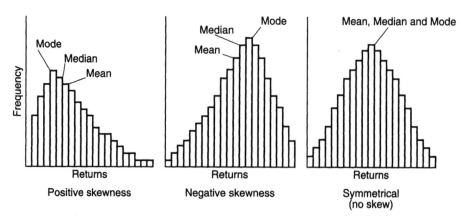

Figure 2.5 Three examples of degrees of skewness.

In the case of the symmetrical distribution, the mean, median and mode all have the same value.

Compounding of periodic returns will give rise to positive skewness. To understand this, consider the terminal value of an asset that showed positive returns of 8% p.a. for two annual periods. If the original investment were 100, the terminal value will be

$$100(1.08)^2 = 116.64$$

Now if the returns were negative by 8% in each year the terminal value would be

$$\frac{100}{(1.08)^2} = 85.73$$

Thus compounding the positive returns gives a gain of 16.64 over the two-year period, whereas the negative returns only lose 14.27 over that period.

There are a number of procedures for quantifying the degree of skewness in data. The **Spearman coefficient of skewness** is calculated as follows:

$$\text{Coefficient of skewness} = \frac{3(\text{mean} - \text{median})}{\text{standard deviation}} \quad (2.15)$$

It is also possible to calculate a coefficient of skewness from quartiles and percentiles. The **quartile coefficient of skewness** is

$$\frac{(Q_3 - Q_2) - (Q_2 - Q_1)}{(Q_3 - Q_1)} = \frac{Q_3 - 2Q_2 + Q_1}{Q_3 - Q_1} \quad (2.16)$$

The **10–90 percentile coefficient of skewness** is

$$\frac{(P_{90} - P_{50}) - (P_{50} - P_{10})}{P_{90} - P_{10}} = \frac{P_{90} - 2P_{50} + P_{10}}{P_{90} - P_{10}} \quad (2.17)$$

However, the measure of skewness that most easily lends itself to spreadsheet calculation is the **moment coefficient of skewness**. This is derived by calculating the third moment about the mean and dividing by the cube of the standard deviation, as follows:

$$\frac{\left[\dfrac{\Sigma\left(X - \bar{X}\right)^3}{n-1}\right]}{\left(\sqrt{\dfrac{\Sigma\left(X - \bar{X}\right)^2}{n-1}}\right)^3} \quad (2.18)$$

Again using the FTSE 100 index returns, refer to Table 2.12 which has applied equation (2.18) to the returns observations.

Table 2.12

FTSE 100	Returns X	$X_i - \overline{X}$	$(X - \overline{X})^3$
2407.5			
2289.2	−5.04	−4.16	−71.99
2160.1	−5.80	−4.93	−119.55
2311.1	6.76	7.64	445.15
2422.7	4.72	5.59	175.10
2345.8	−3.23	−2.35	−12.93
2238.4	−4.69	−3.81	−55.22
2221.6	−0.75	0.13	0.00
2117.9	−4.78	−3.9	−59.40
2371.4	11.31	12.18	1808.77
2372.0	0.03	0.90	0.74
2339.0	−1.40	−0.52	−0.14
2166.6	−7.66	−6.78	−311.37
Σ	−10.54	0.00	1799.17
Mean	−0.88	3rd moment	163.56
		Stand. dev.³	188.71
		Coeff. of skew	0.87

The deviations from the mean are cubed and placed in the fourth column. This column is summed and divided by 11 to give 163.56. This is divided by the cube of the standard deviation, giving a coefficient of skewness of 0.87.

If the distribution of the equity index returns were symmetrical, the coefficient of skewness would be zero. As this calculation shows a positive number, the index returns data is positively skewed.

Kurtosis

Whereas skewness indicates the degree of symmetry in the frequency distribution, kurtosis indicates the peakedness of that distribution. Distributions that are more peaked than the normal distribution (which will be discussed in detail later) are referred to as leptokurtic. Distributions that are less peaked (flatter) are referred to as platykurtic, and those distributions that resemble a normal distribution as referred to as mesokurtic. The general shape of these distributions is given in Fig. 2.6.

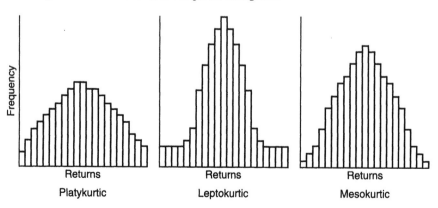

Figure 2.6 Three examples of kurtosis.

Leptokurtic distributions are found in asset returns, when there are periodic jumps in asset prices. Markets where there is discontinuous trading, such as security markets that close over night or at weekends, are more likely to exhibit jumps in asset prices. The reason is that information which has an influence on asset prices but is published when the markets are closed will have an impact on prices when the market reopens, thus causing a jump between the previous closing price and the opening price. This jump in price, which is most noticeable in daily or weekly data, will result in higher frequencies of large negative or positive returns than would be expected if the markets were to trade continuously.

Coefficients of kurtosis can be derived from percentiles and quartiles or from the moments around the mean.

The percentile coefficient of kurtosis. This is calculated as

$$\frac{\frac{1}{2}\left(Q_3 - Q_1\right)}{P_{90} - P_{10}} \tag{2.19}$$

The moment coefficient of kurtosis. The moment coefficient of kurtosis is derived by dividing the fourth moment about the mean by the standard deviation raised to the fourth power. The calculation is as follows:

$$\frac{\dfrac{\Sigma(X - \bar{X})^4}{n-1}}{\left(\sqrt{\dfrac{\Sigma(X - \bar{X})^2}{n-1}}\right)^4} \tag{2.20}$$

The top part of equation (2.20), $\Sigma(X - \bar{X})^4/N$, is the fourth moment around the mean. The lower part is simply the variance squared. An application of equation (2.20) to the equity index returns data used earlier is given in Table 2.13 below.

Table 2.13

FTSE 100	Returns X	$X_i - \bar{X}$	$(X - \bar{X})^4$
2407.5			
2289.2	−5.04	−4.16	299.50
2160.1	−5.80	−4.93	588.91
2311.1	6.76	7.64	3398.97
2422.7	4.72	5.59	979.58
2345.8	−3.23	−2.35	30.34
2238.4	−4.69	−3.81	210.26
2221.6	−0.75	0.13	0.00
2117.9	−4.78	−3.9	231.74
2371.4	11.31	12.18	22038.28
2372.0	0.03	0.90	0.67
2339.0	−1.40	−0.52	0.07
2166.6	−7.66	−6.78	2110.42
Σ	−10.54	0.00	29888.76
Mean	−0.88		
		4th moment	2717.16
		Stand. dev.^{4}	1082.43
		Coeff. of kurtosis	2.51

The fourth moment around the mean is 2717.16, the standard deviation raised to the fourth power is 1082.43. Therefore the moment coefficient of kurtosis is calculated as 2.51.

If the data were normally distributed (i.e. mesokurtic) the moment coefficient of kurtosis would be 3.0. Thus the index returns data in question are less peaked than a normal distribution and therefore are platykurtic. If the data were leptokurtic, i.e. more peaked than in a normal distribution, the moment coefficient of kurtosis would be greater than 3.0.

Measures of association

Earlier it was explained that the variance of a random variable indicated how observations of that variable are distributed around the mean observation. In this section we will introduce the concept of the **covariance**, which **indicates how two random variables behave in relation to each other**. Later in this section we will go on to calculate the **correlation coefficient** which is a more convenient measure of the linear association between two variables.

The covariance

There are many situations in finance and elsewhere when it is important to know how two variables behave in relation to each other. For example, in portfolio management, in order to know the riskiness of a portfolio of assets, it is necessary to know how the price of security X behaves in relation to security Y for all pairs of assets. In other words we need to know the covariance or the correlation between each pair of assets.

If the price of security X generally rises (falls) at the same time that the price of security Y rises (falls), the covariance will be positive. However, if generally while the price of security X rises the price of security Y falls, the covariance will be negative. If there is no pattern to the relationship of the price movements, i.e. the two security prices act totally independently, the covariance will be zero.

The formula for the covariance is

$$\text{cov}_{XY} = \sigma_{XY} = \frac{\sum(X - \bar{X}) * (Y - \bar{Y})}{n - 1} \tag{2.21}$$

Note that the value of the covariance, as calculated, depends on the values of the X and Y observations, and thus a larger covariance could be more to do with having high values to the observations than to a closer association between the variables.

The first step in investigating the relationship between pairs of variables is to plot a scatter diagram of each pair.

Table 2.14 gives data relating the returns to two assets, the S&P 500 index and the FTSE 100 index. For the purposes of this section we will find it more convenient to label these X and Y, respectively. Table 2.14 shows the details of the calculation of the covariance between these two assets.

Equation (2.21) is applied to the data as follows. The deviations of each X or Y observation from its respective mean is placed in the appropriate $X - \bar{X}$ or $Y - \bar{Y}$ column. The corresponding entries in these two columns are multiplied together and placed in the far right-hand column. This column is summed to give a total of 236.491. This figure

Table 2.14

S&P 500 returns	FTSE 100 returns	$X - \bar{X}$	$Y - \bar{Y}$	$(X - \bar{X})(Y - \bar{Y})$
−0.81	−5.04	−0.043	−4.160	0.179
−2.79	−5.80	−2.026	−4.926	9.980
2.73	6.76	3.495	7.635	26.686
0.79	4.72	1.556	5.594	8.703
−7.22	−3.23	−6.449	−2.347	15.137
1.19	−4.69	1.963	−3.808	−7.475
1.78	−0.75	2.544	0.125	0.319
−1.92	−4.78	−1.154	−3.902	4.503
8.90	11.31	9.664	12.184	117.751
−1.00	0.03	−0.233	0.904	−2.211
−1.20	−1.40	−0.356	−0.522	0.186
−9.73	−7.66	−8.961	−6.778	60.733
Mean	Mean		Sum	236.491
−0.77	−0.88		Covariance	21.499
			Correlation	0.793

is divided by $n - 1$ to give a covariance of 21.499. We see that in this example, assets X and Y have a positive covariance between their returns.

It is instructive to consider how the sign of the covariance develops. Consider Fig. 2.7; notice how the plot space has been divided into four by the graticules with axes at $\bar{X}$ and $\bar{Y}$.

Note that for plots in the top left-hand quarter, the values for Y are greater that $\bar{Y}$, thus $Y - \bar{Y}$ is positive. However, any values of X will be below $\bar{X}$. Thus $X - \bar{X}$ will be negative. As a negative figure multiplied by a positive figure results in a negative figure, the contribution of observations in the top left-hand quarter will be negative. By analogous reasoning, the contribution of any observations in the top right-hand corner will be positive, as will the contribution of those in the bottom left-hand corner. The contribution of those observations in the bottom right-hand corner will be negative.

Thus data that predominates in the bottom left and top right quarters will have a positive covariance, whilst data where the observations predominate in the top left and bottom right quarters will have a negative covariance. If the data points are scattered evenly over the four quarters, the covariance will be zero.

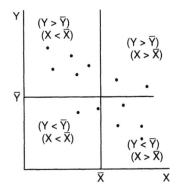 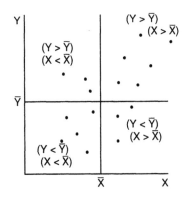

Figure 2.7

The variance–covariance matrix

Often the covariances between several pairs of variables are displayed in a variance–covariance matrix. For example, consider Table 2.15 which shows the covariances between all possible pairs from a group of three assets.

Table 2.15

	(a)				(b)		
	X	*Y*	*Z*		*X*	*Y*	*Z*
X	*xx*	*xy*	*xz*	*X*	*xx*		
Y	*yx*	*yy*	*yz*	*Y*	*yx*	*yy*	
Z	*zx*	*zy*	*zz*	*Z*	*zx*	*zy*	*zz*

The possible pairs of assets in the above three-asset example are *XX*, *XY*, *XZ*, *YX*, *YY*, *YZ*, *ZX*, *ZY* and *ZZ*. In the above tables the covariances are depicted in lower case. Thus the covariance between *X* and *Y* is *xy*. Note that this is the same as the covariance *yx*. Thus all the covariances to the left and below the diagonal are repeated above and to the right of the diagonal. Thus these covariances are entered in the matrix twice. We will illustrate the importance of this point later when we measure portfolio risk.

Note the covariances on this diagonal, entered here in bold. These are the covariances of returns with themselves. The covariance of a variable with itself is the variance of the variable. To understand this consider equation (2.21) above applied to the covariance of asset *X* with itself. The equation would be

$$\text{cov}_{xx} = \frac{\Sigma(X - \bar{X}) * (X - \bar{X})}{n - 1} \tag{2.22}$$

The numerator in equation (2.22) simplifies to $(X - \bar{X})^2$, thus equation (2.22) becomes

$$\text{cov}_{xx} = \frac{\Sigma(X - \bar{X})^2}{n - 1} = \sigma_x^2 \tag{2.23}$$

which is the formula for the variance. Thus a matrix of covariances is often referred to as a **variance–covariance matrix**.

The correlation coefficient

A more widely used measure of the degree of association between two variables is the **correlation coefficient**. It is a unit-free measure of the strength and the direction of a linear relationship between two variables. Consequently, it overcomes one weakness of the covariance, because the size of the correlation coefficient is not influenced by the values of the observations.

The values of the correlation coefficient lie between -1 for a perfectly negative relationship, through zero where the two variables are independent of each other, to $+1$ for a perfectly positive relationship between the variables.

The correlation coefficient, ρ, is calculated by dividing the covariance between X and Y by the product of the standard deviation of X and the standard deviation of Y. This is formally stated as

$$\rho_{xy} = \frac{\sigma_{xy}}{\sigma_x \sigma_y} \tag{2.24}$$

Applying equation (2.24) to the data in Table 2.14 indicates that the correlation coefficient between X and Y is

$$\rho_{XY} = \frac{21.499}{4.73 * 5.74} = 0.793$$

Irrespective of how positive or negative the correlation is, the correlation coefficient is only a measure of statistical association. There is no inference of causality in the statistic. By this we mean that there is no suggestion that a change in variable X causes a change in variable Y, or indeed vice versa.

Thus although the correlation coefficient may indicate how strong the linear association is between two variables, it cannot explain the changes in them. For such an explanation it is first necessary to develop a theory of the causal relationship from a priori reasoning, construct a model that reflects the hypothesized relationship and then test the model statistically. For this regression analysis is used. This is the subject of Chapter 6.

The correlation matrix

Just as covariances are frequently displayed in matrices so correlation coefficients are similarly displayed. For example, consider Table 2.16.

Table 2.16

	(a)				(b)		
	X	Y	Z		X	Y	Z
X	1	xy	xz	X	1		
Y	yx	1	yz	Y	yx	1	
Z	zx	zv	1	Z	zx	zv	1

Note that Tables 2.16(a) and (b) differ from Tables 2.15(a) and (b) in that the diagonal from top left to bottom right must be one by definition. To prove this recall that the numerator in equation (2.23) is the covariance of the variable but from equation (2.22) becomes a variance in these circumstances. Note also that the denominator is the product of the standard deviations of the two variables. In the case of the correlation of a variable with itself, the denominator would be the square of the standard deviation, which is the variance. Thus numerator and denominator would both be the same variance, and the result must be one.

Applications of covariances and correlations

Correlations and covariances have many applications in finance theory and practice. In this section we will see the role that these variables play in the riskiness of portfolios of risky assets. In Chapter 6, we will investigate the role of correlations used in conjunction with regression analysis in hedging risk.

Portfolio risk. The risk in a single asset is measured by the variance or the standard deviation of the returns to that asset, and the risk of a portfolio is measured by the variance or standard deviation of the returns to that portfolio.

However, to measure the risk of the portfolio, we do not only need to know the variability of the returns of individual securities but the degree to which the returns of pairs of securities fluctuate together. We need to know the **covariance** or alternatively the **correlation** of the returns of each pair of securities in the portfolio.

The portfolio risk, as measured by its variance, is calculated as the weighted sum of the covariances between each pair of assets in the portfolio, where each covariance is weighted by the product of the weights of each of the respective assets in the pair, and where the variance of a particular asset is considered as the covariance of the asset with itself.

To demonstrate this consider a portfolio of three assets A, B and C. The returns to each asset are a, b and c. The weights by which each asset is represented in the portfolio are W_A, W_B and W_C.

The covariances of returns from all the possible pairs of assets can be depicted by a covariance matrix, which is the matrix in the box below:

	W_A	W_B	W_C
W_A	**cov (a,a)**	cov (a,b)	cov (a,c)
W_B	cov (b,a)	**cov (b,b)**	cov (b,c)
W_C	cov (c,a)	cov (c,b)	**cov (c,c)**

The portfolio risk can be calculated as

$$\sigma_p^2 = W_A W_A \text{cov}_{aa} + W_A W_B \text{cov}_{ab} + \ldots + W_C W_B \text{cov}_{cb} + W_C W_C \text{cov}_{cc} \qquad (2.25)$$

Note from the covariance matrix that each weighted covariance is effectively included twice, e.g. $W_A W_C$ cov (a, c) is the same as $W_C W_A$ cov(c, a). Recalling also that cov(a, a) is actually the variance of a, equation (2.25) simplifies to

$$\sigma_p^2 = W_A^2 \sigma_a^2 + W_B^2 \sigma_b^2 + W_C^2 \sigma_b^2 + 2 \times W_A W_B \text{ cov}_{ab} + 2 \times W_A W_C \text{cov}_{ac} + 2 \times W_B W_C \text{cov}_{bc} \qquad (2.26)$$

Equation (2.25) can be generalized as follows:

$$\sigma_p^2 = \sum_{i=1}^{N} W_i^2 \sigma_i^2 + 2 \sum_{i=1}^{N} \sum_{j>i} W_i W_j \text{ cov}_{ij} \qquad (2.27)$$

Clearly, the benefits of diversification are derived from adding assets to the portfolio that have low or even negative covariances with other assets in the portfolio, thus reducing the sum of the covariances and therefore the total risk of the portfolio.

As the covariance is an unbounded measure of association, the **correlation coefficient** is frequently used as the measure of association. The advantage of ranking pairs of assets by their correlation coefficients is that it provides a clear system for including assets that enhance the benefits of diversification, and excluding those that do not.

Recalling that the correlation coefficient is calculated as

$$\rho_{ab} = \frac{\sigma_{ab}}{\sigma_a \sigma_b} \qquad (2.28)$$

It follows that the covariance can be calculated as

$$cov_{ab} = \rho_{ab} * \sigma_a \sigma_b \qquad (2.29)$$

Consequently, the variance of a portfolio of two assets, using the correlation coefficients as well as the asset variances is

$$\sigma_p^2 = W_A^2 \sigma_a^2 + W_B^2 \sigma_b^2 + 2W_A W_B (\rho_{ab}\, \sigma_a\, \sigma_b) \qquad (2.30)$$

where

$$\sigma_p^2 = \text{variance of the portfolio}$$
$$\sigma_a^2 \text{ and } \sigma_b^2 = \text{variance of returns of } a \text{ and } b$$
$$\rho_{ab} = \text{correlation of returns of } a \text{ and } b$$
$$\sigma_a \text{ and } \sigma_b = \text{standard deviation of } a \text{ and } b$$
$$(\rho_{ab}\, \sigma_a\, \sigma_b) = \text{the } \mathbf{covariance} \text{ of the returns on assets } a \text{ and } b.$$

The portfolio standard deviation is the square root of the portfolio variance.

The similarity with equation (2.26) will be clear given that $(\rho_{ab}\sigma_a\sigma_b)$ is the covariance between the returns on a and b.

The risk-reducing effects of diversification

To demonstrate the risk-reducing effects of diversification assume that security A has a standard deviation of 15% and that security B has a standard deviation of 14%. Also assume that the assets are equally weighted in the portfolio.

Look first at the special case where the asset returns are perfectly correlated, i.e. the correlation coefficient = +1.0.

The portfolio standard deviation will be as follows:

$$\sigma_p = \sqrt{(0.5)^2(0.14)^2 + (0.5)^2(0.15)^2 + 2(0.5*0.5*1*0.14*0.15)} = 0.145 = 14.5\%$$

This shows that in this special case the portfolio risk is simply the weighted average risk of the individual assets.

Now let us see what happens when the securities have a correlation of only 0.6. The portfolio standard deviation becomes

$$\sigma_p = \sqrt{0.5^2 0.14^2 + 0.5^2 0.15^2 + 2(0.5*0.5*0.6*0.14*0.15)} = 0.1297 = 12.97\%$$

Note that the figure of 12.97 is actually lower than the standard deviations of either of the individual securities.

Next we will look at another special case where the two assets are perfectly negatively correlated. The portfolio standard deviation will be

$$\sigma_p = \sqrt{0.5^2 0.14^2 + 0.5^2 0.15^2 + 2(0.5*0.5*-1*0.14*0.15)} = 0.005 = 0.5\%$$

The portfolio risk is virtually zero because in this special case when one asset rises the other falls by a similar amount and therefore the portfolio value does not fluctuate. This is the basis of hedging transactions, where the negative correlation is achieved by selling a position in an instrument that is highly positively correlated with that to be hedged. However, going short with the hedging instrument creates a negative correlation between the long and short positions.

As security returns are, in general, not perfectly correlated, the standard deviation of a portfolio will be less than the weighted average of the standard deviations of the individual securities. Moreover, the standard deviation of a portfolio falls as the degree of correlation between pairs of assets falls.

Thus effective diversification is not just a case of adding assets to a portfolio, but of adding assets whose returns have the lowest correlations with existing assets in the portfolio.

We saw earlier the effects of the degree of correlation upon diversification. It will be instructive at this point to see what happens in portfolios consisting of many assets.

Consider again equation (2.27) which is repeated here as (2.31):

$$\sigma_p^2 = \sum_{i=1}^{N} W_i^2 \sigma_i^2 + 2\sum_{i=1}^{N} \sum_{j>i} W_i W_j \, \mathrm{cov}_{ij} \tag{2.31}$$

Imagine that there were a very large number of assets available for investment, say an index of 100 or 500 equities. Assume also that all the asset returns are independent. Equation (2.31) would reduce to

$$\sum_{i=1}^{N} W_i^2 \sigma_i^2 \tag{2.32}$$

As the asset returns were assumed to be independent, the covariances are zero. Now assume that an equal amount is invested in each of N assets, the weights become $1/N$. The portfolio variance then becomes

$$\sigma_p^2 = \sum_{i=1}^{N} \left(\frac{1}{N}\right)^2 \sigma_i^2 = \frac{1}{N}\left[\sum_{i=1}^{N} \frac{\sigma_i^2}{N}\right] \tag{2.33}$$

The term in the square brackets is the average variance of the assets in the portfolio. As the number of assets (N) in the portfolio gets larger, $1/N$ gets smaller and the portfolio variance gets smaller and in the extreme approaches zero.

However, in reality asset returns are not all independent, particularly when we are considering assets that come from the same asset class, for example equities or bonds. There will be some degree of correlation between most of the assets. Thus in reality equation (2.32) becomes

$$\sigma_p^2 = \sum_{i=1}^{N} \left(\frac{1}{N}\right)^2 \sigma_i^2 + 2\sum_{i=1}^{N} \sum_{j>i} \left(\frac{1}{N}\right)\left(\frac{1}{N}\right) \mathrm{cov}_{ij} \tag{2.34}$$

This can be expressed as

$$\sigma_p^2 = \left(\frac{1}{N}\right)\sum_{i=1}^{N}\left[\frac{\sigma_i^2}{N}\right] + \frac{(N-1)}{N}2\sum_{i=1}^{N}\sum_{j>i}\left[\frac{\text{cov}_{ij}}{N(N-1)}\right] \qquad (2.35)$$

The first term is the average of the variances met in equation (2.33) above. The second term is also an average: it is the sum of the covariances divided by the number of covariances, $N(N-1)$. Equation (2.35) can therefore be simplified to

$$\sigma_p^2 = \frac{1}{N}\bar{\sigma}_i^2 + \frac{N-1}{N}\overline{\text{cov}}_{ij} \qquad (2.36)$$

This equation helps explain intuitively what happens to portfolio risk when large numbers of assets are included in the portfolio. As the number of assets in the portfolio increases, $1/N$ gets smaller and therefore when multiplied by the average variance the product approaches zero. However, $(N-1)/N$ gets closer to one as N gets larger, so the second part of the right-hand side of equation (2.36) approaches the average covariance. Thus, to summarize, when a portfolio is diversified amongst a large number of assets, the portfolio variance approaches the average covariance of the individual assets.

Thus the total risk of a security held in isolation is more than that of the same security held within a portfolio. The combination of assets with less than perfect correlations reduces the risk of the portfolio, therefore total risk comprises of two parts: (a) that which can be diversified away (**unsystematic risk, also known as specific or residual risk**) and (b) that element that cannot be diversified away (**systematic risk, also known as market risk**).

Index numbers

An index number is a single descriptive statistic that summarizes the relative change in an underlying variable or group of variables. For example, published index numbers summarize, in a single number, the change in price of a large number of retail goods: the **retail price index**, manufacturers' prices: the **producer price index**, share prices: the **FTSE 100 index**, the change in the value of a currency *vis-à-vis* a number of other currencies: the **trade weighted index**.

Index numbers are useful because the single underlying numbers are very large and changes in the absolute number are difficult to comprehend, for example the gross national income of a country. Alternatively, index numbers are used to summarize the collective change in a group of items that are each changing at different rates. For example, the retail price index summarizes the change in price of the basket of goods and services purchased by the "typical" household. The FTSE 100 index summarizes the change in price of 100 shares listed on the London Stock Exchange.

Index numbers as relative indicators

A price index can be construed as a **price relative.** Consider the price of an item today, P_1, compared with the price yesterday, P_0. The relative price of the item, the price relative, is given by P_1/P_0.

If the price had risen the price relative would be greater than one and if the price had fallen the price relative would be less than one. For example, if the price on day 0 is 12 and on day 1 is 14 the price relative is 14/12= 1.1667. If the price had fallen to 11, the price relative would be 11/12 = 0.9167.

Clearly, from the above example it can be seen that a further price relative could be derived to summarize the change in price between day 2 and day 1, i.e. P_2/P_1. However, it is a feature of index numbers that they refer to a **base period**. Thus in our example the index number would refer to the price relative P_2/P_0. Thus the index number can be used to summarize the change in the underlying variable(s) between the base period and the current period.

So far we have considered only price relatives, but quantity relatives or value relatives are equally validly constructed if the underlying data relates to quantities or values rather than prices.

A second notable feature of index numbers is that the base price (or quantity or value) is rescaled to be 100 or sometimes 1000. To do this all we have to do is multiply the price relatives by 100 (or 1000). Thus if $P_0 = 12$ is rescaled to 100 we are effectively dividing by 12 and multiplying by 100. If P_1 is 14 the index number with a base of 100 becomes

$$14/12 * 100 = 116.67$$

This can be generalized to

$$Z\left(\frac{P_t}{P_0}\right) \qquad\qquad (2.37)$$

where

Z = the base value of the index, usually 100 or 1000
P_0 = the value of the variable in the base period
P_t = the current value of the variable.

When Z is set to 100, this format has the advantage of very simply indicating the percentage change in the index since the start date or base period. However, this percentage change does not reflect the change between dates. For example, if the price on day 3, P_3, were 15, the index number on day three would be

$$15/12 * 100 = 125$$

This indicates a 25% increase since the start date. However, the change from day 2 would only be

$$(125 - 116.67)/116.67 = 7.14\%$$

In the examples given above, we have considered the price relative of only one item. Yet in practice index numbers are constructed to summarize the change in a number of underlying variables. This raises two important issues:

1. How to average the many changes so as to arrive at a single index number.
2. How to treat each variable within the group. Are they to be treated as equally important? If not, how is the importance of each to be determined?

Choice of averaging system

So that the changes to a large number of variables can be summarized, the relative values have to be collected and averaged. There are two types of averaging that are applicable to index numbers. These are arithmetic averaging and geometric averaging.

To illustrate arithmetic averaging consider an index consisting of four assets, A, B, C and D. The price relatives are

$$\frac{P_{A_1}}{P_{A_0}}, \frac{P_{B_1}}{P_{B_0}}, \frac{P_{C_1}}{P_{C_0}}, \frac{P_{D_1}}{P_{D_0}}$$

Assuming that all the constituents are equally important, i.e. have equal weighting, the index will be

$$Z \left(\frac{\dfrac{P_{A_1}}{P_{A_0}} + \dfrac{P_{B_1}}{P_{B_0}} + \dfrac{P_{C_1}}{P_{C_0}} + \dfrac{P_{D_1}}{P_{D_0}}}{4} \right) \qquad (2.38)$$

This can be generalized to

$$Z \left(\frac{\displaystyle\sum_{i=1}^{n} \frac{P_{it}}{P_{i0}}}{n} \right) \qquad (2.39)$$

If the same price relatives were included in a geometrically averaged index, it would be constructed as follows:

$$Z \left(\frac{P_{A_1}}{P_{A_0}} * \frac{P_{B_1}}{P_{B_0}} * \frac{P_{C_1}}{P_{C_0}} * \frac{P_{D_1}}{P_{D_0}} \right)^{\frac{1}{4}} \qquad (2.40)$$

This can be generalized to

$$Z \prod_{i=1}^{n} \left(\frac{P_{it}}{P_{i0}} \right)^{\frac{1}{n}} \qquad (2.41)$$

Geometrically averaged indices have an inherent bias in that, except where all the constituents change by the same proportion, the index will understate the absolute magnitude of a rise and overstate the absolute magnitude of a fall. This overestimation is relative to a portfolio of the shares constituting the index. In addition, if one of the constituents of the index falls to zero value, the whole index has only a zero value.

The bias in geometrically averaged indices is illustrated in Table 2.17 where the changes in an arithmetically averaged and geometrically averaged index are compared.

Another problem with geometrically averaged indices is that their behaviour does not replicate that of the underlying constituents. For example, if the index represented

Table 2.17 Illustration of bias in geometrically averaged indices

A	B	C	D	Arithmetic mean	Geometric mean
100	100	100	100	100	100
103	101	105	106	103.75	103.73
110	106	106	108 .	107.5	107.49
115	104	108	112	109.75	109.67
120	113	112	123	117	116.91
110	108	107	115	110	109.96
103	105	104	102	103.5	103.49
88	101	101	90	95	94.81
71	95	96	85	86.75	86.13
0	104	108	80	73	0.00

changes in price of company shares, the returns as suggested by the index would not be the same as the returns actually achieved by a portfolio of those shares constructed to have the same weightings as in the index. However, in the case of an arithmetically averaged index, the returns would match those of a similar underlying portfolio.

Choice of weighting system

So that the index reflects the collective movement in the underlying constituents, each of the price relatives should be weighted to reflect their relative importance. There are four popular forms of weighting:

- weighting the constituents equally
- weighting the constituents by the base period price or quantity
- weighting the constituents by current period price or quantity
- weighting the constituents by their current value, i.e. price * quantity.

Equally weighted indices treat all the items in the group as equally important. This has some drawbacks when the group represented by the index contains items of significantly different degrees of importance. This approach tends to overemphasize the importance of relatively unimportant items relative to important ones. For example, if we equally weighted the change in price of shares of small (by market capitalization) companies we would be overemphasizing the changes in the price of shares of small companies or, alternatively, underemphasizing the change in price of large companies relative to the importance of those companies to average investors.

Equally weighted indices can be constructed as the arithmetic mean or the geometric mean of the price relatives as set out in equations (2.38)–(2.40) above.

If it is decided to use a weighting system that is not "equally weighted", we have to consider how to construct the weights. Are the weights to reflect the relative importance of each variable as at the base period? If so such an index would be a **Laspeyres index** or **base-weighted index**. Alternatively, should the weights reflect the current importance of each variable? Such an index would be a **Paasche index**. Whichever is chosen, it must be used consistently otherwise comparison of the index number over time will not be valid.

Laspeyres index or base-weighted index

Laspeyres indices may be either price or quantity weighted. The common feature is that the weights relate to those derived in the base period and are used without change in calculations relating to subsequent time periods.

The Laspeyres price index measures the current cost of a group of items weighted by their base period quantities relative to the cost of that same basket at the base period. $P_n * Q_0$ is the current cost of the group. $P_0 * Q_0$ is the base period cost of that basket. The index is calculated as

$$\frac{\sum P_n * Q_0}{\sum P_0 * Q_0} \times 100 \tag{2.42}$$

The Laspeyres quantity index measures the quantity of a current group of items weighted at base period prices relative to the base period quantity of that group. The index is calculated as

$$\frac{\sum P_0 * Q_0}{\sum P_0 * Q_0} \times 100 \tag{2.43}$$

One problem with base weighting is that it does not allow for substitution between goods as tastes, technology or relative prices change. For example, over the last 15 years we have seen a dramatic change in the relative price of beef and chicken, with people increasing their consumption of the cheaper chicken relative to the more expensive beef. This change is probably also influenced by changing attitudes to health. A base-weighted index may not take these changes into account, or at least not promptly.

Paasche index or current year weighted index

Paasche indices may also be price or quantity indices. The Paasche price index measures the cost of a group of items weighted by current period quantities relative to the cost of that same group in the base period. The current cost is $P_n * Q_n$, and the base period cost is $P_0 * Q_n$. The index is calculated as

$$\frac{\sum P_n * Q_n}{\sum P_0 * Q_n} \times 100 \tag{2.44}$$

The Paasche quantity index measures the quantity of current output weighted by current period prices relative to the quantity of base period output. The index is calculated as

$$\frac{\sum P_n * Q_n}{\sum P_n * Q_0} \times 100 \tag{2.45}$$

A problem with current weighting the index is that the group of items being compared is not the same between time periods. When tastes or technology bring about rapid change, the comparisons could be rendered meaningless.

Capitalization-weighted indices

Some indices relating to security values, such as most stock exchange indices, weight the constituents by their current market capitalization. This is calculated as **current price** multiplied by the **current quantity** of securities that have been issued.

The weightings change as the relative value of each security issue changes. This change may be due to the changing price of the securities or the changing quantities issued or both. Thus as a share rises in value relative to others in the index, the importance of that share in the index will increase as its weighting (price * quantity) increases relative to the other constituents. Conversely, if a share price falls relative to the others, the importance of that share in the index will decline. In addition if a company issues more shares relative to other constituents, then, provided the price of those securities does not fall, $P * Q$ will increase and the weighting of those securities in the index will increase.

Thus capitalization-weighted indices have a bias in favour of successful companies because these companies have share prices that are rising and/or are able to issue additional shares to investors, thus their weighting increases. Less successful companies on the other hand fall in price and, depending upon the rules for inclusion as an index constituent, ultimately fall out of the index, making way for other more successful companies that originally were not constituents.

A good example of how the market capitalization weighting system affects not only the weights between periods but also the constituents of the index is the takeover of Midland Bank by the Hong Kong and Shanghai Bank. The FTSE 100 index covers the 100 largest companies (by market capitalization) listed on the London Stock Exchange. Individually neither bank was large enough to qualify. However, the combined market capitalization of the two banks was sufficient for the new bank to be included in the index. So the now enlarged Hong Kong and Shanghai Bank was admitted but the smallest of the previously 100 largest had to be removed to make way. In addition, the share price of the Hong Kong and Shanghai Bank rose relative to other constituents, again enhancing its importance in the index.

Calculation of equity indices

There are a large number of equity indices calculated around the world, because there are so many separate stock markets and because many markets have more than one index, each reflecting a different section on the overall market.

The majority of equity market indices are market capitalization weighted, although a few are arithmetic means of price relatives and others are geometric means of equally weighted price relatives. We will illustrate each of these forms below.

Simple aggregate of prices. The most naive form of price index is simply an aggregate of the price of the constituents. As the high-priced items are likely to have larger absolute price changes than low-priced items, changes to high-priced items will have a much greater influence on the index than changes to small-priced items. This weakness is exacerbated because this system takes no account of the size of the unit being measured. For example, a simple aggregate price-weighted index of share prices will be equally sensitive to a 20 pence change in price of one share priced at £10 as it would be with a similar change in price of a £0.25 share.

Sometimes price aggregate indices reflect the average price of the securities in the index, and changes in the index represent the average price change of the securities in the index calculated as

$$\frac{\sum P_i}{n} \tag{2.46}$$

The Dow Jones Industrial Average, the major market index of the US Stock Exchange (MMI) and the Nikkei 225 Stock Average are all of the average price aggregate form of index.

Price aggregate indices give equal weight to all constituents, thus giving greater relative weight to less important constituents. However, in the Dow Jones and the MMI, this influence is negligible because only a few stocks, all relatively large, are included.

Equally weighted geometric mean of price relatives

Some indices are constructed as the geometric mean of price relatives, with each price relative being treated equally. This technique overcomes the problem of giving equal weight to price changes irrespective of their relative size but it does not overcome the problem of not accounting for the relative size of the constituent companies.

The Financial Times Ordinary Index is an equally weighted geometric average of the price relatives of 30 shares.

Market capitalization weighted arithmetic mean of price relatives

This type of index is a **weighted average of prices** because the individual equity prices are weighted to reflect their relative importance within the group of equity prices that constitute the index. As the form of weighting is market capitalization weighting, the relative importance of each equity, and therefore the relative importance of the change in price of each equity, in determining the value of the index is influenced by the relative size of each constituent company.

To illustrate the effects of capitalization weighting, a hypothetical index will be constructed. Assume that we are to construct an index covering the shares of just three companies. Company A has 1000 shares at issue each priced at 100, company B has 10 000 shares at issue priced at 25, company C has 5000 shares at issue priced at 50 and company D has 8,000 shares at issue also priced at 50.

The market capitalization for each of the companies will be as follows:

$$A \quad 1000 * 100 = 100\ 000$$
$$B \quad 10\ 000 * 25 = 250\ 000$$
$$C \quad 5000 * 50 = 250\ 000$$
$$D \quad 8000 * 50 = 400\ 000$$

The total value of the market capitalizations is 1 000 000. This will be the base value of the index, although it will be expressed in terms of some more simple number, say 1000. This is achieved by dividing by a constant divisor; in our example the divisor will be 1000.

To show how the price movements of the constituent shares are reflected in changes in the index, consider the following changed prices after just one day: $A = 105$, $B = 26$, $C = 45$, $D = 55$.

The sum of the market capitalizations will be $105\,000 + 260\,000 + 225\,000 + 440\,000 = 1\,030\,000$. To derive the value of the index this figure must be divided by the constant divisor, i.e. 1000. The new index number is $1\,030\,000/1000 = 1030$, indicating that the overall value of the index has risen since its inception. This is despite the fact that one stock fell in price. The reason is that although the fall in value of company C's share price was as large as the rise in that of companies A and D, company A and D had a much larger combined market capitalization. Therefore, the relative importance of the fall in the share price of C was less than the rise in the share price of A and D.

This example can be generalized to give the construction of a market capitalization weighted index consisting of n equities, as

$$\text{Index}_t = \frac{\sum_{j=1}^{n} m_{j,t} P_{j,t}}{CD} \tag{2.47}$$

where

n = the number of constituent in the index
$m_{j,t}$ = the number of shares outstanding in company j at time t
$P_{j,t}$ = the price of one of those shares at time t
CD = constant divisor.

The constant divisor is the base value of the index, i.e. the sum of $m_{j,0} * P_{j,0}$ divided by the initial value of the index, i.e. 100 or 1000 as appropriate.

Alternatively, the same result can be achieved by multiplying the top part of equation (2.47) by the opening value of the index and dividing the product by the base value of the index instead of the constant divisor. This is shown as

$$\text{Index}_t = \frac{Z \sum_{j=1}^{n} m_{j,t} P_{j,t}}{\text{Base value}} \tag{2.48}$$

where Z is the initial value of the index, i.e. 100 or 1000 as appropriate.

This formula is similar to that used for the calculation of a number of equity indices, for example the FTSE 100 and 250 indices in London, the S&P 100 and 500 indices in the United States and the CAC 40 in France. Only the number of constituents of each index differs.

It is a feature of equity indices that the constituents are changed periodically. Thus in the context of equity indices the base value warrants discussion. Generally in the calculation of a base-weighted price index, the base quantity and the base price are fixed. However, with equity indices that base value must be adjusted periodically to accommodate changes due to companies entering and leaving the index and changes due to the changing number of shares issued by constituent companies.

For example, it has already been explained that the FTSE 100 index consists of the ordinary shares of the largest 100 companies by market capitalization listed on the London Stock Exchange. As relative share prices change over time, some companies will fall out of that classification and new ones will enter. The base value of the index must be adjusted to reflect these changes, otherwise changes in the index level would reflect not only relative price movements, but also changes in constituents and changes in the capital structure of those constituents.

Thus a mechanism is required that allows for changes in constituents but provides continuity of the index series. This is achieved by a process known as chain linking.

The chain linking process requires that the base value of the index be re-calculated to take account of the changes to the constituents and/or the capital structure of the constituents. Although there may be complex capital changes to an index as constituents are added and deleted, the net change in the aggregate market capitalization of the constituents is given as

$$CD_t = CD_{t-1} \left[1 + \frac{C_t}{\sum_{i=1}^{n} N_{i(t-1)} \times P_{i(t-1)}} \right] \tag{2.49}$$

where

CD = constant divisor
C_t = the change in the capitalization
N_i = number of shares in company i at issue at time $t - 1$
P_i = price of shares in company i at time $t - 1$.

Price indices and total rate of return

It should be noted that most published indices of financial markets are price only indices. The movements in the index only reflect movement in the prices of the underlying instruments, they take no account of the dividends or other income received by investors and, therefore, do not calculate the total rate of return.

The exclusion of dividends or other income payments results in the return as measured by movements in the index understating the actual returns to holding a portfolio that replicates that index. To see how this occurs, imagine that, using our earlier hypothetical example of an index, company C paid a dividend of 5. On the ex-dividend date the share price will fall by 5 (assuming no tax effects) to 45. The index will fall to 975. Yet the investor gets the 5 in dividend and the overall value of the portfolio has not changed. Thus when comparing the returns to an index with the returns to a portfolio, the influence of dividends upon the index must be taken into account.

Choice of base period

With any index number methodology a base period must be chosen. There are no hard and fast rules as to what constitutes the "right" base period. It may be a single date. It may be a period of more or less length. The important consideration is that it must not be an "unusual" period. In other words it must be representative of the periods that have gone before. Unfortunately, of course, that is no guarantee that it will be representative of what follows.

In addition it is considered that the base period should not be so far in the past that it is difficult for contemporary observers to relate to conditions in the base period.

Choice of constituents

This choice must depend on the function of the index and the experience of those constructing the index. In some cases, survey work may be required to determine the most appropriate constituents. For example the retail price index constituents and weighting are determined by the Household Expenditure Survey.

Within the realms of financial markets, the choice of constituents will be dictated by the sector which the index is intended to reflect. Thus an index reflecting the major constituents of the stock market will be dominated by the larger companies. However, an index intended to reflect the lowest capitalization stocks will not by definition contain large companies. It is important in these circumstances that the criteria for inclusion in an index are clearly established and widely publicized. As companies are dynamic entities, the constituents of equity indices change, some companies leaving the index as they no longer meet the criteria for inclusion and others joining the index as they meet those same criteria for the first time.

Examples of security market indices

FTSE 100. This is a capitalization weighted index consisting of the top 100 stocks listed on the London Stock Exchange. The base value is 1000, and the base date is 31 December 1984. It is calculated each minute of the trading day. As the constituents are classified by market capitalization, they will change as relative capitalizations change. Additions and deletions to the constituency are made quarterly. This index is calculated as price-only and total rate of return forms, but the price-only form is the type most frequently quoted.

This index is based upon the average of the bid–offer quotes and not from actual transaction prices.

FTSE 250. This index is conceptually similar to the FTSE 100 index except that it contains the 250 largest London stocks immediately below the top 100. There is also a FTSE 350 index which is a combination of the FTSE 100 and 250 indices.

Standard and Poors 100 and 500 indices. These are capitalization-weighted indices covering the top 100 and 500 most actively traded stocks in the USA. The S&P 500 covers the largest 456 stocks on the New York Stock Exchange, 36 stocks traded on the NASDAQ and eight traded on the American Stock Exchange.

New York Stock Exchange composite index. This is a market capitalization weighted index which covers all shares traded on the New York Stock Exchange (about 1700). It is computed every 15 seconds.

DAX (Deutscher Aktienindex). This is an arithmetic capitalization-weighted index of 30 German companies listed on the Frankfurt Stock Exchange. This index is noteable for ignoring ex-dividend and ex-rights adjustments. Consequently, the market capitalization of each constituent (i.e. price * number of shares) is not quite the same as in, say, the FTSE 100 index. The DAX behaves more like a total rate of return index with dividends and bonus shares reinvested.

CAC (Compagnie des Agents de Change) 40 index. This index is a market capitalization weighted index of 40 shares drawn from the top 100 largest companies listed on the Paris Bourse. It is recalculated every 30 seconds.

Exercises

1. Using the following ungrouped data relating to 20 observations of the daily returns to the FTSE 100 index calculate:
    (a) the median daily return
    (b) the mode
    (c) the arithmetic mean
    (d) the geometric mean.

Ungrouped data: 20 observations of daily FTSE index returns:

−0.43, −0.13, −0.38, −0.50, −0.68, 0.84, −0.05, −0.53, −0.04, −0.35,
  1.07,   0.58, −0.75,   0.26, −0.04, 0.68, −0.51,   0.71, −0.12 ,   0.02.

2. Assuming that the above returns are consecutive daily observations calculate the rate of growth of £1000 invested for the 20-day period in question. Compare this with the rates of growth suggested by your arithmetic and geometric means.

3. From the grouped returns data below, calculate the median and arithmetic mean.

   Grouped data

Returns	Frequency
up to −0.05	1
more than −0.05 and up to −0.04	2
more than −0.04 and up to −0.03	5
more than −0.03 and up to −0.02	24
more than −0.02 and up to −0.01	191
more than −0.01 and up to 0.00	714
more than 0.00 and up to 0.01	816
more than 0.01 and up to 0.02	235
more than 0.02 and up to 0.03	25
more than 0.03 and up to 0.04	4
more than 0.04 and up to 0.05	1
Total	2018

4. From the ungrouped data given in question 1, calculate the variance, standard deviation and quartile deviation.

5. From the grouped data provided in question 3 calculate the variance, standard deviation and quartile deviation.

6. Explain the circumstances where the semi-variance and the semi-deviation might be used as measures of dispersion.

7. Explain what is meant by skewness. Calculate the moment coefficient of skewness from the ungrouped data in question 1 and interpret your result.

8. Explain what is meant by kurtosis. Calculate the moment coefficient of kurtosis from the ungrouped data in question 1 and interpret your result.

9. From the following data relating to the FTSE 100 index and the S&P 500 index, calculate the covariance and the correlation coefficient.

FTSE 100	S&P 500
3491.8	481.61
3328.1	467.14
3086.4	445.77
3125.3	450.91
2970.5	456.5
2919.2	444.27
3082.6	458.25
3251.3	475.49
3026.3	462.69
3097.4	472.35
3081.4	453.69
3065.5	459.27
2991.6	470.42
3009.3	487.39
3137.9	500.71
3216.7	514.71

10. (a) From the following data construct a market capitalization weighted index with a base period value of 1000.

Company	Number of shares at issue	Share price
A	1,000,000	2.50
B	5,000,000	1.75
C	10,000,000	0.80
D	8,000,000	1.60
E	7,500,000	3.00

(b) Calculate the value of the index given the following new share prices.

A = 2.70, B = 1.30, C = 1.20, D = 1.40, E = 2.70.

Answers to selected questions

1. Median = −0.085
   Mode = −0.04
   Arithmentic mean = −0.0175
   Geometric mean = 0.000189

3. Median = 0.000888
   Mean = 0.000525

4. Var = 0.2905
   Std = 0.5390
   QD = 0.39375

5. Var = 0.0000914
   Std = 0.00956

7. Skew = 0.560761

8. Kurt = 2.0645

9. Cov = 1092.14445
   Cor = 0.4092

10. 965.17

Further reading

Bowers, D. (1991) *Statistics for Economics and Business*. Macmillan, London.
Curwin, J. and Slater, R. (1993) *Quantitive Methods for Business Decisions*, 2nd edn. Chapman & Hall, London.
Silver, M. (1992) *Business Statistics*. McGraw-Hill, London.

Appendix: the sample standard deviation – why the $n - 1$ divisor?

Consider a random sample of size n drawn from a population with mean μ and standard deviation σ, these paraameters being unknown.

The sample mean is

$$\frac{1}{n}\sum_{i=1}^{n} X_i = \bar{X}$$

We take this as an estimate for μ.

To estimate the variance we would like to compute

$$\frac{1}{n}\sum_{i=1}^{n}(X_i - \mu)^2$$

However, we do not know μ and we must use our estimate of it.

$$\frac{1}{n}\sum_{i=1}^{n}(X_i - \bar{X})^2$$

To see what effect this has consider

$$\sum_{i=1}^{n}(X_i - \bar{X})^2 = \sum_{i=1}^{n}(X_i - \mu + \mu - \bar{X})^2 = \sum_{i=1}^{n}((X_i - \mu) + (\mu - \bar{X}))^2$$

$$= \sum_{i=1}^{n}((X_i - \mu)^2 + 2(X_i - \mu)(\mu - \bar{X}) + (\mu - \bar{X})^2)$$

$$= \sum_{i=1}^{n}(X_i - \mu)^2 + 2(\mu - \bar{X})\sum_{i=1}^{n}(X_i - \mu) + n(\mu - \bar{X})^2$$

$$= \sum_{i=1}^{n}(X_i - \mu)^2 + 2(\mu - \bar{X})\left(\sum_{i=1}^{n} X_i - n\mu\right) + n(\mu - \bar{X})^2$$

$$= \sum_{i=1}^{n}(X_i - \mu)^2 + 2(\mu - \bar{X})(n\bar{X} - n\mu) + n(\mu - \bar{X})^2$$

$$= \sum_{i=1}^{n}(X_i - \mu)^2 - 2n(\bar{X} - \mu)^2 + n(\mu - \bar{X})^2$$

$$= \sum_{i=1}^{n}(X_i - \mu)^2 - n(\bar{X} - \mu)^2$$

Now by definition the expected value of

$$\sum_{i=1}^{n}(X_i - \mu)^2 \text{ is } n\sigma^2$$

Furthermore, as a result of later work on the variance of the sample mean, we know that the expected value of $(\bar{X} - \mu)^2$ is σ^2/n (see Chapter 5).

Thus the expected value of

$$\sum_{i=1}^{n}(X_i - \bar{X})^2 \text{ is } n\sigma^2 - n\frac{\sigma^2}{n} = (n-1)\sigma^2$$

So for an unbiased estimator of σ^2 we need to compute

$$\frac{1}{n-1}\sum_{i=1}^{n}(X_i - \bar{X})^2$$

3

Calculus applied to finance

Introduction

Calculus is that branch of mathematics which studies change. It is divided into two main sub-branches: differential calculus and integral calculus. Differential calculus calculates how, or at what rate, a given variable is changing. In particular it calculates how one variable will change for a given small change in another variable. Integral calculus is used to calculate the areas and volumes of shapes that are bounded by curved lines or curved surfaces.

In this chapter we will introduce the concepts of calculus from first principles and then apply the techniques to specific financial problems. Differential calculus will be applied to calculating the way bond prices fluctuate in response to changes in yield. Integral calculus will be applied to calculating the area under a curve. Finding the area under a curve will be applied in subsequent chapters when we need to determine the probability of a financial variable taking on a value within a particular range of values.

Differential calculus

Differential calculus provides the methodology to determine the amount by which one variable changes in response to a change in one or more other variables, and whether that rate of change is increasing, decreasing or constant. For example, we can calculate the speed of a vehicle from knowledge of the distance travelled and the time travelled. In the realms of finance the interest rate sensitivity of bond prices can be derived.

The rate of change of one variable, Y, in response to a change in another variable, X, is known as the first derivative of Y with respect to X. Whether that rate of change is increasing or decreasing or constant, i.e. whether Y is accelerating, decelerating or changing at a constant rate, is given by the second derivative. We will now look at these concepts in detail.

The first derivative – the rate of change

In this section we will investigate the rate of change of one variable, Y, in response to a change in a single other variable, X. Later in this chapter we will investigate how Y changes when only one of several other variables changes, and when all of several other variables change together.

Let us first consider the following linear functions: $Y = 12 + 3X$, and $Y = 3X$. The first function says that when X is zero Y will have a value of 12, and that otherwise the value of Y will be 12 plus three times the value of X. The second function says that the value of Y is three times the value of X. Below we give values of Y for possible values of X for both functions

$Y = 12 + 3X$	X	0	1	2	3	4	5	6	7	8	9	
	Y	12	15	18	21	24	27	30	33	36	39	

$Y = 3X$	X	0	1	2	3	4	5	6	7	8	9	10
	Y	0	3	6	9	12	15	18	21	24	27	30

Both of these functions are straight lines and are plotted in Fig. 3.1.

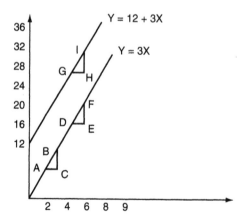

Figure 3.1

The rate of change of Y resulting from a change in X is given by the slope of the line. The slope of each of these lines is given by the ratio of ΔY, the change in Y, over ΔX, the change in X; this ratio is $\Delta Y/\Delta X$. If we observe the right-angled triangles where the hypotenuse is given by part of the lines plotted from the above data, we will note that the ratio of the vertical over the horizontal is, in these examples, always three, irrespective of on which part of each line we draw the triangles, and also irrespective of which of the two lines.

We could discern the same result by just looking at the two functions. We know that the change in Y will be three times the change in X, thus in our examples $\Delta Y/\Delta X$ will be three. Clearly as the function in question is linear, the slope of the line or the rate of change of Y resulting from a change in X will always be three, thus $\Delta Y/\Delta X$ will be constant. In addition, the constant term, 12, in the first function does not influence the slope of the line, only its position. Thus the constant does not influence the rate of change of Y.

However, what happens when the function is not linear. For example, consider the function $Y = 4 + 2X^2$. The values of Y for various values of X are given below:

X	Y
0	4
1	6
2	12
3	22
4	36

This function can be plotted as shown in Fig. 3.2.

The important feature of this line is that it gets steeper as X increases. In other words Y increases at an increasing rate as X increases, even though X is increasing at a constant rate. Thus the gradient or rate of change is not constant.

Clearly it is not possible to form a right-angled triangle with the hypotenuse being formed by the curve. However, consider forming a right-angled triangle between two points on the non-linear curve in Fig. 3.2. The ratio $\Delta Y/\Delta X$ no longer gives the actual or instantaneous rate of change of Y for a given change in X at any point along the curve. Instead it gives only an average rate of change. This average will be less representative

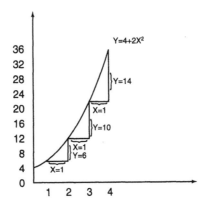

Figure 3.2

of the instantaneous rate change the larger the distance between the two points on the curve. Conversely, the smaller the distance between the two points the more accurate will be the measure of the rate of change of Y. As the distance between the two points gets smaller and smaller it becomes infinitesimally small or close to zero – mathematicians say it is approaching its limiting value, or the limit. The limit is the proportional change in Y when the change in X approaches zero. To indicate such a limiting change, mathematicians alter their notation, thus $\Delta Y/\Delta X$ when X approaches its limit is given as dY/dX.

Clearly then, to find the instantaneous rate of change of Y we have to calculate dY/dX. But how do we find the value of dY/dX? To understand this, consider the function $Y = X^2$. Now assume that X increases by a very small amount, δX. In order to maintain the equality, Y must also grow, actually by δY. Thus, given the function $Y = X^2$,

$$Y + \delta Y = (X + \delta X)^2 \tag{3.1}$$

Expanding the right-hand side of this function gives

$$(X + \delta X)^2 = (X + \delta X)(X + \delta X) =$$
$$X * X + X * \delta X + X * \delta X + \delta X * \delta X = X^2 + 2(X * \delta X) + \delta X^2$$

Thus

$$Y + \delta Y = X^2 + 2(X + \delta X) + (\delta X)^2 \tag{3.2}$$

Recall that we want to find the value of dY, and also recall that the original function was $Y = X^2$. We can keep equation (3.1) in equilibrium by deducting Y from the left-hand side and X^2 from the right-hand side. The result is

$$\delta Y = 2(X * \delta X) + (\delta X)^2 \tag{3.3}$$

Dividing through by δX

$$\frac{\delta Y}{\delta X} = 2X + \delta X \tag{3.4}$$

Letting δX tend to zero gives, in the limit

$$\frac{dY}{dX} = 2X \tag{3.5}$$

As the original function was $Y = X^2$, so dY/dX has become $2X$. Thus to derive dY/dX we have simply multiplied the X variable by a number equal to the power to which X was raised in the original function, and then reduced that power by one. Thus X^2 becomes $2 * X^1$. Remembering that $X^1 = X$, $dY/dX = 2X$.

This can be generalized as follows. When the function is $Y = X^n$

$$dY/dX = nX^{n-1} \tag{3.6}$$

Extending this to the earlier function $Y = 4 + 2X^2$

$$dY/dX = 2 \times 2X^1 = 4X \tag{3.7}$$

Note that constants are eliminated in differentiation. The logic of this was shown right at the beginning of this chapter when it was shown that the slopes of both of the linear functions in Fig. 3.1 were the same, i.e. they were unaffected by the constant. Clearly the rate of change of Y is not affected by a constant value.

Having looked at the basic principles behind differentiation, we can now summarize some basic rules for differentiation.

Differentiating constants

1. Given the function $Y = a$

$$dY/dX = 0 \tag{3.8}$$

In this case Y is a constant "a", which by definition cannot change.

2. Given the function $Y = bX$

$$dY/dX = b \tag{3.9}$$

In this case Y increases at a constant rate of b times X, so the rate of increase in Y per unit increase in X is a constant b. $Y = bx$ is in fact a function representing a straight line.

3. Given the function $Y = a + bX$

$$dY/dX = b \tag{3.10}$$

The reasoning is that the constant, a, is ignored as in rule (1) above, and Y grows at a constant rate bX as in rule (2) above. Again, $Y = a + bX$ is also a function for a straight line.

Differentiating when X is raised to a power

4. Given the function $Y = X^n$

$$dY/dX = nX^{n-1} \tag{3.11}$$

5. There is a special case of rule (4), and that is when X is raised to a negative power, i.e. $Y = X^{-n}$. In this case we multiply X by n as in rule (4) and we make the power n to which the X is raised more negative by one. For example, if $Y = X^{-2}$ (remembering that a negative power indicates a reciprocal, so that X^{-2} is $1/X^2$)

$$dY/dX = -2X^{-3} \tag{3.12}$$

This case is particularly important in bond portfolio management because it is used to determine the interest rate sensitivity of bonds. This will be demonstrated later in this chapter.

6. There is a second special case of rule (4) and that applies when the power to which X is raised is a fraction, i.e. $Y = X^{1/n}$; $X^{1/n}$ is the nth root of X; $X^{p/n}$ is the nth root of X raised to the power p.

When $Y = X^{1/n}$

$$dY/dX = \frac{1}{n} \times X^{\frac{1}{n}-1}$$

(3.13)

For example, if $Y = X^{1/3}$ (this is the same as $\sqrt[3]{X}$)

$$dY/dX = \frac{1}{3} \times X^{\left(\frac{1}{3}-1\right)} = \frac{1}{3} \times X^{-\frac{2}{3}} = \frac{1}{3\sqrt[3]{X^2}}$$

Differentiating the sum of two functions of X

7. Given the function $Y = u + v$, and both u and v are functions of X

$$dY/dX = du/dX + dv/dX$$

(3.14)

For example, if $Y = 2X^4 + 3X^2$

$$dY/dX = 8X^3 + 6X$$

8. Given the function $Y = u - v$, and both u and v are functions X

$$dY/dX = du/dX - dv/dX$$

(3.15)

For example, if $Y = 3X^3 - 2X$

$$dY/dX = 9X^2 - 2$$

Differentiating the product of two functions of X

9. When Y is a product of two functions of X, dY/dX is found by multiplying each function by the derivative of the other function and adding the two products together. Thus given the function $Y = u * v$

$$dY/dX = v(du/dX) + u(dv/dX)$$

(3.16)

For example, if $Y = (5X + 2)(4X^2)$, we can express the right-hand side as $u * v$, where $u = 5X + 2$ and $v = 4X^2$.

The first step is to find du/dX and dv/dX

$$du/dX = 5$$

$$dv/dX = 8X$$

The second step is to calculate $v(du/dX)$, which in this example is

$$4X^2(5) = 20X^2$$

The third step is to calculate $u(dv/dX)$, which is

$$(5X + 2)8X = 40X^2 + 16X$$

Thus

$$dY/dX = 20X^2 + 40X^2 + 16X = 60X^2 + 16X$$

Differentiating the quotient of two functions of X

10. Given the function $Y = u/v$

$$dY/dX = [v(du/dX) - u(dv/dX)]/v^2 \qquad (3.17)$$

For example, assume again that $u = 5X + 2$ and $v = 4X^2$

$$\frac{du}{dX} = 5, \quad \frac{dv}{dX} = 8X$$

$$\frac{dY}{dX} = \frac{4X^2\left(\dfrac{du}{dX}\right) - (5X + 2)\left(\dfrac{dv}{dX}\right)}{(4X^2)^2}$$

$$\frac{dY}{dX} = \frac{4X^2(5) - (5x + 2)(8x)}{(4X^2)^2}$$

$$\left(\text{This can be further simplified to } \frac{-5x - 4}{4x^3}\right)$$

Differentiating a function of a function

11. Given the function $Y = f(u)$, where u is itself a function of another variable, for example $u = y(x)$

$$dY/dX = (dY/du)(du/dX) \qquad (3.18)$$

This is called the chain rule.

To illustrate this, assume that $Y = (2X^3 + 3)^6$. Y would be equal to u^6, where $u = 2X^3 + 3$

$$du/dX = 6X^2 \quad \text{and} \quad dY/du = 6u^5$$

Consequently, $dY/dX = 6u^5 * 6X^2$. Therefore

$$\frac{dY}{dX} = 6(2X^3 + 3)^5(6X^2)$$

Differentiating an exponential function

The exponential function is a particularly important function in calculus because, except for the zero function, this is the only function that differentiates to itself.

12. Take for example the function $Y = e^X$

$$\mathbf{dY/dX = e^X} \tag{3.19}$$

13. If $Y = e^{aX}$, then

$$\mathbf{dY/dX = a\ e^{aX}} \tag{3.20}$$

For example, if $Y = e^{3X}$, then $dY/dX = 3e^{3X}$.

14. If e is raised by a more complex function, say

$$Y = e^{X^3 + 2X^2}$$

we apply the chain rule of equation (3.18) above. For example, let the power of e be represented by u, thus $Y = e^u$. We then find du/dX and then dY/dX as $dY/du * du/dX$. dY/du is e^u from rule (12) above, and

$$\frac{du}{dX} = 3X^2 + 4X$$

So

$$\frac{dY}{dX} = e^u * \frac{du}{dX} \tag{3.21}$$

This becomes

$$(3X^2 + 4X)e^{(X^3 + 2X^2)}$$

Differentiating a natural logarithmic function

15. If $Y = \log_e X$ then

$$\mathbf{dY/dX = 1/X} \tag{3.22}$$

Thus the first derivative of the natural log of a number, is the reciprocal of that number.

16. If $Y = \log_e$ of a more complex function of X, say

$$\mathbf{\log_e (X^3 + 2X^2)}$$

then let $X^3 + 2X^2$ be represented by u. Thus $Y = \log_e (u)$. We then find du/dX and then dY/dX as $dY/du * du/dX$.
  dY/du is given as

$$\frac{dY}{du} = \frac{1}{u} = \frac{1}{X^3 + 2X^2} \tag{3.23}$$

We know that du/dX is 3X^2 + 4X, so

$$\frac{dY}{dX} = \frac{dY}{du} * \frac{du}{dX} = \frac{1}{u}(3X^2 + 4X) = \frac{3X^2 + 4X}{X^3 + 2X^2}$$

Second derivatives

The differentiating that has been explained so far results in the first derivative of the function, the rate of change. To determine whether the variable is accelerating, decelerating or changing at a constant rate we need the second derivative of the function. This is designated **d^{2}Y/dX^2**.

To derive the second derivative of a function one simply has to differentiate the first derivative of that function. For example, consider the function $Y = 4 + 2X^2$. The first derivative is

$$\frac{dY}{dX} = 4X$$

and the second derivative is found by differentiating 4X, i.e. 4.

Taylor series expansions

It is often useful to approximate to a function by using simpler functions. Taylor series expansions provide a methodology for doing this. For example, consider the price of a two-year bond paying annual coupons. The price P as a function of the yield to maturity, y, is

$$P = f(y) = \frac{CF_1}{(1+y)} + \frac{CF_2}{(1+y)^2}$$

The graphical representation of such a function is known as a price yield curve and is represented schematically in Fig. 3.3. The function $f(y)$ is quite complex, and would get more so the greater the number of cash flows due under the bond. It would be easier if we had an approximation to $f(y)$. Taylor series approximations are polynomial expressions, combinations of a number and powers of a variable which we will call h. We can construct constant, linear, quadratic, cubic, degree 4, degree 5, ... , etc. approximations.

By their nature approximations are usually valid over a limited range of values of the underlying variable (which we shall take to be y in this section). We can choose to approximate to f about any point we like, so let us indicate that point by y. Having chosen this point we will fix it (so that y is a sort of fixed variable!) and consider points that entail a small change in y, say $y + h$. Here h is taken to be small, and it can be negative. So h is the variable.

We list below Taylor series approximations for $f(y)$ about y, The constant approximation we have called P_0. The linear approximation we have called P_1, the quadratic P_2, etc. and so on.

$$P_0(y+h) = f(y)$$

$$P_1(y+h) = f(y) + f'(y) \times h$$

$$P_2(y+h) = f(y) + f'(y) \times h + \frac{f''(y)}{2} \times h^2$$

$$(3.24)$$

$$P_3(y+h) = f(y) + f'(y) \times h + \frac{f''(y)}{2} \times h^2 + \frac{f'''(y)}{6} \times h^3$$

$$P_4(y+h) = f(y) + f'(y) \times h + \frac{f''(y)}{2} \times h^2 + \frac{f'''(y)}{6} \times h^3 + \frac{f^{iv}(y)}{24} \times h^4$$

Note that $f'''(y)$ means the third derivative of $f(y)$ with respect to y evaluated at $y = y$, and that $f^{iv}(y)$ is the fourth derivative evaluated at $y = y$.

Note also that the divisors, 2, 6, 24, etc. are factorial numbers, $2! = (2 \times 1)$, $3! = (3 \times 2 \times 1)$, $4! = (4 \times 3 \times 2 \times 1)$, etc.

There is no sense in which these approximations claim to be the "best" linear, or "best" quadratic, etc., approximations. They are constructed so that P_0 has the **same value** as f at $y = y$, P_1 has the **same value** and the **same first derivative** as f at $y = y$, P_2 has the **same value**, **same first derivative** and **same second derivative**, etc.

How do we interpret these approximations? Consider Fig. 3.3 below. The height of $f(y)$ gives a point on the line equal to P. Now to approximate $P = f(y)$ when y changes by a small amount we could add a linear approximation. This would be $P = f(y + h) = f(y) + hf'(y)$. Being a linear approximation the change in the value of P would be approximated by moving along the straight line, not along the curve.

Thus, for any but the smallest of changes in y, the approximation will be inaccurate. We can improve the approximation by making a quadratic approximation, which entails adding the second derivative to the linear approximation, thus $P = f(y) + hf'(y) + (h^2/2!)f''(y)$.

For some functions, although not usually for bond analysis, it may be necessary to use the cubic and higher order approximations.

To illustrate the application of the linear and quadratic approximations we will apply the principles first to estimating changes in bond prices and secondly to deriving bond price volatility.

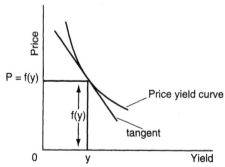

Figure 3.3 Price yield curve of a bond.

Applications of differential calculus

Applying Taylor series to estimate bond price changes

Figure 3.3 is the schematic representation of the relationship between the yield and the price of a bond. The curve, known as the price yield curve of a bond, is non-linear and negatively sloped. To model the change in the price resulting from a change in the yield may be very complex. However, from our understanding of Taylor series expansions, we should be able to approximate the price yield function by some stages of a Taylor series. We could, for example, apply the first derivative of the bond price with respect to yield, and the second derivative and the third derivative, and so on. In fact we will see below that applying just the first two stages of a Taylor series provides a very good measure of a change in price of a bond resulting from small changes in yield. Moreover, if we divide the various elements of the Taylor series by the bond price, we can derive a very useful measure of bond price volatility.

To illustrate how the Taylor series may be applied to approximating bond price changes, take the simple example of a one-year zero coupon bond paying 100 one year hence. If the yield-to-maturity is 10% the current price will be 90.9090, thus $P = f(0.10) = 90.9090$. Now what would be the price of the bond if y changed from 0.10 to 0.11? We can make a linear approximation of this change by adding the first derivative multiplied by the change in yield to the constant giving

$$90.9090 + h\frac{dP}{dy}$$

Recall that $P = f(y) = P = 100/(1 + y)$. Thus

$$\frac{dP}{dy} = \frac{-100}{(1+y)^2} \tag{3.25}$$

so that, when $y = 0.10$

$$\frac{dP}{dy} = \frac{-100}{1.10^2} = 82.6446$$

The linear approximation to P with $y = 0.11$, i.e. $P = f(0.11)$, is

$$90.9090 - (0.01 * 82.6446) = 90.0826$$

Actually if the yield-to-maturity does rise instantaneously to 11%, the bond price would only fall to 90.09. Thus, the linear approximation is inaccurate. Indeed, we can see from Fig. 3.3 that the linear approximation overstates the downward movement of the bond price.

We can improve on this approximation by making a quadratic approximation which is given by

$$P(y + h) = f(y) + \frac{dP}{dy}h + \frac{1}{2}\frac{d^2P}{dy}h^2 \tag{3.26}$$

To proceed further we must revise our understanding of negative exponents. $1/y$ is equal to y^{-1}, $1/y^2 = y^{-2}$, and this can be generalized to

$$1/y^n = y^{-n} \tag{3.27}$$

Accordingly

$$1/(1 + y)^1 = (1 + y)^{-1}; \ 1/(1 + y)^2 = (1 + y)^{-2} \text{ and } 1/(1 + y)^T = (1 + y_T)^{-T}$$

and so on.

We must apply this rule in the derivation of the second derivative. As the second derivative is simply differentiation of the first derivative

$$\frac{dP}{dy} = \frac{-100}{(1+y)^2} = -100 * \frac{1}{(1+y)^2}$$

This is the same as

$$-100 * (1 + y)^{-2}$$

which, when differentiated, becomes

$$-2 * -100 * (1 + y)^{-3}$$

Thus, the second derivative of our zero coupon bond price with respect to yield is given by

$$\frac{d^2P}{dy^2} = \frac{200}{(1+y)^3}$$

which, when $y = 0.10$, gives

$$\frac{d^2P}{dy^2} = \frac{200}{1.10^3} = 150.2630$$

Thus, the quadratic approximation becomes

$$90.0826 + \tfrac{1}{2}(150.2630 * 0.01^2) = 90.0826 + (0.5 * 0.0150) = 90.0901$$

This is equal to the true value, to four decimal places.

Applying calculus to measure bond price risk

The first element of a Taylor series divided by the bond price is known as **modified duration**; the second element divided by the bond price is known as **convexity**. Higher order elements are usually considered to be insignificant in measuring bond price sensitivity.

Volatility of bonds

We will now use some of the concepts learned so far in this chapter to measure the price volatility of bonds. In Chapter 1 we explained that the yield to maturity of a bond is the annualized internal rate of return that equates the value of future cash flows to the current bond price.

The current value of the coupon paying bond is calculated as

$$P_{CB} = \frac{CF_1}{(1+y)} + \frac{CF_2}{(1+y)^2} + \frac{CF_3}{(1+y)^3} + \dots + \frac{CF_n}{(1+y)^n} \qquad (3.28)$$

where y is the periodic yield to maturity, i.e. the internal rate of return reflecting the periodicity of the cash flows.

The slope of the tangent of the curve in Fig. 3.3 is the first derivative of the price with respect to yield. The first derivative divided by price gives a measure of the percentage change in the bond price for a 1% change in yield, which is known as modified duration.

In order to understand how to calculate the first derivative dP/dy for the whole bond, we will first calculate it for the present value of a single cash flow

$$P_z = CF_T * \frac{1}{(1+y)^T} \qquad (3.29)$$

where $1/(1+y)^T$ is the discount factor. It is the present value of one currency unit discounted at the appropriate rate of interest for T time periods. Clearly multiplying the present value of one currency unit by the number of units actually due in the future will give the present value of that expected future cash flow.

We can invoke the fifth rule for differentiating given above, but first we must transform equation (3.29) above into

$$P = CF_T * (1+y)^{-T} \qquad (3.30)$$

and dP/dy is

$$\frac{dP}{dy} = (-T)CF_T(1+y)^{-(T+1)} \qquad (3.31)$$

Given that it is normal to transform the exponents into positive form whenever possible, this becomes

$$\frac{(-T)CF_T}{(1+y)^{T+1}} \qquad (3.32)$$

This application can be extended to finding the first derivative of price with respect to yield of a coupon-paying bond by invoking rule (7) above and recalling that a coupon-paying bond can be treated as a portfolio of zero coupon bonds. Thus, rule (5) is used to find dP/dy for each cash flow and then, according to rule (7), these first derivatives are combined to give dP/dy for the whole bond. For example, using the equation for the coupon-paying bond which is repeated here

$$P_{CB} = \frac{CF_1}{(1+y)} + \frac{CF_2}{(1+y)^2} + \frac{CF_3}{(1+y)^3} + \dots + \frac{CF_n}{(1+y)^n}$$

this could be expressed as

$$P_{CB} = CF_1 * (1+y)^{-1} + CF_2 * (1+y)^{-2} + CF_3 \times (1+y)^{-3} + \dots + CF_n * (1+y)^{-n}$$

Then, recalling the way we calculated dY/dy for the zero coupon bond, dP/dy for the individual cash flows will be

$$\frac{dP_1}{dy} \text{ for } CF_1 = \frac{(-1)CF_1}{(1+y)^2}$$

$$\frac{dP_2}{dy} \text{ for } CF_2 = \frac{(-2)CF_2}{(1+y)^3}$$

$$\frac{dP_3}{dy} \text{ for } CF_3 = \frac{(-3)CF_3}{(1+y)^4} \tag{3.33}$$

$$\frac{dP_n}{dy} \text{ for } CF_n = \frac{(-n)CF_n}{(1+y)^{n+1}}$$

dP/dy is the sum of $dP_1/dy + dP_2/dy + \ldots + dP_n/dy$. Therefore collecting all the individual derivatives together as per rule (7) gives

$$\frac{dP}{dy} = \frac{(-1)CF_1}{(1+y)^2} + \frac{(-2)CF_2}{(1+y)^3} + \frac{(-3)CF_3}{(1+y)^4} + \ldots + \frac{(-n)CF_n}{(1+y)^{n+1}} \tag{3.34}$$

Rearranging equation (3.34) and dividing both sides by P (actually multiplying by $1/P$) we get

$$\frac{dP}{dy}\frac{1}{P} = \frac{1}{(1+y)}\left[\frac{(-1)CF_1}{(1+y)^1} + \frac{(-2)CF_2}{(1+y)^2} + \frac{(-3)CF_3}{(1+y)^3} + \ldots + \frac{(-n)CF_n}{(1+y)^n}\right]\frac{1}{P} \tag{3.35}$$

The expression in the square brackets multiplied by $1/P$ is known as Macaulay's duration after Macaulay (1938). The right-hand side of equation (3.35) is known as the **modified duration** and is used by bond market practitioners as an indicator of the interest rate risk in bonds. The modified duration may be interpreted as approximating the percentage change in the price of a bond resulting from a 1% change in yield over small changes in yield in the next instant in time.

Note that in calculating duration, the yield to maturity is used as the discount rate. This assumes that the term structure of interest rates is flat, because the same rate is used to discount all cashflows irrespective of their timing. It follows from this that the term structure is assumed only to shift in parallel. These assumptions are not supported by empirical evidence. These points are reviewed again in Chapter 11.

Numerical example of modified duration

To illustrate the use of differentiation to calculate modified duration, we will apply equation (3.35). Consider a two-year bond that pays 5 semi-annually and the yield to maturity is 8% p.a. The price and cash flow pattern will be

$$P = \frac{5}{(1.04)} + \frac{5}{(1.04)^2} + \frac{5}{(1.04)^3} + \frac{105}{(1.04)^4}$$

$$= 4.8077 + 4.6228 + 4.4450 + 89.7544 = 103.6299$$

First we have to value the arguments within the square brackets in equation (3.35)

$$\frac{-1*5}{(1.04)} + \frac{-2*5}{(1.04)^2} + \frac{-3*5}{(1.04)^3} + \frac{-4*105}{(1.04)^4}$$

$$= -4.8077 + (-9.2456) + (-13.335) + (-359.0175) = -386.4058 \qquad (3.36)$$

next we multiply this by $1/(1 + y)$

$$\frac{1}{1.04} * -386.4058 = \frac{-386.4058}{1.04} = -371.5440$$

Finally, we divide by the price of the bond 103.6299

$$\frac{-371.5440}{103.6299} = -3.5853$$

This modified duration relates to the cash flow periods. It is the bond market convention to quote duration in years and omit mention of the negative sign. As the bond in question pays two cash flows per year, duration in years is arrived at by dividing the calculated figure by two, thus 3.5853/2 = 1.7926.

Modified duration can be interpreted as indicating the relative interest rate sensitivity of the bond. In particular, the percentage change in the bond price will be given as minus the modified duration multiplied by the change in the yield to maturity. In the case of the bond analysed above, for a 0.5% rise in yield the bond price will fall by 0.5×1.7926 = 0.8963%.

The second derivative – the rate of change of the rate of change

If we refer back to Figs 3.2 and 3.3, we will see that in each diagram the slope of the tangent at any two points is different. Indeed the slope of that tangent changes continuously as we move up or down the curve. This is important because the change in the Y variable as indicated by dY/dX only applies for the given value of X. Thus it is also important to find out how dY/dX changes as X itself changes. This change in dY/dX is represented by the second derivative of Y with respect to X, and indicated by $\mathbf{d^2Y/dX^2}$.

Intuitively, if dY/dX represents the rate of change, d^2Y/dX^2 indicates whether that rate of change is accelerating, constant or decreasing.

To calculate the second derivative, one simply differentiates the first derivative. Thus if $\mathbf{Y = X^2}$ and $\mathbf{dY/dX = 2X}$

$$\mathbf{d^2Y/dX^2 = 2} \qquad (3.37)$$

Indeed the third derivative is found by differentiating the second derivative and so on.

Application of the second derivative: bond convexity

One application of the second derivative is to improve on our measure of bond price sensitivity given by modified duration. We can use our knowledge of the second derivative of a bond price with respect to the yield to maturity to calculate what is called bond convexity.

To illustrate this, we will again begin by relating the analysis to a single future cash flow. Recall the dP/dy for a single cash flow can be expressed as

$$\frac{dP}{dy} = (-T)CF_T(1+y)^{-(T+1)} \tag{3.38}$$

To get the second derivative we simply differentiate the first derivative, thus

$$\frac{d^2P}{dy^2} = -(T+1)*(-T)CF_T(1+y)^{-(T+2)} \tag{3.39}$$

Rearranging, this becomes

$$\frac{d^2P}{dy^2} = \frac{-(T+1)*(-T)CF_T}{(1+y)^{(T+2)}} \tag{3.40}$$

The second derivatives for each of the cash flows will be

$$\frac{d^2P_1}{dy^2} \text{ for } CF_1 = (-2)(-1)CF_1(1+y)^{-3}$$

$$\frac{d^2P_2}{dy^2} \text{ for } CF_2 = (-3)(-2)CF_2(1+y)^{-4}$$

$$\frac{d^2P_3}{dy^2} \text{ for } CF_3 = (-4)(-3)CF_3(1+y)^{-5}$$

$$\frac{d^2P_n}{dy^2} \text{ for } CF_n = (-(n+1))(-n)CF_n(1+y)^{-(n+2)}$$

Rearranging, this becomes

$$\frac{d^2P_1}{dy^2} \text{ for } CF_1 = \frac{(-2)(-1)CF_1}{(1+y)^3} = \frac{+2CF_1}{(1+y)^3}$$

$$\frac{d^2P_2}{dy^2} \text{ for } CF_2 = \frac{(-3)(-2)CF_2}{(1+y)^4} = \frac{+6CF_2}{(1+y)^4}$$

$$\frac{d^2P_3}{dy^2} \text{ for } CF_3 = \frac{(-4)(-3)CF_3}{(1+y)^5} = \frac{+12CF_3}{(1+y)^5}$$

$$\frac{d^2P_n}{dY^2} \text{ for } CF_n = \frac{(-(n+1)(-n)CF_n}{(1+y)^{n+2}} = \frac{(n+1)nCF_n}{(1+y)^{n+2}}$$

Collecting all the individual derivatives together as per rule (7) gives

$$\frac{d^2P}{dy^2} = \frac{(2)CF_1}{(1+y)^3} + \frac{(6)CF_2}{(1+y)^4} + \frac{(12)CF_3}{(1+y)^5} + \ldots + \frac{n*(n+1)CF_n}{(1+y)^{n+2}}$$

Numerical example of convexity

We will illustrate the use of the second derivative in bond risk management by calculating the convexity of the two-year bond we looked at earlier. Convexity is half the second derivative of the bond price with respect to yield, divided by the bond price.

The second derivative of the bond price with respect to yield is given as

$$\frac{2*5}{(1.04)^3} + \frac{6*5}{(1.04)^4} + \frac{12*5}{(1.04)^5} + \frac{20*105}{(1.04)^6} =$$

$$8.890 + 25.644 + 49.316 + 1659.661 = 1743.510$$

Given that the bond price is 103.6299, convexity is

$$\frac{1}{2} * \frac{1743.510}{103.6299} = 8.4122$$

As with modified duration, it is bond market practice to quote convexity in years squared. It is therefore necessary to divide the above figure by the square of the number of cash flows per year. In this example there are two cash flows per year, so we must divide by $2^2 = 4$. Thus the convexity of this bond is 2.1031.

We can also use calculus to derive closed-form solutions of modified duration and convexity of coupon-paying bonds, if we assume the appropriate discount rate is the gross redemption yield.

The coupon cash flows under the bond can be considered as an annuity plus a redemption payment. The present value of an annuity paying C for n years is given as

$$P = C\left[\frac{1 - [1/(1+y)^n]}{y}\right] \tag{3.41}$$

Thus, the price of a bond paying C for n years plus a redemption payment of 100 would be

$$P = C\left[\frac{1 - [1/(1+y)^n]}{y}\right] + \frac{100}{(1+y)^n} \tag{3.42}$$

We can derive modified duration by calculating the first derivative and dividing by P. The second derivative is used to calculate convexity. We will begin with the first derivative.

We will start by rewriting equation (3.42) in a way that makes differentiation easier, for example

$$P = C\left(1 - \frac{1}{(1+y)^n}\right)y^{-1} + 100(1+y)^{-n} \tag{3.43}$$

The first derivative, dP/dY, is

$$\frac{dP}{dy} = C\left[-\left(1 - \frac{1}{(1+y)^n}\right)y^{-2} + \left(\frac{n}{(1+y)^{n+1}}\right)y^{-1}\right] - \frac{100n}{(1+y)^{n+1}}$$

$$= C\left[\frac{1}{y^2}\left(\frac{1}{(1+y)^n} - 1\right) + \frac{n}{y(1+y)^{n+1}}\right] - \frac{100n}{(1+y)^{n+1}}$$

$$= \frac{C}{y^2}\left(\frac{1}{(1+y)^n} - 1\right) + \frac{n}{(1+y)^{n+1}}\left(\frac{C}{Y} - 100\right) \qquad (3.44)$$

To arrive at modified duration, the first derivative must be divided by the dirty price of the bond, P, thus

$$\text{MOD DUR} = \frac{\left[\frac{C}{y^2}\left(\frac{1}{(1+y)^n} - 1\right) + \frac{n}{(1+y)^{n+1}}\left(\frac{C}{Y} - 100\right)\right]}{P} \qquad (3.45)$$

The closed-form solution for convexity can be derived in a similar manner.
   The first derivative of equation (3.44) is

$$\frac{d^2P}{dY^2} = -\frac{2C}{y^3}\left(\frac{1}{(1+y)^n} - 1\right) + \frac{C}{y^2}\left(-\frac{n}{(1+y)^{n+1}}\right) - \frac{n(n+1)}{(1+y)^{n+2}}\left(\frac{C}{y} - 100\right) + \frac{n}{(1+y)^{n+1}}\left(-\frac{C}{y^2}\right)$$

Collecting the arguments together and dividing by twice the dirty price of the bond gives

$$\frac{1}{2}\frac{\left[-\frac{2C}{y^3}\left(\frac{1}{(1+y)^n} - 1\right) - \frac{2C}{y^2}\left(\frac{n}{(1+y)^{n+1}}\right) - \frac{n(n+1)}{(1+y)^{n+2}}\left(\frac{C}{y} - 100\right)\right]}{P} \qquad (3.46)$$

Maxima and minima

There are many situations in finance where we need to know the minimum or maximum value of a function. For example, we may wish to know which combination of assets will give the minimum risk, yet achieve a given level of return. Alternatively, there are situations in which we wish to maximize the probability that the estimated parameters of a model are the true parameters. The first type of problem is elaborated on in Chapter 9 which covers optimization. The second type of problem is explained in the Appendix to Chapter 7 which covers maximum likelihood estimation and particularly applies to time series analysis. Calculus is used to identify the minimum and maximum values of these and many other functions.

   The combination of the first derivative and the second derivative can be used to determine whether a particular point on a curve is at a peak (maxima) or the bottom of a trough (minima) or whether there is a change in the slope from concave to convex or vice versa. For an example of peaks and troughs, consider Fig. 3.4.

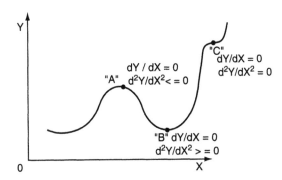

Figure 3.4 Maxima and minima.

From Fig. 3.4 we can clearly see that point A is a local maximum. In terms of calculus the curve is neither rising (positive change), nor falling (negative change), so the first derivative is zero. Point B is clearly at the bottom of a trough, it is a local minimum, and in terms of calculus the first derivative is also zero. Point C is a stationary point of inflection. It is neither a maximum nor a minimum, but it also has a first derivative equal to zero. Thus, the local maximum, the local minimum and the stationary point of inflection all have one thing in common: they are **stationary points**, thus $dy/dx = 0$ in each case.

If the local maximum, the local minimum and stationary points of inflection are all indicated by the first derivative equalling zero, how would we know which is which if we did not have the advantage of the diagram?

The answer lies in the second derivative. When $dY/dX = 0$ and the second derivative is greater than zero, $d^2Y/dX^2 > 0$, the point is a local minimum. If on the other hand $dY/dX = 0$ and the second derivative is less than zero, i.e. $d^2Y/dX^2 < 0$, the point is at a local maximum. To understand this recall that the second derivative indicates the change in the first derivative. Now consider what is happening to the first derivative at point A. To the left of point A, the first derivative is positive. To the right of point A, the first derivative is negative. Thus at point A, the first derivative changes from positive to negative, a change that itself is negative, and thus the second derivative is negative.

Now consider what happens at point B. To the left, the first derivative is negative, and to the right it is positive, a change that is itself positive. Thus the second derivative is positive.

If the first derivative is zero and the second derivative is also zero, we may have a stationary point of inflection. To understand this, again consider Fig. 3.4. At point C the first derivative is zero. To the left of point C the second derivative is negative, whilst to the right it is positive, but at point C it is zero. Thus points of inflection are points where both the first and second derivatives are zero but the second derivative is changing sign.

We can summarize a number of rules for determining, minima, maxima and stationary points of inflection.

1. At any point where $dY/dX = 0$, that point is either a local minimum or a local maximum point or a stationary point of inflection.
2. If at that point d^2Y/dX^2 is negative, that point is a local maximum point.
3. If at that point d^2Y/dX^2 is positive, that point is a local minimum point.
4. If at that point d^2Y/dX^2 is equal to zero and is also changing in sign, that point is a stationary point of inflection.

Finding minimum and maximum values of a function

We can use our understanding of minimum and maximum points to find the minimum or maximum values of a function.

In order to find the maximum value we must set the first derivative of the function to equal zero. This will indicate the position of the turning points. Next, the value of the second derivative is calculated. If it is negative, the turning point will be a maximum point.

A minimum value of a function is found in a similar way, except that if the second derivative has a positive value then the turning point is at a minimum.

These points can be illustrated as follows.

- Find the maximum point of the function $Y = 4X^3 - 2X$.
  The first step is to set dY/dX equal to zero, thus

$$\frac{dY}{dX} = 12X^2 - 2 = 0$$

$$\therefore 12X^2 = 2$$

$$\frac{2}{12} = X^2$$

$$X = \pm\sqrt{\frac{1}{6}} = \pm 0.4082$$

  Thus, one turning point is at $X = 0.4082$, the other is at $X = -0.4082$.
- The second step is to check the sign of the second derivative. The second derivative itself is

$$\frac{d^2Y}{dX^2} = 24X$$

At $X = +0.4082$ the second derivative is positive so the turning point is at a minimum. At $X = -0.4082$, the second derivative is negative and thus the turning point is a maximum. The maximum value of Y is found by substituting $X = -0.4082$ into the original function, $Y = 4X^3 - 2X$, i.e.

$$Y = 4(-0.4082)^3 - 2(-0.4082) = -0.2721 + 0.8164 = +0.5443$$

Differentiating functions of more than one variable

So far we have considered differentiating a function where one variable is a function of only one other variable. In finance, as in other aspects of economics, one variable is often a function of several other variables. When we differentiate a function of several variables with respect to only one of those variables, we engage in **partial differentiation**.

Equations that contain partial differentials are known as **partial differential**

equations and are particularly important in the valuation of derivative instruments. In Chapter 10 we will apply continuous-time partial differential equations to the pricing of options.

Partial differentiation

Consider the function $Y = f(X, Z)$, where X and Z are both independent of each other. Such a function can be differentiated with respect to one independent variable, the others being assumed to be fixed. This form of differentiation is known as partial differentiation.

With respect to partial differentiation the notation changes slightly, the dY/dX becomes $\partial Y/\partial X$.

In order to partially differentiate a function of more than one independent variable, that function is differentiated with respect to one variable while all the other functions are treated as constants. Thus to get the partial derivatives of

$$Y = X^2 + 2Z^3 + Z^2 \tag{3.47}$$

we find $\partial Y/\partial X$, treating the Z variable as a constant, and $\partial Y/\partial Z$, treating the X variable as a constant. For example

$$\frac{\partial Y}{\partial X} = 2X$$

$$\frac{\partial Y}{\partial Z} = 6Z^2 + 2Z$$

Now consider the following function with three independent variables

$$Y = X^4 + W^3 X + XZ - 4Z^3$$

$$\frac{\partial Y}{\partial X} = 4X^3 + W^3 + Z$$

$$\frac{\partial Y}{\partial W} = 3W^2 X \tag{3.48}$$

$$\frac{\partial Y}{\partial Z} = X - 12Z^2$$

To interpret these results $\partial Y/\partial X$ indicates that as X is increased, Y will increase by $(4X^3 + W^3 + Z) * \delta X$. If W is increased, Y will increase by $3W^2 X * \delta W$, and if Z is increased, Y will increase by $(X - 12Z^2) \times \delta Z$.

The second derivatives $\partial^2 Y/\partial X^2$, $\partial^2 Y/\partial W^2$ and $\partial^2 Y/\partial Z^2$ tell us how the marginal changes in Y behave when X, W or Z are changed but the other two variables remain fixed.

If we wish to know how $\partial Y/\partial W$ changes as X changes, we simply differentiate $\partial Y/\partial W$ with respect to ∂X. This is written $\partial^2 Y/\partial X\partial W$, indicating that we want to find the change in $\partial Y/\partial W$ as X changes. Likewise, if we wanted to find the effect on $\partial Y/\partial W$ of a change in Z, we would find $\partial^2 Y / \partial Z \partial W$. This notation indicates that we are differentiating $\partial Y / \partial W$ with respect to Z, i.e. ∂Z.

If we use the notation $\partial / \partial Z$ to indicate that we wish to differentiate with respect to Z

$$\frac{\partial^2 Y}{\partial Z \partial W} = \frac{\partial}{\partial Z}\left(\frac{\partial Y}{\partial W}\right)$$

Total differentiation

Total differentiation explains how the dependent variable, Y, will change when all the independent variables change. Consider again $Y = f(W, X, Z)$. For small changes in the independent variables

$$\Delta Y \approx \frac{\partial Y}{\partial W}(\Delta W)$$

$$\Delta Y \approx \frac{\partial Y}{\partial X}(\Delta X)$$

$$\Delta Y \approx \frac{\partial Y}{\partial Z}(\Delta Z)$$

Note that either δ or Δ is used to indicate a small change.

To indicate how Y will change in response to simultaneous small changes in each of the independent variables we simply add together the products of each partial derivative multiplied by the small change in its respective variable. The complete equation is usually given by

$$\Delta Y = \frac{\partial Y}{\partial W}(\Delta W) + \frac{\partial Y}{\partial X}(\Delta X) + \frac{\partial Y}{\partial Z}(\Delta Z) \tag{3.49}$$

where ΔY is known as the total differential. To illustrate this, consider the following function again

$$Y = X^4 + W^3 X + XZ - 4Z^3$$

$$\frac{\partial Y}{\partial X} = 4X^3 + W^3 + Z$$

$$\frac{\partial Y}{\partial W} = 3W^2 X$$

$$\frac{\partial Y}{\partial Z} = X - 12Z^2$$

The completely differentiated function would be

$$\Delta Y = + (4X^3 + W^3 + Z)(\Delta X) + (3W^2 X)(\Delta W) + (X - 12Z^2)(\Delta Z)$$

Maxima and minima of functions of more than one variable

We will limit our discussion of this topic to the two variable cases. Let the function of interest be $f(X, Y)$. Then we can find two partial derivatives and four partial second derivatives. These are

$$\frac{\partial f}{\partial X}, \frac{\partial f}{\partial Y} \tag{3.50}$$

and

$$\frac{\partial^2 f}{\partial X^2}, \frac{\partial^2 f}{\partial Y^2}, \frac{\partial^2 f}{\partial Y \partial X}, \frac{\partial^2 f}{\partial X \partial Y} \tag{3.51}$$

However, we should note that

$$\frac{\partial^2 f}{\partial Y \partial X} = \frac{\partial^2 f}{\partial X \partial Y} \tag{3.52}$$

In the two-variable situation, there are three types of stationary point, local maxima, local minima and saddle points. The latter are like the high points of mountain passes – mountain peaks on each side and valleys in front and behind.

The conditions for so-called "strong local extrema" are as follows.

Local maxima

The criteria for local maxima are that both first derivatives are equal to zero, i.e.

$$\frac{\partial f}{\partial X} = \frac{\partial f}{\partial Y} = 0 \tag{3.53}$$

and in addition we have

$$\frac{\partial^2 f}{\partial X^2} < 0, \qquad \frac{\partial^2 f}{\partial X^2} \frac{\partial^2 f}{\partial Y^2} > \left(\frac{\partial^2 f}{\partial X \partial Y}\right)^2 \tag{3.54}$$

Local minima

The criteria for local minima are that both first derivatives are equal to zero, i.e.

$$\frac{\partial f}{\partial X} = \frac{\partial f}{\partial Y} = 0 \tag{3.55}$$

and in addition we have

$$\frac{\partial^2 f}{\partial X^2} > 0, \qquad \frac{\partial^2 f}{\partial X^2} \frac{\partial^2 f}{\partial Y^2} > \left(\frac{\partial^2 f}{\partial X \partial Y}\right)^2 \tag{3.56}$$

The above conditions relate to "strong" local extrema. Weak local extrema, which refer to ridges instead of peaks, or valley bottoms instead of bottoms of bowls, are also provided for simply by changing the $>$ and $<$ to $\geq$ and $\leq$.

Maximization and minimization subject to constraints: Lagrangian multipliers

There are many examples in business and finance where we wish to find the maximum or minimum of a function subject to a constraint. For example, in portfolio management we frequently wish to know the expected return subject to a maximum level of risk. We use what is known as a Lagrangian multiplier.

To illustrate this, suppose that we wish to maximize the following function

$$R = 5X + 2X^2 - 4Y \qquad (3.57)$$

Subject to the constraint

$$2X + Y = 20 \qquad (3.58)$$

The first step is to transform the constraint function so that it equals zero, i.e.

$$20 - 2X - Y = 0$$

This form of the constraint is then multiplied by an unspecified variable, known as lambda and designated λ, the Lagrangian multiplier

$$\lambda(20 - 2X - Y) \qquad (3.59)$$

This is subtracted from the original function to form a new function, which then becomes

$$L(X, Y, \lambda) = 5X + 2X^2 - 4Y - \lambda(20 - 2X - Y) \qquad (3.60)$$

This new function is known as the Lagrangian; it is differentiated for all of the variables, including the lambda, thus

$$\frac{\partial L}{\partial X} = 5 + 4X + 2\lambda = 0$$

giving $5 + 4X = -2\lambda$

$$\frac{\partial L}{\partial Y} = -4 + \lambda = 0$$

giving $\lambda = 4$

$$\frac{\partial L}{\partial \lambda} = 20 - 2X - Y = 0$$

giving $2X + Y = 20$

Solving

$$\lambda = 4$$

$$5 + 4X = -2\lambda$$

$$2\lambda = 8$$

$$\therefore 5 + 4X = -8$$

$$4X = -13$$

$$\therefore X = -\frac{13}{4} = -3.25$$

$$2X + Y = 20$$

$$2(-3.25) + Y = 20$$

$$-7.5 + Y = 20$$

$$Y = 27.5$$

Integral calculus or integration

There are two forms of integration. The first is actually the reverse of the differentiation process and is known as the **indefinite integral**. Thus from information about the derivative, we can determine the form of the original function. The second type of integration is the process used to find areas bounded by curves, and is known as the **definite integral**. We will show that these two processes are essentially the same. (This is known as the fundamental theorem of calculus.)

The instruction to integrate is given in the form of an elongated s, $\int$, and is always accompanied an indication of the variable with respect to which the original function was differentiated. Thus, $\int X^2 \, dX$ is interpreted as "find the integral of X^2 with respect to X".

The indefinite integral

To illustrate the indefinite integral, otherwise known as reverse differentiation, consider the function $Y = aX^n$. We know that $dY/dX = naX^{n-1}$. By reversing the differentiation process we get the original function. To illustrate this consider

$$Y = \int naX^{n-1} \, dX \qquad (3.61)$$

This simply says, find the integral of naX^{n-1} with respect to X. The answer is actually

$$Y = \int naX^{n-1} * \mathrm{d}X = aX^n \qquad (3.62)$$

To find the indefinite integral we need only follow some simple rules:

1. To integrate $Y = X^n$ we add one to the exponent of X and divide by the new exponent. For example, consider the function $4X$, add 1 to the exponent, i.e. $4X^{1+1}$ and divide by $1 + 1 = 2$. Therefore the indefinite integral is

$$Y = \frac{4X^{1+1}}{1+1} = \frac{4X^2}{2} = 2X^2$$

2. Recall that any constants in the original function $Y = f(X)$ were lost on differentiation. Consequently in integration we have to allow for the constants, although we may not know what they are. Thus, C, the constant of integration, is customarily added to the integration result. It is because the value of C is unspecified that this form of integration is known as indefinite integration.

Thus the rule for the integration of a power can be refined as follows: add one to the exponent, divide by the new exponent and add a constant, C, i.e.

$$\int X^n \, \mathrm{d}X = \frac{X^{n+1}}{n+1} + C \qquad (3.63)$$

Finding areas under curves

We can use what we have learned about finding the **indefinite integral** to enable us to find areas under curves. However, before we begin this process we must introduce a new concept, known as a **primitive**.

A primitive is the function F whose first derivative is the function f. Thus $2X$ is the first derivative of X^2, so X^2 is a primitive of $2X$. We explained above how to find the indefinite integral. We can use the same process to find the primitive of a function, because the indefinite integral of a function is a primitive of that function. Thus a primitive of $2X$ is

$$\frac{2X^{1+1}}{2} = X^2$$

and a primitive of X^2 is

$$\frac{X^{2+1}}{2+1} = \frac{X^3}{3}$$

Keep this knowledge in reserve as we will draw upon it very soon.

Now consider Fig. 3.5. The diagonal line has the equation $Y = X$. The area of the square $0abX_1$ is equal to X^2, where $X = X_1$. The area of the triangle $0bX_1$ must be equal to half that area, i.e. $X^2/2$. For example, if $X = 2$ the area under the curve between 0 and $X = 2$ must be $2^2/2 = 4/2 = 2$.

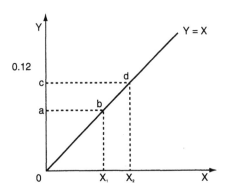

Figure 3.5

Next consider the rectangle $0cdX_2$. The area must again be X^2, for the value of $X = X_2$. The area of the triangle must be $X^2/2$.

Now consider the area of the polygon X_1bdX_2. This area is the difference between the two triangles, i.e. $0dX_2 - 0bX_1$. We could calculate the area of the polygon by calculating the area of each triangle and deducting the smaller from the larger. However, there is an easier way and one which is also applicable when the upper curve is not linear. This easier method uses the concept of the primitive, introduced earlier.

You will have noted that the function for the straight line was $Y = X$, and that irrespective of the magnitude of X, the area of the two triangles was a function of $X^2/2$. It so happens that $X^2/2$ is a primitive of X. Recalling that $X = X^1$, we can show this as follows:

$$\text{primitive of } X = \frac{X^{1+1}}{1+1} = \frac{X^2}{2}$$

Thus, we are able to find the area under the curve between X_1 and X_2 by finding the primitive of X, evaluating the primitive for values of X_1 and X_2, and deducting the smaller value from the larger value.

The instruction to evaluate the definitive integral using the primitive of the function $f(X)$ is formally given as follows:

$$\int_j^k f(x)\ dX = [F(X)]_j^k \tag{3.64}$$

where j and k are the lower and upper values of X in the function being integrated.

Now assume that $j = 2$ and $k = 4$. We know that the primitive of X is $X^2/2$, so to find the area under the curve $Y = X$ between values of $X = 2$ and $X = 4$ we evaluate the following definite integral by substituting the two values of X into the primitive

$$\int_2^4 X\ dX = \left[\frac{X^2}{2} \right]_2^4$$

Firstly, by substituting 2 for X we get

$$\frac{2^2}{2} = \frac{4}{2} = 2$$

Secondly, by substituting 4 for X we get

$$\frac{4^2}{2} = \frac{16}{2} = 8$$

Therefore, the area under the curve $Y = X$ between $X = 2$ and $X = 4$ is

8 − 2 = 6

We can check this by inspecting Fig. 3.5. The area of the triangle $0dX_2$ if $X = 4$, is $4^2/2$ = 8. The area of the triangle $0bX_1$ if $X = 2$ we already know is 2. Subtracting the area of triangle $0bX_1$ from the area of triangle $0dX_2$ gives 8 − 2 = 6, the same result as found using the primitive of the function.

Now we will use another linear function, this time $Y = 4X$. This is illustrated in Fig. 3.6.

We wish to find the area of the polygon $JLMK$. To do this we need to find the area of the triangles $0LJ$ and $0MK$ and deduct the former from the latter. This is achieved by evaluating the definite integral

$$\int_J^K 4X \ dX = [F(X)]_j^k$$

where F is the primitive of $4X$.

First we find the primitive of $4X$, i.e.

$$\text{Primitive of } 4X = \frac{4X^{1+1}}{2} = 2X^2$$

To find the area under the curve between, say $J = 2$ and $K = 5$, the values of 2 and 5 are substituted for X and the difference calculated, for example

$$\int_2^5 4X \ dX = [2X^2]_2^5$$

which is

$$2(5)^2 - 2(2)^2 = 2 * 25 - 2 * 4 = 50 - 8 = 42$$

Thus, the area under the curve bounded by $X = 2$ and $X = 5$ is 42 square units.

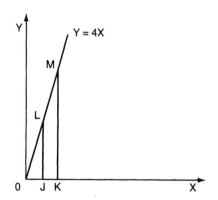

Figure 3.6

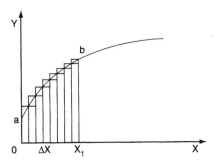

Figure 3.7

This is an application of the fundamental theorem of calculus which says to integrate the area under a curve, $Y = f(x)$ we simply evaluate a primitive of f between points j and k.

The above examples related to situations where Y was a linear function of X and therefore the upper boundary of the area was a straight line. However, this same process can be used to find the area under curves where Y is a non-linear function of X. To show this we will again resort to the use of triangles. Consider Fig. 3.7.

We wish to calculate the area under the curve between a and b. We can begin by drawing many rectangles with a width ΔX and height equal to the curve at the upper boundary of ΔX. You will notice that if we sum the areas of each of these rectangles we will overstate the area under the curve because a small triangle-like part of each rectangle is above the curve. Note that this shows an interval over which the function in question is increasing. A similar argument will hold for intervals for which the function is decreasing.

Now let us look at one of these rectangles in more detail as depicted in Fig. 3.8(a).

Notice that within the left-hand rectangular column we can actually observe three smaller rectangles. One has a width ΔX and a height Y, corresponding to point J on the curve. The second has a height $Y + \Delta Y$ corresponding to the height at point K. The third, the small one at the top, has a height ΔY. Clearly the area of the rectangle $\Delta X(Y + \Delta Y)$ overstated the area under the curve between points J and K, while the rectangle ΔXY will understate that same area.

If we call the whole area under the curve $0abX_1$, "A" and the true area under the curve pertaining to ΔX as ΔA, this relationship can be formally stated as

$$(Y + \Delta Y)(\Delta X) > \Delta A > (Y)(\Delta X) \tag{3.65}$$

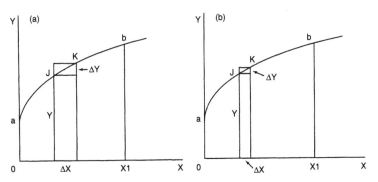

Figure 3.8

If we divide this expression through by ΔX, we get

$$Y + \Delta Y > \frac{\Delta A}{\Delta X} > Y \qquad (3.66)$$

However, consider what happens as ΔX gets smaller and smaller. We can see from Fig. 3.8(b) that ΔX is smaller than in Fig. 3.8(a) and the size of the rectangle $\Delta Y * \Delta X$ gets smaller. Accordingly, the overstatement and the understatement also get smaller. Now consider what happens as ΔX approaches its limit of zero. In that case

$$\frac{\Delta A}{\Delta X} \rightarrow \frac{dA}{dX} \qquad (3.67)$$

and equation (3.66) becomes

$$Y + \Delta Y > \frac{dA}{dX} > Y \qquad (3.68)$$

so

$$\frac{dA}{dX} \rightarrow Y \qquad (3.69)$$

However, Y is a function of X, so $dA/dX = Y = f(X)$. Thus dA/dX equals the height of the curve.

To derive the area under the curve between a and b, we simply integrate dA/dX for values of X between a and b. We do this using the primitive of the Y function.

Thus we have shown that whether the function $Y = f(X)$ is linear or non-linear, the area under that curve can be found by evaluating the definite integral using the primitive of the function

$$\int_a^b f(X) \; dX = F(b) - F(a) \qquad (3.70)$$

Using the earlier notation this is expressed as

$$\int_a^b f(X) \; dX = [F(X)]_a^b \qquad (3.71)$$

where $F(X)$ is a primitive of the function $Y = f(X)$.

We will illustrate this using the function $Y = X^4 + 3X - 3$. We wish to calculate the area under the curve between $X = 3$ and $X = 6$. More formally

$$\int_3^6 (X^4 + 3X - 3) \; dX \qquad (3.72)$$

A primitive of this function is

$$\left[\frac{X^5}{5} + \frac{3X^2}{2} - 3X \right] \qquad (3.73)$$

and to evaluate this function between $X = 3$ and $X = 6$ we have

$$\int_3^6 (X^4 + 3X - 3) dX = \left[\frac{X^5}{5} + \frac{3X^2}{2} - 3X \right]_3^6 \qquad (3.74)$$

Now substitute 6 for X in the right-hand side of equation (3.74)

$$\frac{6^5}{5} + \frac{3(6)^2}{2} - 3(6) =$$

$$\frac{7776}{5} + \frac{108}{2} - 18 = 1591.2$$

(3.75)

Now substitute 3 for X in the right-hand side of equation (3.74)

$$\frac{3^5}{5} + \frac{3(3)^2}{2} - 3(3)$$

$$\frac{243}{5} + \frac{27}{2} - 9 = 53.1$$

(3.76)

The area under the curve is given by equation (3.75) minus equation (3.76), i.e. $1591.2 - 53.1 = 1538.1$ square units.

Exercises

1. Calculate the first derivative of Y with respect to X of the following:

$$Y = X, \qquad Y = CX,$$
$$Y = X^2, \qquad Y = X^4$$
$$Y = X^{1/4}, \qquad Y = X^{1/2}$$

2. Calculate the first derivative of Y with respect to X of the following:

$$Y = 3X^3 + 4X^5, \qquad Y = 5X + 4X^2$$
$$Y = 3X^5 - 2X^2, \qquad Y = X^3 - 3X^2$$

3. Calculate the first derivative of Y with respect to X of the following:

$$Y = X^2(X + 2)^3, \qquad Y = (3X + 2)2X^3$$
$$Y = e^{2X}, \qquad Y = X^3e^X$$
$$Y = X^3e^{3X}, \qquad Y = X\ln(X)$$
$$Y = X^2(X + 2)^3 e^{2X}$$

Note: first revise your understanding of differentiating exponential functions and natural log functions.

4. Calculate the first derivative of Y with respect to X of the following

$$Y = \frac{3X}{X + 2}, \qquad Y = \frac{\ln(X)}{X^2(X + 2)^3}$$

5. Calculate the first derivative of Y with respect to X of the following

$$Y = (X^2 + 2)^3, \qquad Y = (3X^2 + 5)^4$$
$$Y = e^{X^2(X + 2)^3}, \qquad Y = \ln(3X)$$

Convince yourself of the last result by sketching the graphs of $Y = \ln(X)$ and $Y = \ln(3X)$.

6. Calculate the first derivative of Y with respect to X of the following

$$\frac{1}{(1+X)}, \quad \frac{1}{(1+X)^2}$$

$$\frac{1}{(1+X)^{10}}, \quad \frac{2X}{(1+X)^3}$$

7. Calculate the second derivative of Y with respect to X for each of the exercises in questions 1 and 2.

8. Calculate the value of a three-year zero coupon bond that has a yield to maturity (spot rate) of 7.5% p.a. Use a Taylor series approximation to determine the value of the bond if the spot rate changes to 7.6%

9. A three-year bond paying annual coupons of 10 and a final redemption payment of 100, has a current yield to maturity of 8% p.a. Apply your understanding of Taylor series approximation to determine the value of the bond if the yield to maturity changed to 8.1%.

10. Using the cash flows of the bond in question 9, apply your understanding of calculus to calculate the modified duration and the convexity of that bond.

11. Find the minimum point of the function $Y = 2X^2 + 3X$.

12. Totally differentiate the function

$$Y = X^3 + 4X^2 + Z^2$$

13. Maximize the following function

$$R = 3X + X^2 - 2Y$$

subject to

$$2X + Y = 5$$

14. Find primitives for the following functions

$$X^3, \quad e^{2X}$$

$$\frac{1}{(1+X)^2}, \quad \frac{\ln X}{X}$$

$$X^2, \quad e^{X^3}, \quad Xe^{X^2}$$

For part 4 try differentiating $(\ln x)^2$.
For the functions in the last row, try differentiating e^{X^2}.

Answers to selected questions

1.

$$1; \quad C; \quad 2X; \quad -4X^{-5}$$

$$\frac{1}{4}X^{-\frac{3}{4}} = \frac{1}{4\sqrt[4]{X^3}}$$

$$\frac{1}{2}X^{-\frac{1}{2}} = \frac{1}{2\sqrt{X}}$$

2.

$$9X^2 + 20X^4; \quad 5 + 8X; \quad 15X^2 - 4X; \quad 3X^2 - 6X$$

3.

$$2X(X+2)^3 + 3X^2(X+2)^2$$
$$6X^2(3X+2) + 6X^3$$
$$2e^{2X}$$
$$3X^2e^x + X^3e^x$$
$$\ln(X) + 1$$
$$2X(X+2)^3e^{2X} + 3X^2(X+2)^2e^{2X} + 2X^2(X+2)^3e^{2X}$$

4.

$$\frac{3(X+2)-3X}{(X+2)^2}$$

$$\frac{X(X+2)^3 - \ln(X)(2X(X+2)^3 + 3X^2(X+2)^2)}{X^4(X+2)^6}$$

5.

$$6X(X^2+2)^2; \quad 24X(3X^2+5)^3$$

$$(2X(X+2)^3 + 3X^2(X+2)^2)e^{x^2(x+2)^3}$$

$$\frac{1}{X}$$

6.

$$\frac{-1}{(1+X)^2}; \quad \frac{-2}{(1+X)^3}; \quad \frac{-10}{(1+X)^{11}}$$

$$\frac{2(1+X)^3 - 6X(1+X)^2}{(1+X)^6}$$

7. For question 1:

$$0; \quad 2; \quad 20X^{-6}; \quad \frac{-3}{16}X^{-\frac{7}{4}}; \quad \frac{-1}{4}X^{-\frac{3}{2}}$$

For question 2:

$$18X + 80X^3; \quad 8; \quad 30X - 4; \quad 6X - 6$$

8. Value:

$$\frac{100}{(1.075)^3} = 80.496$$

This is the value of

$$\frac{100}{(1+X)^3}$$

when $X = 0.075$
if

$$Y = \frac{100}{(1+X)^3}, \quad \frac{dY}{dX} = \frac{-300}{(1+X)^4}, \quad \frac{d^2Y}{dX^2} = \frac{1200}{(1+X)^5}$$

$$\frac{100}{(1+0.075+0.001)} \approx$$

$$80.496 + 0.001\left(\frac{-300}{(1.075)^4}\right) + \frac{0.001^2}{2}\left(\frac{1200}{(1.075)^5}\right) =$$

$$80.496 - 0.225 + 0.000 = 80.271$$

9.

$$f(X) = \frac{10}{1+X} + \frac{10}{(1+X)^2} + \frac{110}{(1+X)^3}$$

$$f'(X) = \frac{-10}{(1+X)^2} - \frac{20}{(1+X)^3} - \frac{330}{(1+X)^4}$$

$$f''(X) = \frac{20}{(1+X)^3} + \frac{60}{(1+X)^4} + \frac{1320}{(1+X)^5}$$

so

$$f(0.081) \approx \frac{10}{1.08} + \frac{10}{(1.08)^2} + \frac{110}{(1.08)^3}$$

$$+0.001\left(\frac{-10}{(1.08)^2} - \frac{20}{(1.08)^3} - \frac{330}{(1.08)^4}\right)$$

$$+\frac{0.001^2}{2}\left(\frac{20}{(1.08)^3} + \frac{60}{(1.08)^4} + \frac{1320}{(1.08)^5}\right)$$

$$= 105.154 - 0.267 + 0.000 = 104.89$$

10.

$$MD = \frac{1}{105.154}\left(\frac{-10}{1.08^2} + \frac{-20}{1.08^3} + \frac{-330}{1.08^4}\right) = -2.54$$

$$CONV = \frac{\dfrac{1}{105.154}\left(\dfrac{20}{1.08^3} + \dfrac{60}{1.08^4} + \dfrac{1320}{1.08^5}\right)}{2} = 4.56$$

11.

$$Y = 2X^2 + 3X$$

$$\frac{dY}{dX} = 4X + 3 = 0 \quad \text{when } X = -\frac{3}{4}$$

$$X = -\frac{3}{4} \Rightarrow Y = 2\left(\frac{9}{16}\right) + 3\left(-\frac{3}{4}\right)$$

$$= \frac{9}{8} - \frac{18}{8} = -\frac{9}{8}$$

$$\frac{d^2Y}{dX^2} = 4 > 0 \therefore \text{MIN}$$

12.

$$\frac{dY}{dX} = 3X^2 + 8X; \qquad \frac{dY}{dZ} = 2Z$$

$$\therefore \Delta Y = (3X^2 + 8X)dX + 2Z\Delta Z$$

13.

$$L(x, y; \lambda) = 3X + X^2 - 2Y - \lambda(2X + Y - 5)$$

$$\frac{\partial L}{\partial X} = 3 + 2X - 2\lambda$$

$$\frac{\partial L}{\partial Y} = -2 - \lambda$$

so require $\lambda = -2$

$$3 + 2X - 2(-2) = 0 \Rightarrow X = -3.5$$
$$2X + Y = 5 \Rightarrow Y = 12$$

Maximum value of $R =$

$$3(-3.5) + (-3.5)^2 - (2 * 12) = -22.25$$

14.

$$\int X^3 dX = \frac{X^4}{4}; \qquad \int e^{2X} dX = \frac{1}{2}e^{2X}; \qquad \int \frac{1}{(1+X)^2}dX = \frac{-1}{1+X}$$

$$\int \frac{\ln X}{X}dX = \frac{1}{2}(\ln X)^2; \quad \int X^2 e^{X^3}dX = \frac{1}{3}e^{X^3}; \qquad \int Xe^{X^2}dX = \frac{1}{2}e^{X^2}$$

References and further reading

Hunt, R. (1994) *Calculus of a Single Variable*. HarperCollins, New York.

Larson, R. E., Hostetler, R. P. and Edwards, B. H. (1994) *Calculus*. D. C. Heath, Lexington, MA.

Macaulay, F. R. (1938) *Some Theoretical Problems Suggested by Movements in Interest Rates, Bond Yields and Stock Prices in the US since 1856*. National Bureau of Economic Research, New York.

Watsham, T. J. (1993) *International Portfolio Management: A Modern Approach*. Longman, London.

4

Probability distributions: applications to asset returns

Introduction to probability

Probability is a measure of the possibility of an event happening. This is measured on a scale between zero (certainly will not happen) and one (certainly will happen). Probability distributions are mathematical models of the possibility (probability) of uncertain events happening.

Probability has a substantial role to play in financial analysis because, almost without exception, the outcomes of financial decisions are uncertain. For example, most people would expect the future value of a share listed on a stock exchange to be uncertain, because it is accepted that share prices fluctuate day by day. We may have a view of what the future price should be, but we accept that the actual price may be different. At the other extreme, the interest earned on a bank or building society deposit quoting a fixed rate of interest is also uncertain, because the bank or building society may fail and thus we would not receive the interest due. We presumably would not expect the bank or building society to fail, otherwise we would not place our money there. However, experience tells us that these types of institutions do sometimes fail, and therefore there must be some degree of uncertainty surrounding the amount of interest that we will receive.

Probability distributions are akin to the frequency distributions we met in Chapter 2. They describe the way the probabilities associated with a random variable are distributed across all the possible values of that variable. They are important in decision making because they enable us to evaluate the amount of uncertainty surrounding all types of events and, therefore, can be applied to financial decision making.

In this book we are interested in two groups of probability distributions. The first group, which is the subject of this chapter, comprises those probability distributions that are useful for describing the way asset returns behave. A knowledge of these distributions enables us to evaluate the risk in portfolios of financial assets and also to price correctly derivative instruments such as options. These distributions include the lognormal, normal, binomial, Poisson and Pareto–Levy distributions.

The second group consists of the probability distributions (known as sampling distributions) of certain descriptive statistics used to test the significance of hypotheses that have been developed. These distributions include the Student-t, χ-squared and F distributions. These are applied in Chapters 5 and 6.

It is important to get to grips with some jargon before proceeding too deeply into a new subject. In this chapter we have to understand the terms "experiment", "sample space" and "event". An experiment is any action which has a defined set of outcomes. The array of possible outcomes is the sample space, and the specific outcomes, or combination of outcomes, are known as events.

The objective of this chapter is to provide a basic introduction to the theory of probability and then introduce some models of probability which can be applied to the distribution of asset returns. With regard to the theory, we will explain three approaches, the **classical** or *a priori* approach, the **empirical** approach and the **subjective** approach. Next we will introduce some rules for calculating probabilities. We will then proceed to discuss the algebra of random variables and finally we will discuss a number of probability distributions that are widely used in finance, together with examples of their application.

The classical or *a priori* approach to probability

This theory is applicable when the range of possible uncertain outcomes is known and equally likely. Logic can then be applied to determine the probabilities of each outcome.

Consider the tossing of a fair coin, i.e. one where the result, which must be either a head or a tail, is equally likely. The range of possible outcomes is limited to two (a head or a tail), by the construction of the coin. The probability of the result being a head must be 0.5 and the probability of the result being a tail must also be 0.5. The tossing of the coin is the **experiment**, the **sample space** refers to the two possible outcomes, the head and the tail, and the **event** is whether the outcome is a head or a tail.

Similarly, in the case of throwing a fair six-sided die, i.e. one where each side is equally likely, the experiment consists of the throwing of the die, the six possible faces constitute the sample space and the event is the face that shows uppermost. As the die is assumed to be fair the probability of each face showing must be 1/6 or 0.166666.

In these cases the probability of each outcome has been determined from the construction of the coin or the die. This is referred to as the "*a priori*" or "classical" theory of probability.

In such circumstances the probability of an event occurring is simply

$$P(A) = \frac{\text{No. of equally likely outcomes associated with the event}}{\text{Total number of equally likely outcomes}} \qquad (4.1)$$

where $P(A)$ = the probability of A occurring.

The probability of A not occurring is $P(\text{not } A) = 1 - P(A)$

The empirical approach to probability

However, in finance, as in many other fields, we cannot rely on the exactness of a process to determine the probabilities. For example, a scientist may have to repeat an experiment many times in order to determine the probability of possible outcomes. The range of outcomes of the returns of financial assets is virtually unlimited, thus financial analysts have to observe many movements in asset prices in order to determine the probability of future price changes of a given magnitude.

In such situations the probability of a given outcome Z, $P(Z)$, is calculated as the ratio of the number of times that Z occurs divided by the number of experiments, i.e. the number of times the experiment is conducted. Put formally

$$P(Z) = \frac{\text{No. of } Z \text{ occurrences}}{\text{No. of experiments}} \qquad (4.2)$$

The reader may recognize this as an analysis of the relative frequency of observations. This approach analyses historical data of events to determine the probabilities that can be assigned to future events.

Consider, for example, a sample of 100 consecutive daily movements in a share price, where each movement is analogous to a single experiment, and thus a total of 100 experiments are conducted. Each movement is an event, and the sample space is the array of price changes of a given magnitude actually experienced. This approach is known as the **relative frequency or empirical approach to probability**.

It is this approach that is of particular interest in this chapter because it is from the historical behaviour of asset returns that we are able to hypothesize the probability distribution that will describe future asset returns. To illustrate this, assume that of the 100 absolute movements, five movements were 0.5 pence each, 15 were 1 pence each, 20 were 1.5 pence, 30 were 2 pence each, 20 were 2.5 pence and 10 were 3 pence each. The probability that on a randomly choosen day in the period a price change would be 1 pence is 15/100 or 0.15. Similarly, the probability of a single price change being 3 pence is 10/100 or 0.1.

We can formally state this as $P(1 \text{ pence}) = 0.15$ and $P(3 \text{ pence}) = 0.10$.

Adding up all the probabilities for all the price changes

$$0.05 + 0.15 + 0.20 + 0.30 + 0.20 + 0.10 = 1.0$$

Thus it is a feature of probability that the probability of a single outcome lies somewhere between zero and one, and that all the probabilities of a given set of outcomes, for example price changes of our 100 shares, must sum to one.

The subjective approach to probability

There is a third approach to probability, known as **subjective probability**. Probability under this approach is simply defined as the strength of belief that an event will occur.

Subjective probability is applied to many business problems because the probabilities cannot be derived from logic, nor are there sufficient empirical observations upon which to base probability estimates. For example, subjective probability is incorporated in the forecasting of company profits by investment analysts. It is also incorporated in some methods used to calculate expected returns of financial investments.

Basic rules of probability

Irrespective of the approach to probability there are a number of formal rules which are to be applied. Which of these rules is applicable will depend on whether:

1. we are concerned with a single event, in which case the outcomes relate only to that event
2. we are concerned with combinations of several events, for example the changes in the FTSE 100 index and the changes in the S&P 500 index
3. the combined events are independent or mutually exclusive.

These rules are:

- the addition rule
- the multiplication rule.

The addition rule applies where we are concerned with A **or** B happening, and then we need to know whether or not A and B are mutually exclusive.

The multiplication rule applies when we want the probability of A **and** B occurring. Then we need to know whether or not A and B are independent of each other.

The addition rule applied to mutually exclusive events

If two events are mutually exclusive, for example that the outcome can be either A or B but not both, we invoke the **specific rule of addition** and **add the probabilities** of each event in order to determine the probability of one or the other event occurring.

If events A and B are mutually exclusive then

$$P(A \text{ or } B) = P(A) + P(B) \qquad\qquad (4.3)$$

This rule will be demonstrated later with respect to binomial representations of asset price movements.

The addition rule applied to non-mutually exclusive events

If the results of an experiment are non-mutually exclusive we apply the **general rule of addition**. In such circumstances $P(A \text{ or } B)$ is given by

$$P(A \text{ or } B) = P(A) + P(B) - P(A \text{ and } B) \qquad\qquad (4.4)$$

The reasoning behind this is that some events could have an outcome A, some an outcome B, but because A and B are not mutually exclusive, some events could have both outcomes. Thus, because we want to know the probabilities of outcomes A or B, we must deduct from the total outcomes of A and B those events that achieve both outcomes simultaneously, otherwise the intersection would be counted twice, once within A and once within B.

Note that by $P(A \text{ or } B)$ we refer to the probability of A occurring, or of B occurring, or both. The correct notation for this is $P(A \vee B)$, but the notation $P(A \cup B)$ is often used, "$\cup$" standing for the union of two sets.

Let us consider the example of the probabilities of two equity indices rising or falling. The events are not mutually exclusive because clearly both may rise together, both may fall together, one may fall and the other rise or vice a versa. Assume that the FTSE 100 index may rise with a probability of 0.55 and fall with a probability of 0.45. Also assume that in a particular time interval the S&P 500 index may rise with a probability of 0.35 and fall with a probability of 0.65. There is also a probability of 0.3 that both indices may rise together. What is the probability of either the FTSE 100 index or the S&P 500 index rising?

$$P(\text{FTSE } 100 \uparrow) = 0.55 \qquad P(\text{S\&P } 500 \uparrow) = 0.35$$

$$P(\text{FTSE } 100 \downarrow) = 0.45 \qquad P(\text{S\&P } 500 \downarrow) = 0.65$$

$$P(\text{FTSE } 100 \uparrow \text{ and S\&P } 500 \uparrow) = 0.30$$

$$P(\text{FTSE } 100 \uparrow \text{ or S\&P } 500 \uparrow) = 0.55 + 0.35 - 0.30 = 0.60$$

These probabilities are reflected diagramatically in Fig. 4.1

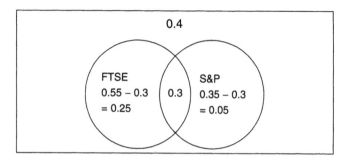

Figure 4.1

The multiplication rule applied to independent events

Two events are said to be independent in terms of probability when knowledge of the outcome of A does not affect the probability of B occurring. Thus

$$P(A \mid B) = P(A) \tag{4.5}$$

A | B means "A given B".

If the events are independent the probability of A and B occurring is given by

$$P(A \textbf{ and } B) = P(A) * P(B) \tag{4.6}$$

Again, the correct notation for A and B is $A \wedge B$, but $A \cap B$ is often used, "$\cap$" standing for the intersection of sets.

Independence is defined as being when the two variables have zero covariance with each other. For example, if two stock market indices had no influence on each other with regard to movements, their covariance would be zero and the two indices would be "independent". However, it should be noted that the covariance between major financial markets is usually not zero.

Imagine that our equity indices were actually independent of each other. What would be the probability of the FTSE 100 rising and the S&P 500 rising?

$$P(\text{FTSE } 100 \uparrow) = 0.55, \quad P(\text{S\&P } 500 \uparrow) = 0.35$$

So the probability of both indices rising is

$$P(\text{FTSE } 100 \uparrow \text{ and S\&P } 500 \uparrow) + 0.55 \times 0.35 = 0.1925$$

The multiplication rule applied to non-independent events

If the events are not independent, the probability of A and B occurring is given by the product of the probability of A, P(A), and the **conditional probability** of B occurring given that A has occurred. This conditional probability is given as $P(B \mid A)$. The vertical line should be read as meaning "given", so $P(B \mid A)$ represents the probability of B occurring given that A has occurred.

Thus the probability of *A* and *B* occurring when *A* and *B* are not independent is given as

$$P(A \text{ and } B) = P(A) * P(B \mid A)$$

i.e.

$$P(B \mid A) = P(A \text{ and } B)/P(A) \qquad (4.7)$$

This probability rule would be appropriate in the case of the two stock markets if there was a positive or negative covariance between the market movements. Each movement in a market would be an event, and the conditional probability would be the probability of one market rising or falling given that the other market has risen or fallen.

We know that in fact the FTSE 100 and the S&P 500 are not independent of each other. Indeed we know from Chapter 2 that the correlation coefficient is about 0.793.

In our example the probability of the FTSE 100 rising over the next time period **and** the S&P 500 rising is 0.30. We know that the probability of the FTSE 100 rising is 0.55. We can thus deduce the probability of the S&P 500 rising given that the FTSE 100 has risen

$$P(\text{FTSE 100} \uparrow \text{ and S\&P 500} \uparrow) = P(\text{FTSE 100} \uparrow) * P(\text{S\&P 500} \uparrow \mid \text{FTSE 100} \uparrow)$$

i.e. $0.30 = 0.55 * P(\text{S\&P 500} \uparrow \mid \text{FTSE 100} \uparrow)$

Thus the conditional probability $P(\text{S\&P 500} \uparrow \mid \text{FTSE 100} \uparrow) =$

$$0.30/0.55 = 6/11 = 0.5454 \dots$$

Discrete and continuous random variables

A random variable is a variable that behaves in an uncertain manner. As this behaviour is uncertain we can only assign probabilities to the possible values of these variables. Thus the random variable is defined by its probability distribution and possible outcomes. In Chapter 2, we classified data as discrete or continuous; in a similar vein we can classify random variables as discrete or continuous. Just as there are two types of random variables, so there are two types of probability distributions: **continuous distributions** and **discrete distributions**. With discrete probability distributions we are concerned with the probability of a variable having only certain discrete values. For example, the throw of dice can only give a discrete number of outcomes one to six. Continuous probability distributions, on the other hand, relate to distributions where the variable can take on any value within a specified range.

Discrete random variables

Discrete random variables are those that have only a finite number of possible outcomes. Consider again the situation when a six-sided die is thrown. Each of the possible outcomes has a probability associated with it; if the die is unbiased, each of those six probabilities is 1/6. This process is modelled mathematically as a discrete random variable.

In this case we could call the random variable Z and define it by the possibility of Z taking values between one and six and the probabilities associated with each outcome, for example

Thus a discrete random variable (X) is defined by its **probability frequency function**. The possibilities, together with their associated probabilities, constitute the probability frequency function, e.g.

Possibilities	$r =$	1	2	3	4	5	6
Probability that	$Z = r$	1/6	1/6	1/6	1/6	1/6	1/6

Note that the probabilities, irrespective of how many possibilities there are, must always sum to one, i.e.

$$\sum_{-\infty}^{+\infty} P(X) = 1 \tag{4.8}$$

Examples of discrete distributions are the binomial distribution and the trinomial distribution. The throwing of a coin results in a binomial distribution of outcomes because the result can either be a head or a tail. Asset prices that can rise, fall or remain unchanged give rise to trinomial distributions because there are three types of possible outcome, up, down or no change, from each event.

Continuous random variables

Continuous random variables are those that can be subdivided into an infinite number of subunits for measurement. Distance, speed, time and asset returns are examples. The unit of measurement can be increased or decreased by infinitesimally small amounts. Consider, for example, the return from holding a security. As we noted above, this is a continuous random variable. The number of possible outcomes for this particular random variable is uncountably infinite. For example a movement in an asset price from 105 to 109 will give a return of 3.8% or 3.81% or 3.8095% depending upon how many decimal places one chooses to measure the return. Under these circumstances it makes no sense to consider the possibility of the random variable taking the value of (say) 3.81%! It only makes sense to consider the possibility of the continuous random variable taking values between two limits, for example the possibility that the return will be between 3.8% and 3.82%.

Clearly, if we hoped to derive the expected value of a continuous random variable by summing, as in the discrete case, we would have difficulty because of the need to sum over infinitely many possibilities. To overcome this practical problem, we must define our continuous random variable not by summing a probability frequency function, which gives individual probabilities, but by integrating what is known as a **probability density function (pdf)**.

Thus for a continuous random variable (X) we have

$$\int_{-\infty}^{+\infty} f(X)\, dX = 1 \tag{4.9}$$

where f is the pdf. From Chapter 3, you will recognize this as the area under a curve. In fact a probability density function is the function representing probability. Figure 4.2 is an example of such a graph. It is the area under the curve which represents probability. (Note that the total area beneath the probability density function must equal one.)

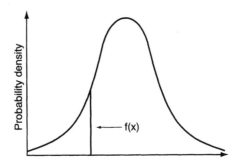

Figure 4.2

The algebra of discrete random variables

Multiplication of a random variable

When multiplying a random variable by a fixed number, the probabilities remain unchanged but the possibilities are multiplied by that fixed number. For example, if X is a discrete random variable, then $2X$ is defined by the same probability distribution as is X, except that the possibilities are all doubled (the probabilities remain the same).

Thus if X is defined by:

Possibility (r)	0	1	2
Probability that $X = r$	1/4	1/2	1/4

then $2X$ is defined by:

Possibility (r)	0	2	4
Probability that $X = r$	1/4	1/2	1/4

Adding two independent random variables

Now we can consider adding a random variable Y to X. Suppose that

X is defined by:

Possibility (r)	0	1	2
Probability that $X = r$	1/4	1/2	1/4

Y is defined by:

Possibility (r)	4	5
Probability that $Y = r$	1/2	1/2

and that X and Y are independent. By checking possibilities and their associated probabilities, it can be seen that $X + Y$ is given by:

Possibility (r)	4	5	6	7
Probability that $X + Y = r$	1/8	3/8	3/8	1/8

Lets us see why.

First the possibilities:

$$0 + 4 = 4, \quad 0 + 5 = 5 \quad 1 + 4 = 5, \quad 1 + 5 = 6, \quad 2 + 4 = 6, \quad 2 + 5 = 7$$

Thus the possibilities are: four (one-way), five (two-ways), six (two-ways) and seven (one-way).

One possibility is four, i.e. $0 + 4$. This has a probability of $1/4 * 1/2 = 1/8$.

There is the possibility of five, achieved in two ways, $0 + 5$ with probability $1/4 * 1/2 = 1/8$, or $1 + 4$ with probability $1/2 * 1/2 = 1/4$. Adding the probabilities for each possibility gives a probability of $1/8 + 1/4 = 3/8$.

In a similar manner, the possibility of six can be achieved two ways with probabilities $= 3/8$.

The possibility of seven can be achieved in one way with probabilities $= 1/8$.

Note that, for random variables, $2X$ is not the same as $X + X$! To understand this take the random variable X given above. It has possibilities 0, 1, 2. Now add a copy of that random variable to the original variable, i.e. $X + X$. The possible outcomes are $0 + 0 = 0, 1 + 0 = 1, 0 + 1 = 1, 1 + 1 = 2, 0 + 2 = 2, 2 + 0 = 2, 1 + 2 = 3, 2 + 1 = 3$ and $2 + 2 = 4$. The probabilities will be:

0	1	2	3	4
1/16	1/4	3/8	1/4	1/16

Expected value and variance of a discrete random variable

The expectation, or probabilistically weighted mean

The expected value (or expectation) of a discrete random variable is defined by

$$E(X) = \sum_r X * \text{prob} \, (X = r) \qquad (4.10)$$

where E is the expectation operator.

For example, our random variable X defined one page 129 has an expected value of

$$E(X) = 0 * \frac{1}{4} + 1 * \frac{1}{2} + 2 * \frac{1}{4} = 1$$

We can see that the expected value of a random variable is the probabilistically weighted mean of various expectations of our random variable. However, it should be noted that the expected value need not be a member of the set of possibilities. For example, consider the six-sided die used earlier. The expected value of a single outcome is

$$\left(1 * \frac{1}{6}\right) + \left(2 * \frac{1}{6}\right) + \left(3 * \frac{1}{6}\right) + \left(4 * \frac{1}{6}\right) + \left(5 * \frac{1}{6}\right) + \left(6 * \frac{1}{6}\right) = 3.5$$

This is because the expectation is the probabilistically weighted mean of the possibilities and like the arithmetic mean (refer to Chapter 2) can give a result that does not coincide with any particular possibility.

Note that the expectation of a group of random variables is a linear combination of the expectation of each random variable. For example, note that

$$E(X + Y) = E(X) + E(Y) \tag{4.11}$$

In a similar way there is a property of a random variable which corresponds to what we would expect to be the variance of a large number of realizations. It is called the variance [sic] of the random variable, and is defined by

$$\text{var}(X) = E(X - E(X))^2 \tag{4.12}$$

To illustrate this with a numerical example, consider the expected return from a particular asset. Assume that an investment manager considers that over a given future period one of three economic scenarios, or states of the world, will apply: high growth, no growth or a recession. The probabilities of each scenario occurring are estimated to be 0.25, 0.5 and 0.25, respectively. Readers will recognize this as an example of subjective probability.

The investment manager expects a particular asset to earn 20% if high growth prevails, 10% if no growth prevails and –4% if there is a recession.

As the expected value of a random variable is the value of each expected outcome multiplied by the respective probability, the expected return of the asset in question would be

$$E(r) = (0.20 * 0.25) + (0.10 * 0.50) + (0.04 * -0.25) = 0.09 = 9\%$$

This may be generalized to the following equation

$$E(r) = \sum_{i=1}^{n} r_i P_i \tag{4.13}$$

As the variance of a random variable is the sum of the products of the squared deviation of the expected outcomes from the mean expected outcome, multiplied by their respective probabilities, thus

$$\text{var}(r) = \sum_{i=1}^{n} (r_i - \bar{r})^2 P_i \tag{4.14}$$

Thus the numerical example of the variance of the above expected returns is calculated as

$$\text{var}(r) = (0.20 - 0.09)^2 0.25 + (0.10 - 0.09)^2 0.50 + (-0.04 - 0.09)^2 0.25$$
$$= 0.003025 + 0.00005 + 0.004225 = 0.0073$$

This outcome is in percentages squared which is not intuitively very appealing, so the square root of the variance, the standard deviation, is usually calculated. The square root of 0.0073 is

0.085 or 8.5%

Thus we see that if asset returns are a random variable, the expected return is the probabilistically weighted mean of the expectation of returns, and the risk, as measured by the variability of those expectations, is described by the variance or standard deviation.

To compute the variance of a linear combination of random variables we need to take into account whether or not the variables are independent of each other, so we need to look at the way they interact as represented by their covariance. The result is

$$\text{var } (aX + bY) = a^2\text{var } (X) + b^2\text{var } (Y) + 2ab \text{ cov } (X, Y) \tag{4.15}$$

where cov XY is the covariance of X and Y, and is given by

$$\text{cov } (X, Y) = E[(X - E(X))(Y - E(Y))] = E(XY) - E(X)E(Y) \tag{4.16}$$

Two random variables are said to be uncorrelated when their covariance is zero. Under these circumstances var $(X + Y) = $ var $(X) + $ var (Y).

We will now illustrate the expected value and expected variance with an example of the expected return on a portfolio of assets and the variance and standard deviation of that portfolio. An example using a two-asset portfolio follows.

Application of discrete random variables: calculating the returns and standard deviation of a portfolio

The portfolio returns

The portfolio return is simply the weighted average of the returns to the individual assets. For example, in a two-asset portfolio the portfolio return would be calculated as follows:

$$E(R_p) = W_aE(r_a) + W_bE(r_b) \tag{4.17}$$

where:

$E(R_p)$ = the expected portfolio return
W_a = weighting of security A
$E(r_a)$ = the expected return to security A
W_b = weighting to security B
$E(r_b)$ = the expected return to security B

The variables $E(r_a)$ and $E(r_b)$ are actually the expected returns of the respective assets. The weights are the proportion of the total value of the portfolio invested in each of the assets.

To illustrate this with a numerical example, assume that the expected returns to A and B are 10% and 12% respectively, and a portfolio is constructed of 60% A and 40% B. The expected return to the portfolio will be

$$\mathbf{0.6 * 0.1 + 0.4 * 0.12 = 0.108 = 10.8\%}$$

Equation (4.17) can be generalized to the following

$$E(R_p) = \sum_{i=1}^{n} W_iE(r_i) \tag{4.18}$$

The portfolio standard deviation

The risk of the portfolio cannot be measured just by the weighted average of the variance of returns of each security. The reason is that, in measuring the risk of the portfolio, we are not only concerned with the variability of the returns of individual securities but the degree to which the returns of pairs of securities fluctuate together – the degree of interaction or covariability. This covariability is measured by the **covariance** or alternatively the **correlation** of the returns of pairs of securities.

Recall from Chapter 2 that the **correlation coefficient** has a value between $+1$ (perfect positive correlation) and -1 (perfect negative correlation). Positive correlation means that the returns to each asset in the respective pair generally change in the same direction, that relationship being stronger the nearer the correlation coefficient is to $+1$. Negative correlation indicates that those returns move in the opposite direction, with the relationship being stronger the closer the coefficient is to -1.0.

The importance of correlations in portfolio construction can be explained by the fact that if the expected returns were so highly correlated that they exhibited perfect positive correlation, there would be no benefit to diversification because the returns from all securities would fluctuate in the same direction at the same time and by the same degree. On the other hand, if the returns on a pair of securities were perfectly negatively correlated, i.e. moved in opposite directions at the same time and by the same degree, the variability of the returns to a portfolio of the two securities would be close to zero.

In Chapter 2 we saw that the standard deviation of a portfolio of two assets is

$$\sigma_p = \sqrt{W_a^2 \sigma_a^2 + W_b^2 \sigma_b^2 + 2W_a W_b (\rho_{ab} \sigma_a \sigma_b)} \qquad (4.19)$$

where

$$\sigma_p = \text{standard deviation of the portfolio}$$
$$W_a \text{ and } W_b = \text{weights of } a \text{ and } b \text{ in portfolio}$$
$$\sigma_a^2 \text{ and } \sigma_b^2 = \text{variance of returns of } a \text{ and } b$$
$$\rho_{ab} = \text{correlation of returns of } a \text{ and } b$$
$$\sigma_a \text{ and } \sigma_b = \text{standard deviation of the returns of } a \text{ and } b$$
$$(\rho_{ab} \sigma_a \sigma_b) = \text{the } \textbf{covariance} \text{ of the returns of } a \text{ and } b.$$

Note that the correlation coefficient is defined as $\text{cov}_{ab}/\sigma_a \sigma_b$. Respecifying equation (4.19) in terms of expectations, it becomes

$$\sigma^2_{(port)} = W_a^2 E((r_a - E(r_a))^2)$$

$$+ W_b^2 E((r_b - E(r_b))^2) \qquad (4.20)$$

$$+ 2W_a W_b \rho_{ab} \sqrt{(E((r_a - E(r_a))^2)E((r_b - E(r_b))^2))}$$

The portfolio standard deviation will simply be the square root of the variance.

Expectation and variance of continuous random variables

The expected value of a continuous random variable is given as

$$E(X) = \mu = \int\limits_{-\infty}^{+\infty} Xf(X)\,dX \qquad (4.21)$$

If X is limited to a range of possible values, say between possibilities y and z, the expected value would be

$$E(X) = \mu = \int\limits_{y}^{z} Xf(X)\,dX \qquad (4.22)$$

The variance and standard deviation of a continuous random variable are

$$\mathrm{var}(X) = \int\limits_{-\infty}^{+\infty} (X - \mu)^2 f(X) \qquad (4.23)$$

$$\mathrm{SD}(X) = \sqrt{\int\limits_{-\infty}^{+\infty} (X - \mu)^2 f(X)} \qquad (4.24)$$

Whilst the distinction between the discrete and the continuous is vital in building and understanding theoretical models, we can sometimes use a continuous random variable to model a discrete situation (and vice versa). For instance, consider the price of a particular share on the stock exchange at midday on the next trading day. Clearly there **are** only a discrete set of possibilities (shares are priced in pounds sterling, pence and occasionally fractions of pence). Nevertheless it may well be that a continuous random variable may provide the best model of the behaviour of that share price.

Important probability distributions in finance

The remainder of this chapter analyses a number of probability distributions that may be applicable to the behaviour of asset returns under the appropriate assumptions. We start with two continuous distributions, the normal and the lognormal. We then discuss two discrete distributions, the binomial and the Poisson. Finally we discuss a group of other continuous distributions, the stable Pareto–Levy distributions. However, before we proceed to discuss these distributions in detail, it is appropriate to explain the charac-teristics of a probability distribution that are desirable from the financial analyst's point of view.

Generally financial analysts are searching for probability distributions as a basis for forecasting the future distribution of asset prices or asset returns. Consequently it is very desirable that the distributions should have the following characteristics:

- stationarity
- stability
- finite variance.

By stationarity we mean that the parameters of the probability distribution are unchanging through time.

Much of our interest in financial assets centres upon their price. However, probability distributions of asset prices are of little use in analysis because they tend not to be stationary. The reason for this is that over time asset prices change collectively – equities, for example, grow over time. Thus the mean price and the standard deviation of prices will generally be higher in one year than in a previous year. Moreover, as asset prices can theoretically rise to infinity but cannot fall below zero, as asset prices rise over time there will be a tendency for the distribution of prices to be skewed to the right.

The probability distributions of changes in prices are also unlikely to be stationary because the magnitude of an absolute change in the asset price is also likely to change as the asset price changes. However, percentage changes in asset prices, percentage asset returns, are likely to be independent of the level of asset prices. Consequently, the percentage return will not necessarily change between time periods simply because the asset price has changed.

Stationary probability distributions of asset returns enable analysts to make probabilistic estimates of future returns. In addition, the parameters of the past history of returns are the basis for estimates of the uncertainty surrounding future returns, and thus can be a basis for measuring future risk. Stationarity is discussed in more detail in Chapter 7.

It is also desirable to consider distributions which have the property that linear combinations of them result in the same type of distribution. For example, adding two independent normal distributions produces a normal distribution, albeit with different parameters from either of the original two. Such distributions are called stable distributions. To illustrate the importance of stability, consider the probability distribution of asset returns on two separate one-day periods. It is desirable that the distribution of returns over the combined two-day period is of the same type.

A finite variance is desirable because without it efficient parameter estimators constructed from samples will not converge to the true population statistics as the sample size is increased. Furthermore, we have adopted the standard deviation as our measure of risk. This will not be defined unless the variance is finite. Evidence that we may not have a finite variance may come from data with significant outliers. The presence of a significant number of outliers will lead to parameter estimates which differ significantly from sample to sample.

This leaves us only to identify those probability distributions which best describe the probability distribution of asset returns.

The normal distribution

We noted in Chapter 2 that the most widely occurring frequency distribution is the **normal or Gaussian distribution.** It follows, therefore, that the most widely used probability distribution is the normal distribution. It is a continuous distribution, but is widely used to model discrete random variables.

The shape of the normal distribution is a symmetrical bell shape as shown in Fig. 4.3.

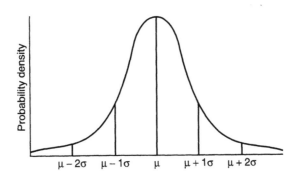

Figure 4.3

This distribution has the desirable statistical characteristics of being completely defined by the **mean and the standard deviation of the distribution.** The mean indicates the position of the centre of the bell, and the standard deviation indicates how spread out the bell is.

If a variable is normally distributed, 68.27% of the observations will fall within plus or minus one standard deviation from the mean. Moreover, 95.45% of the observations will fall within plus or minus two standard deviations, and 99.73% of the observations will fall within plus or minus three standard deviations from the mean.

Recall that the normal curve is actually plotting the height of the **normal density function** for any point on the horizontal axis. The equation for the normal density function is

$$Y = \frac{1}{\sigma\sqrt{2\pi}} e^{-(x-\mu)^2/2\sigma^2}$$

(4.25)

where

μ = the mean of the distribution
σ = the standard deviation
π = 3.14159
e = 2.71828

Thus if this expression is evaluated separately for all possible values of X, and the results plotted, the outcome would be a normal curve. We will develop this point further when we examine the standardized normal probability density function.

The basis of normality: the central limit theorem

The intellectual basis of the assumption that random variables are normally distributed is based on the **central limit theorem**. This says that the mean of a large number of independent samples of a variable will be normally distributed irrespective of the original distribution of the data, provided that distribution has a finite variance.

To illustrate the central limit theorem, consider the minute-by-minute changes in the FTSE 100 index. Assume that over the next minute in time the index will rise or fall one point with equal probability. This is described as follows:

Possibility	−1	+1
Probability	0.5	0.5

In the following minute the index could rise by one or fall by one, each with equal probability. Thus the possible outcomes for a two-minute period would be

Possibility	−2	0	+2
Probability	0.25	0.5	0.25

In the third minute the index will change as follows. If it was at −2, it could fall to −3 or rise to −1. If it was at 0, it could fall to −1 or rise to +1. If it was at +2, it could rise to +3 or fall to +1. Thus the possible outcomes would be

Possibility	−3	−1	+1	+3
Probability	1/8	3/8	3/8	1/8

In the fourth time period the possible outcomes are

Possibility	−4	−2	0	+2	+4
Probability	1/16	1/4	3/8	1/4	1/16

It can be seen that after only four intervals, the probability of getting extreme values, i.e. −4 or +4 is relatively low, and the probability of getting a value around the mean is relatively high. If we were to continue this process many more times we will see that the probability function will conform to the symmetric bell shape of the normal distribution.

Note that the above example merely illustrates the tendency of accumulated data to cluster. The central limit process describes the behaviour of **means** of accumulated data.

The standardized normal probability density function

A **standardized variable**, usually depicted as z, is one that has a mean of zero and a standard deviation of 1.0, and is calculated as follows:

$$z = \frac{(x - \mu)}{\sigma} \tag{4.26}$$

If the random variable is normally distributed, the standardized variable, z, will have a standard normal distribution, i.e. a mean of zero and a standard deviation of 1. As the normal distribution of a standardized variable is a probability distribution, the area under the normal curve will equal 1.0. The curve is depicted in Fig. 4.4 below.

The height under the curve at any point is the **standardized normal density function**. The equation for the **density function** of the standardized normal distribution is given by

$$Y = \frac{1}{\sqrt{2\pi}} e^{\frac{-z^2}{2}} \tag{4.27}$$

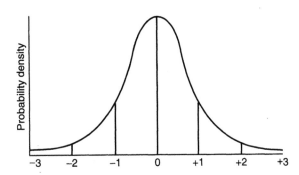

Figure 4.4

The association with the non-standardized normal density function given in equation (4.25) will be clear.

The **area of any vertical segment under the curve represents the probability of the normally distributed random variable having a value within the appropriate range bounded by the segment**. This is shown in Fig. 4.5 below.

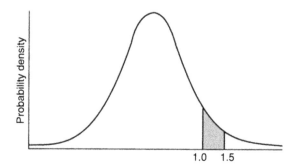

Figure 4.5

Finding areas under a normal curve: using tables

Tables give areas beneath the standard normal curve. This is called the standard normal distribution. Precisely what area is tabulated is usually indicated in a diagram given with the tables, but in the absence of a diagram an examination of the table will quickly reveal whether areas to the left or areas to the right are tabulated for various values of X. Tables will generally only give areas for values of X from zero upwards; other values can be deduced by using the symmetry of the curve. Areas (i.e. probabilities) can be found by computing the difference between the tabulated values at $X = a$ and at $X = b$.

In order to determine the probability of the value of a normally distributed random variable being within a given range, firstly it is necessary to calculate the value of z for the lower and upper bounds of that range.

Secondly, using z tables, find the area under the curve between $z = 0$ and $z = $ the lower bound of the range.

Thirdly, find the area under the curve between $z = 0$ and $z = $ the upper bound of the range. The difference between these two areas will give the probability of the variable being within the given range.

This is illustrated in the following example.

We wish to know the probability of a given asset, which is assumed to have normally distributed returns, providing a return of between 4.9% and 5%. The mean of the returns on that asset to date is 4%, and the standard deviation is 1%. Remember that with continuous probability distributions we have to consider the probability of the variable having a value within a given range, not a value at a given point.

The z values for yields of 4.9% and 5%, respectively, are

$$z_{4.9} = \frac{4.9 - 4}{1.0} = 0.9$$

$$z_5 = \frac{5 - 4}{1.0} = 1.0$$

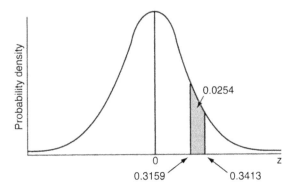

Figure 4.6

From the standard normal distribution table we find that the area between $z = 0$ and $z = 1.0$ is 0.3413, and the area between $z = 0$ and $z = 0.9$ is 0.3159. Thus the area of the standard normal distribution between $z = 5$ and $z = 4.9$ is given by $0.3413 - 0.3159 = 0.0254$. This can be interpreted as indicating that there is a 2.54% probability of the asset providing a yield of between 4.9% and 5.0%. This example is depicted in Fig. 4.6.

We can use the same approach to determine the probability of a random variable, such as an asset price, being above or below a certain figure. Assume, for example, that we know that the daily return to a particular security is normally distributed with a mean of 0.5% and a standard deviation of 0.1%, and we wish to know the probability of the daily return being above 0.525%. We begin by calculating the z value as

$$z = \frac{0.525 - 0.5}{0.1} = 0.25$$

With the value of z calculated as 0.25, we then proceed to the z tables and determine that the area under the curve between $z = 0$ and $z = 0.25$ is 0.0987. Recalling the symmetry of the normal distribution, the probability of z being below zero is 0.5. Therefore, there is a 59.87% probability that the daily returns will be below 0.525%. Thus there is a 40.13% probability of the returns being above 0.525%.

Usually the area under a curve is found by integration. Unfortunately, there is no primitive for the normal pdf. In times before computers this would have posed a difficulty, were it not for the fact that the standardization process leaves areas unchanged, so that it is possible to find areas for any normal distribution by referring to tabulated values for just one standard distribution. With a computer several convenient approaches are possible.

There are two approaches to using a computer to find an area under a normal curve. The first is to use a **numerical integration** approach such as the trapezium rule or Simpson's rule. An alternative approach is to fit a polynomial function which approximates to the function defined by the area under the curve. Both of these approaches are illustrated in Chapter 8 which covers numerical methods.

The lognormal distribution

We noted earlier that the central limit theorem provides the mechanism whereby additive processes tend to produce normally distributed random variables. What then of multiplicative processes?

Take, for example, the price relative of a security over a time interval Δt. Let $S(t)$ be the price at time t, and $S(t + \Delta t)$ be the price at time $t + \Delta t$, then the price relative over the time interval Δt is

$$\frac{S(t + \Delta t)}{S(t)} \tag{4.28}$$

But if we observe the changes in price over a number (n) of smaller intervals of time, δt, $n\delta t = \Delta t$, then we can construct price relatives for each of those smaller intervals. We can then see that

$$\frac{S(t + \Delta t)}{S(t)} = \frac{S(t + \delta t)}{S(t)} * \frac{S(t + 2\delta t)}{S(t + \delta t)} * \frac{S(t + 3\delta t)}{S(t + 2\delta t)} * \cdots * \frac{S(t + n\delta t)}{S(t + (n-1)\delta t)} \tag{4.29}$$

Thus the price relative is the result of a multiplicative process. However, it can be converted to an additive by taking the natural logarithms of the price relatives as follows:

$$\ln\frac{S(t + \Delta t)}{S(t)} = \ln\frac{S(t + \delta t)}{S(t)} + \ln\frac{S(t + 2\delta t)}{S(t + \delta t)} + \ln\frac{S(t + 3\delta t)}{S(t + 2\delta t)} + \cdots + \ln\frac{S(t + n\delta t)}{S(t + (n-1)\delta t)} \tag{4.30}$$

Suppose that each of those price relatives for the short dt time periods were random variables, independent and identically distributed (IID) – let's say X_1, X_2, and so on, where each X_i is an identical copy of a random variable X. Then $S_{(t+\Delta t)}/S_{(t)}$ will also be a random variable, Y, say.

Assuming that $\ln(X_1)$, $\ln(X_2)$, etc. are IID, we can invoke the **central limit theorem** in order to be able to assume that $\ln(S_{(t+\Delta t)}/S_t)$ is approximately normally distributed. Recall that the **central limit theorem** states that if we take a large random sample of a variable, then the mean of that sample will be normally distributed. Thus when time is divided into a large number of sub-periods, i.e. >30, which we do when we work in continuous time, the sum of the natural logarithms on the right-hand side of equation (4.30) will be normally distributed, and so $\ln(S_{(t+\Delta t)}/S_t)$ will be normally distributed.

A variable is said to be **lognormally** distributed if the **natural logarithm** of the variable is **normally** distributed. Consequently, if $\ln S_{(t+\Delta t)}/S_{(t)}$ is normally distributed as indicated above, $S_{(t+\Delta t)}/S_{(t)}$ must be lognormally distributed.

This is an intuitively appealing model of the distribution of security price relatives because, if the price rises, the price relative will be greater than one, whereas if the price falls, the price relative will be less than one, but it cannot be negative. Now consider the lognormal probability density function in Fig. 4.7 below. It shows that the lognormal

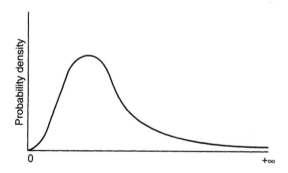

Figure 4.7

distribution is skewed to the right but has no negative numbers. This is compatible with the possible distribution of security prices relatives because they cannot fall below zero but a few observed values could be very high.

The variable cannot take negative values, and the probability of extremely high values is asymptotically zero, as one would expect for a random variable describing, among other things, security price relatives. The lognormal distribution is often used to model variables of this nature, and a multiplicative process may often be found giving rise to such variables.

We have noted in Chapter 1 that the natural logarithm of the security price relative, $\ln (S_t/S_{t-1})$, is the continuously compounded rate of return to holding security S over the time period $t - (t - 1)$ where $S_t/S_{t-1} = S_{(t+\Delta t)}/S_t$ If security price relatives are lognormally distributed, the continuously compounded rate of return, $\ln (S_t/S_{t-1})$, will be normally distributed.

We shall return to this distribution in Chapter 10, where we use it in the context of option pricing. The following results are proved in the appendix to this chapter.

If

$$\ln (X) \sim N(\mu, \sigma^2) \tag{4.31}$$

then

$$E(X) = e^{\mu + \frac{\sigma^2}{2}} \tag{4.32}$$

Note that for ease of printing this is often written as

$$\exp\left(\mu + \frac{\sigma^2}{2}\right) \tag{4.33}$$

$$\text{var} (X) = e^{2\mu + \sigma^2}(e^{\sigma^2} - 1) \tag{4.34}$$

The binomial distribution

One of the most important **discrete distributions** in finance is the binomial distribution. To form a **binomial distribution,** the discrete random variable must satisfy the four following conditions:

1. Only two possible values or outcomes can be taken on by the variable in a given time period or a given event. Each of these time periods or events is known as a binomial **trial**. The two possible outcomes are sometimes referred to as a **"success"** if the outcome is favourable, and a **"failure"** if the outcome is unfavourable.
2. For each of a succession of trials the probability of each of the two outcomes is constant.
3. Each binomial trial is identical.
4. Each trial is independent.

The binomial random variable, X, is the number of successes recorded when a number, n, of independent trials are performed. The outcome of each trial is recorded as success or failure, and the probability of success on each trial is a constant, p. We say $X \sim$ Binomial(n, p). X, the number of successes in n trials, can take values 0, 1, 2, ..., n. Thus

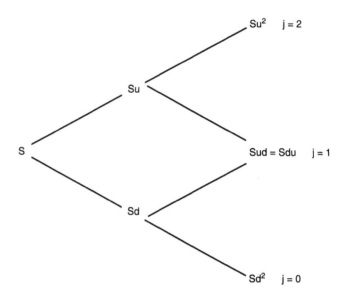

Figure 4.8

with n trials there are $n + 1$ possible outcomes. One outcome will have n successes, another will have $n - 1$ successes and so on, and one will have no successes.

This is illustrated in Fig. 4.8 where the variable S can move up by a factor of $u(u > 1)$ or down by a factor of d $(0 < d < 1)$. In this example there are two binomial trials and three outcomes. The outcome Su^2 is the result of two successes, i.e. $j = 2$. The outcome $Sud = Sdu$, i.e. $j = 1$, is the result of one success. The outcome Sd^2 is the result of zero successes, i.e. $j = 0$.

The binomial distribution gives the probabilities of X taking each of these outcomes. The probability of achieving each outcome depends on:

1. the probability of achieving a success, i.e. p
2. the total number of ways of achieving that outcome.

To explain the probabilities associated with this tree, consider the probability of getting two consecutive successes. Assume that the probability of a success is 0.5 and recalling that the trials are independent, the probability of two consecutive successes $P(Su^2)$ is given as $0.5 * 0.5 = 0.25$. By similar reasoning the probability of a success followed by a failure is also 0.25.

The probability of a success followed by a failure is the same as that of a failure followed by a success. Therefore there are two ways to achieve an outcome with one success. Each way has a probability of 0.25, and so there is a 0.5 probability of the outcome $j = 1$.

To derive each probability in more complex examples, e.g. where the up and the down probabilities are different, or where there are many more trials, we need to know the number of outcomes that have a specific number of successes. In our particular example we need to know how many outcomes resulted from two successes, how many resulted from one success and how many result from no successes.

We can calculate the number, j, of successes for a given number, n, of binomial trials by using the following formula

$$^nC_j = \frac{n!}{j!(n - j)!}$$

(4.35)

where

^nC_j = the number of ways of choosing j successes from n trials, often referred to as n choose j

n = the number of binomial trials

j = the number of successes

The ! after the n and the j is the factorial notation. This means that in the case of n, for example, n is multiplied by $(n-1)(n-2)(n-3)$, etc. For example if $n = 4$, $n!$ would be

$$4 * 3 * 2 * 1 = 24.$$

$j!$ would be treated in a similar manner.

To illustrate the usage, we can determine the number of outcomes that have one success each

$$\frac{2!}{1!(2-1)!} = \frac{2*1}{1*1} = 2$$

This result can easily be checked. In our example, Sud and Sdu were each achieved with one success.

Now that we know the number of outcomes with j successes, we need to know the probability of j successes in n trials as this will give us the probability associated with the number of outcomes that have j successes. This is given by the following formula

$$p(x = j) = \frac{n!}{j!(n-j)!} p^j (1-p)^{(n-j)} \qquad (4.36)$$

where p is the probability of success and $1-p$ is the probability of failure.

To illustrate this, recall that our example assumes $p = 0.5$ and $(1-p) = 0.5$. The probability of a particular outcome (Su^2) after two successes in two trials is therefore

$$p(Su^2) = \frac{2!}{2!(2-2)!} 0.5^2(1-0.5)^{(2-2)} = \frac{2}{2} 0.25 = 0.25$$

Thus there is just one outcome with $j = 2$ successes and that outcome has a probability of 0.25.

We will now calculate the probability of arriving at Sud and Sd^2.

For Sud

$$p(Sud) = \frac{2!}{1!(2-1)!} 0.5^1(1-0.5)^{2-1} = 0.5$$

In fact there are two possible outcomes that have $j = 1$ success, each with a probability of 0.25. Thus two such outcomes have a probability of 0.5.

For Sd^2

$$p(Sd^2) = \frac{2!}{0!(n-0)!} 0.5^0(1-0.5)^{2-0} = 0.25$$

Summing these probabilities gives one as we would expect.

The expectation, variance and standard deviation of the distribution of X are given by

$$E(X) = np$$
$$\text{Var}(X) = np(1 - p)$$
$$\text{SD}(X) = \sqrt{(np(1 - p))}$$

When there are a large number of observations of a binomial random variable, and where the expected value is neither close to zero nor n, the probability distribution approximates a normal distribution, with mean and variance

$$E(x) = np$$
$$\sigma_x^2 = np(1 - p) \tag{4.37}$$

A binomial tree of asset prices

The most common application of the binomial distribution in finance is in relation to security price changes, where it is assumed that over the next small interval of time security prices will either rise ("a success") by a given amount or fall ("a failure"), by a given amount. This binomial distribution is an assumption in some option pricing models. There are three stages in developing the expected value of an asset price using a binomial model. Firstly create a binomial lattice of the possible values of the asset. Secondly, determine the probabilities of each possible outcome. Finally multiply each possible outcome by the appropriate probability and sum the products to arrive at the expected value.

To illustrate this process, consider that we wish to know the value of an asset after two time periods, i.e. two binomial trials. Assume that in each of the time periods the asset may rise (a success) with a probability of 0.5 (i.e. $p = 0.5$), or it may fall (a failure) with a probability of 0.5 (i.e. $1 - p = 0.5$). Assume also that the asset price movement in one time period is independent of that in the other time period. The prospective price movements can be depicted by the binomial tree in Fig. 4.9.

The current time is T_0. At the end of the first time period, T_1, the asset will have risen to Su or have fallen to Sd, each with equal probability. At the end of the second time period, T_2, if the asset had previously risen by a factor of u it would either rise again by u to Su^2 or would fall by d to Sud. Alternatively, if the asset had previously fallen to Sd, it could rise by u to Sud or fall further to Sd^2. Thus after two time periods, or binomial trials, there are three outcomes, Su^2, Sud and Sd^2. To get to Su^2 the trials need two successes, to get to Sud, the trials need one success and to get to Sd^2 the trials need no successes.

To put figures to this process, suppose that $u = 1.10$, $d = 1/1.10$ and S is 50.

Thus at the end of the first time period T_1, S will have risen to $S * 1.1 = 55$, or S will have fallen to $S * (1/1.10) = 45.45$. At the end of the second time period T_2, $Su^2 = 60.50$, $Sud = 50$ and $Sd^2 = 41.32$.

To derive the expected value of the asset at the end of the second time period we must multiply each possible outcome by the probability corresponding to each outcome and sum the products. We know the three possible outcomes: Su^2, Sud and Sd^2. Now we need to know the probabilities associated with each of these outcomes. These are found by applying equation (4.36) above. Given the similarity of this example with the earlier one, it will be no surprise to know that they are 0.25, 0.50 and 0.25, respectively.

We can therefore calculate the expected value of the asset depicted in the binomial lattice by multiplying the various possible outcomes after two time periods by the

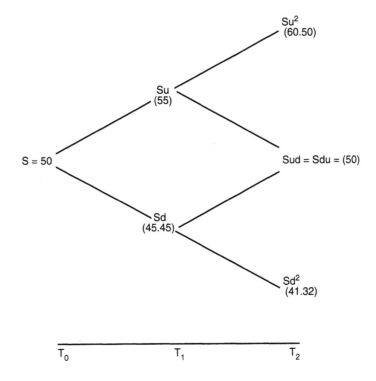

Figure 4.9

probabilities of each outcome. Thus the possible outcomes, are $Su^2 = 60.50$ with a probability of 0.25, $Sud = 50$ with a probability of 0.5, and $Sd^2 = 41.32$ with a probability of 0.25. The expected value is calculated as

$$(60.50 * 0.25) + (50.0 * 0.50) + (41.32 * 0.25) = 50.46$$

The variance is

$$(60.5 - 50.46)^2 * 0.25 + (50 - 50.46)^2 * 0.50 + (41.32 - 50.46)^2 * 0.25 = 46.18$$

It might be thought that such a model that allows only for up movements or down movements is impossibly unrealistic, but the fact is that there is sufficient flexibility in the choice of the parameters (i.e. p, u and d) for it to be a powerful tool in modelling the behaviour of asset prices. Consequently, the binomial lattice is also useful in modelling the pricing of derivative securities, particularly when constructing numerical approximations. This is illustrated in Chapter 8 on numerical methods.

The Poisson distribution

To understand the circumstances when the Poisson distribution is applied, suppose that information which causes market prices to move significantly arrives at a market in discrete pieces, independently, randomly and at a rate of (say) 10 pieces per minute. We might ask the question "What is the probability of only eight pieces of information arriving in the next minute?"

We could attempt to model this process with a binomial distribution with $n = 60$ and $p = 1/6$. This would be the result of envisaging the process as one in which in every second either a single piece of information arrives, or there is no information arriving.

This binomial (60, 1/6) distribution has an expected value of 10 (given by $60 \times 1/6$). There are 61 possible outcomes ranging from 0 (representing no information arriving throughout the 60 seconds) to 60 (representing one piece arriving in each and every one-second period). Each of those possibilities has an associated probability. For the possibility $j = 8$ the probability is given by

$$^{60}C_8\left(\frac{1}{6}\right)^8\left(\frac{5}{6}\right)^{52} = \frac{60!}{8!\,52!}\left(\frac{1}{6}\right)^8\left(\frac{5}{6}\right)^{52} \approx 0.1162$$

Note that $^{60}C_8$ is the number of ways in which eight pieces of information can arrive (for example one in each of the first 8 seconds, one in each of the last 8 seconds, or in many other combinations in between).

The model is subject to some obvious criticisms, not the least being that it is conceivable, in this era of high-speed communications, that more than one piece of information could arrive in any given one-second period. We could improve our model by considering (say) half-second intervals. The appropriate binomial model would then be binomial (120, 1/12). Under this model the probability of obtaining an outcome of eight is given as 0.1145

$$^{120}C_8\left(\frac{1}{12}\right)^8\left(\frac{11}{12}\right)^{112} = \frac{120!}{8!\,112!}\left(\frac{1}{12}\right)^8\left(\frac{11}{12}\right)^{112} \approx 0.1145$$

As we decrease the size of our time intervals the computations become increasingly difficult to handle due to the large numbers involved. Furthermore the additional work has not generated much of a change in the final answer, which might lead us to suspect that we are approaching a limit. This is in fact the case. We can demonstrate this by noting that the binomial (240, 1/24) model has a probability of 0.113534 for an outcome of eight, whereas a (480, 1/44) model has a probability of 0.113067 and a (960, 1/96) model has a probability of 0.112834 for an outcome of eight.

The Poisson distribution is actually the limiting case of the binomial distribution. It is thus applicable where the number of trials (n) is tending towards infinity, the probability of success (p) is tending towards zero and the mean $\lambda = np$ is constant. The formula is

$$p(X) = \frac{e^{-\lambda}\lambda^X}{X!} \tag{4.38}$$

The Poisson has one parameter, λ, the mean rate at which incidents are occurring. In our example this is 10 (i.e. an average of 10 pieces of information arriving within the one-minute period in which we are interested). There are infinitely many possibilities – 0, 1, 2, 3, 4, 5, 6, 7, 8, 9, 10, 11, 12, ... , each with its probability.

For a Poisson with a mean of 10 the probability of an occurrence of eight events is given by

$$\frac{e^{-10}10^8}{8!} = 0.1126$$

Note that the Poisson distribution is fairly symmetric if the distribution has a large enough mean, and under these circumstances the distribution may be approximated by a normal distribution. This is helpful when working out groups of probabilities and under circumstances when the intermediate calculations are again becoming unmanageable.

The Poisson has its variance equal to its mean, so we can approximate to a Poisson (λ) by using a normal (λ, λ) provided that λ is large enough. However, it must be noted that the normal distribution is a continuous distribution, whereas the Poisson is discrete. Thus a **continuity correction** is needed in the approximation.

To demonstrate this we can compute that if $X \sim$ Poisson (30) the probability that X is between 25 and 30 inclusive is 0.391 (this figure is accurate to three decimal places).

We write $P(25 \leqslant X \leqslant 30) = 0.391$.

We can now approximate to X by $Y \sim N(30, 30)$.

To check our approximation we evaluate $P(24.5 < Y < 30.5)$, the 0.5 values representing the continuity correction because the discrete value of 25 on the Poisson distribution is approximated by the range 24.5–25.5 on the continuous normal distribution

$$P = (24.5 < Y < 30.5) = P\left(\frac{24.5 - 30}{\sqrt{30}} < z < \frac{30.5}{\sqrt{30}} \right) = P(1.00 < z < 0.09) = 0.377 \qquad (4.39)$$

where $z \sim N(0, 1)$

The approximation will be better for larger λ – when it will be needed.

Thus we see that the Poisson distribution is applied in similar situations to the binomial distribution, except that the number of successes, j, is very small, and the number of trials, n, is very large. We will now demonstrate its use in modelling the probability of large jumps in the daily returns of the FTSE 100 index.

Suppose that we are interested in modelling a process such as the FTSE 100 index daily changes being in excess of 1%. We might imagine that, over a given period, these particular changes are random, but occur (say) at a rate of about one a month, i.e. 12 per year. How many such changes could we expect in the next six months? Clearly we might expect six, but we wouldn't be unduly surprised if we were to see four, five or eight, or any other number close to six.

As a first attempt we might try to model this by splitting our monthly period into weekly sub-periods. We could then imagine such a change occurring or not occurring in each interval, each with probability 12/52 (52 weeks per year). Supposing these events are independent, this would give us a binomial (26, 12/52) model. Whilst this is a good starting point, it is inadequate as a theoretical model of the physical situation because it only allows for zero or one arrival in each weekly sub-interval. We could divide our time intervals into days and develop a binomial (183, 12/365), but as we noted at the beginning of this section, it would be much more efficient to assume that our random variable, the changes, have a Poisson distribution.

The Poisson distribution is described as follows:

$$\text{Prob}(X = r) = \frac{e^{-\lambda} \lambda^r}{r!} \qquad (4.40)$$

We will illustrate the Poisson distribution by actually investigating the probability of there being more than three jumps in excess of 1% in the next six-month period.

Analysis of our daily data of the FTSE 100 index from 3/1/84 to 3/4/92 show that over that 8.25 year period the average number of daily changes in excess of 1% occurring in each six-month period was 5. The probability of having one such change is given as

$$P(X = 1) = \frac{e^{-5} 5^1}{1!} = 0.0337$$

To find the probability of at least three such changes in the next six months we must find the probability of $X = 0$, $X = 1$ and $X = 2$, sum these and subtract from one. This is illustrated as follows:

$$P(X = 0) = \frac{e^{-5}5^0}{0!} = 0.0067$$

$$P(X = 1) = \frac{e^{-5}5^1}{1!} = 0.0337$$

$$P(X = 2) = \frac{e^{-5}5^2}{2!} = 0.00842$$

Thus

$$P(X \geq 3) = 1 - [P(X = 0) + P(X = 1) + P(X = 2)]$$
$$= 1 - 0.1247 = 0.875$$

So there is an 87.5% chance that there will be at least three jumps of 1% or more in the daily closing of the FTSE 100 index in the six-month period following that data period.

Stable Paretian or Pareto–Levy distributions

The normal distribution is not a panacea for all modelling problems. In particular its use in modelling price relatives of securities is open to question in at least two respects:

1. the assumption of the independence of price relatives over time – they actually demonstrate autocorrelation
2. the likelihood of occurrence of extreme values – more likely in reality than would be implied by a normal model.

These factors point to the need for a probability distribution which, under the log transformation, is more "peaked" than the normal (as a consequence of the autocorrelation), and which has "fatter tails" (as a consequence of the frequency of extreme values).

Certain such symmetric distributions have their coefficient of kurtosis greater than zero (the measure has value zero for the normal distribution). Distributions with a coefficient greater than zero are called leptokurtic (see Fig. 4.10). (Those with the coefficient less than zero are called platykurtic.)

Whilst previous distributions have been derived from the physical circumstances to which they pertain, it is sometimes useful to select a theoretical distribution which most closely matches observed phenomena, even if there are no clear physical links between the two. In such circumstances families of distributions with characteristics which depend on a number of parameters are particularly helpful.

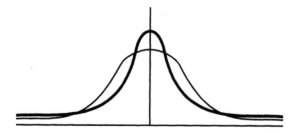

Figure 4.10

Such a family is the family of stable distributions, so-called because an important method of combining distributions (multiplying together linear combinations of their characteristic functions), when applied to two members of the family, produces another member of the family. These stable distributions are, in turn, built from underlying distributions. Those built from the Pareto distribution (which has probability density function $f(x) = \alpha/x^{\alpha+1}$ for $x > 1$) have the characteristics required (symmetry, high peak and fat tails) for particular choices of their four defining parameters. Those four parameters are as follows:

α which determines the heights of the tails
β which determines the skewness
γ determining the horizontal scale of the graph of the pdf
δ which determines location.

Symmetric distributions are given by choosing $\beta = 0$. The maximum possible value of α is two, in which special case we have the normal distribution (in which case $\gamma = \sigma^2/2$). Empirical evidence indicates that appropriate values for a for modelling distributions of log(rates of return) are less than two, possibly between 1.7 and 1.9, although such distributions do suffer from the disadvantage of having infinite variance!

It is not possible to give explicit expressions for the pdfs of such distributions, and associated probabilities must be computed numerically.

Exercises

Basic definitions

1. A standard six-sided die is tossed. What is the probability that the score obtained is:

   (a) three
   (b) not a three
   (c) less than three
   (d) more than three
   (e) seven.

2. Two standard six-sided dice are tossed together and their scores noted. List the elements of the sample space. The scores on the two dice are added. What is the probability that the total score is 6?

Addition rule

3. If S is the set of positive integers below 13, classify the following pairs of subsets of S as mutually exclusive and/or exhaustive:

   (a) {even integers}, {odd integers}
   (b) {multiples of 3}, {multiples of 4}
   (c) {1,2,3}, {integers $\geqslant$ 3}
   (d) {1,2,3}, {4,5,6}.

   Calculate the probability that a randomly-selected member of S belongs to each of the eight subsets given in (a)–(d).

Multiplication rule/conditional probability/independence

4. A number is randomly selected from {1, 2, 3, 4, 5, 6, 7, 8, 9}.

   (a) Calculate the probability that the selected number is greater than 5.
   (b) Calculate the probability that the selected number is an even number greater than 5.
   (c) Using your answers to (a) and (b) calculate the probability that the selected number is even, given that it is greater than 5. Confirm your result by another method.

5. A number is randomly selected from the set **S** = {1, 2, 3, 4, 5, 6, 7}.
   Consider the following subsets of events: **A** = {1, 2, 3}, **B** = {2, 4, 6}.
   Decide whether or not **A** and **B** are statistically independent.

6. Two standard six-sided dice, one red and one blue, are tossed. Let **A** be the event "The score on the blue die is 6", and let **B** be the event "The sum of the scores on the two dice is 7". Are events **A** and **B** statistically independent?

7. In a group of 1000 people, 452 have current accounts, 336 have deposit accounts and 302 have both. Are the attributes "possessing a current account" and "possessing a deposit account" statistically independent?

8. A security moves up in value by 1% over a period of a month with probability 0.6. It moves down in value by 1% with probability 0.4. Assuming that the monthly changes in value are independent, calculate:

   (a) the probability that it attains a value equal to $(1.01)^3$ times its initial value over a three-month period
   (b) the probability that it attains a value equal to $0.99(1.01)^2$ times its initial value over a three-month period.

   Justify your answer to (b) by reference to rules for adding and multiplying probabilities.

9. A study of the daily price changes in two financial markets revealed the following:

market 1

		prices up	prices down
market 2	prices up	165	26
	prices down	32	137

Calculate the probabilities on a given day of:

   (a) the prices in market 1 moving upwards
   (b) the prices in market 1 moving upwards given that the prices in market 2 have moved upwards
   (c) the prices in market 2 moving upwards given that the prices in market 1 have moved upwards.

10. A market analysis system is developed to decide when to buy securities. It is known from historical studies that 5% of a particular market will turn out to be poor investments. The proposed system, applied to past data, correctly identifies 98% of poor investments as potential poor investments, but also identifies 15% of good investments as potential poor investments. Given that a security is scored as a potential poor investment, what is the probability that the security will turn out to be a poor investment?

Comment on the appropriateness of the test for the purposes of making investment decisions

Random variables

11. The random variable **X** takes the value 0 with probability 0.5, 1 with probability 0.3 and 2 with probability 0.2. Calculate:

   (a) $E[\mathbf{X}]$, (b) $E[3 + \mathbf{X}]$, (c) $E[3\mathbf{X}]$, (d) $E[\mathbf{X}^2]$, (e) $Var(\mathbf{X})$.

12. A review of the accounts of 400 stock exchange investors revealed the following information regarding the number of transactions during the past quarter:

X (transactions)	0	1	2	3	4	5	6	7	8	9	10
Number of holders	146	97	73	34	23	10	6	3	4	2	2

   (a) Construct and graph the probability distribution of X, the number of transactions.
   (b) Find the probability that a randomly-selected investor will have had:

   (i)   no transactions
   (ii) at least one transaction
   (iii) more than five transactions
   (iv) fewer than six transactions.

   (c) Find the mean and variance of the number of transactions.

13. Asset A has an expected return of 8% with a standard deviation of 7%. Asset B has an expected return of 11% with a standard deviation of 10%. The correlation of the returns on the two assets is 0.7.

Find the expected return and the standard deviation of the return on a portfolio consisting of 35% of asset A and 65% of asset B.

The normal distribution

14. If Z has a standard normal distribution (i.e. $m = 0$, $s = 1$) find the following:

   (a) $P(Z > 1.2)$ (b) $P(Z \geqslant 1.34)$ (c) $P(Z \leqslant 1.01)$
   (d) $P(Z > 0.85)$ (e) $P(Z \leqslant 2.14)$ (f) $P(Z \leqslant 0.07)$
   (g) $P(Z < 1.37)$ (h) $P(Z \geqslant -2.03)$ (i) $P(Z \leqslant -0.17)$
   (j) $P(Z \geqslant -1.36)$ (k) $P(0.34 \leqslant Z \leqslant 1.29)$ (l) $P(-2.01 \leqslant Z \leqslant 1.52)$.
   (m) $P(-1.21 \leqslant Z \leqslant -0.34)$

   (A sketch for each might be helpful.)

15. If $X \sim N(5,36)$ (i.e. X has normal distribution with $m = 5$ and $s = 6$) find the following:

    (a) $P(X \geqslant 14)$ (b) $P(X \leqslant 9.5)$ (c) $P(X \geqslant 3.5)$
    (d) $P(2 \leqslant X \leqslant 12.5)$.

16. The current share price of a particular company is modelled approximately as a random variable having normal distribution with mean £15.28 and standard deviation £0.12. Find the probabilities that the current price is

    (a) at least £15.50
    (b) at most £ 15.00
    (c) between £15.10 and £15.40
    (d) between £15.05 and £15.15.

17. The price of a particular security is approximately normally distributed. In the last year on about 20% of working days the price was below 20. On 75% of working days the price was above 25. Find the mean and standard deviation of the price.

    Criticize the use of the normal distribution as a model for this situation. Explain how the lognormal distribution is used to overcome the problem.

The binomial distribution

18. A coin is biased so that $P(\text{head}) = 0.8$. If the coin is tossed five times find the probability of getting:

    (a) exactly 2 heads
    (b) exactly 4 heads
    (c) at least 2 heads.

19. Given that the security in question 8 is currently priced at £10, find the probability that it is priced at £10.40 in a year's time.

20. A broker has a bonus scheme to encourage profitable trading. Under the rules of the scheme any trader who drops below his daily target more than three times in a two-week period (10 working days) will forfeit his bonus at the end of that period. If the probability that an employee will be below target on any one day is 0.15, how many bonuses will be lost by 100 traders in a 50-week year?

    What assumptions did you make in deriving your answer? Are these likely to hold?

The Poisson distribution

21. The arrival of significant pieces of information on a market floor during a busy trading period follows a Poisson process with mean 3.5 per minute. Find the probabilities that in any given minute:

    (a) no significant information arrives
    (b) at least one significant piece of information arrives
    (c) two pieces arrive
    (d) four pieces arrive.

Find the probability that more than 20 pieces of information arrive in a given 5-minute period, and examine how close to this is the answer provided by the appropriate normal approximation.

Answers

1.

$$\text{(a) } \frac{1}{6}, \quad \text{(b) } \frac{5}{6}, \quad \text{(c) } \frac{2}{6} = \frac{1}{3}, \quad \text{(d) } \frac{3}{6} = \frac{1}{2}, \quad \text{(e) } 0$$

2. Each cell of the table represents an outcome. The number in each cell gives the total score corresponding to that outcome.

score on second die

	1	2	3	4	5	6
1	2	3	4	5	6	7
2	3	4	5	6	7	8
3	4	5	6	7	8	9
4	5	6	7	8	9	10
5	6	7	8	9	10	11
6	7	8	9	10	11	12

score on first die (left of table, rows 1–6)

$$\text{Probability (total score} = 6) = \frac{5}{36}$$

3. (a) mutually exclusive and exhaustive
   (b) neither
   (c) exhaustive but not mutually exclusive
   (d) mutually exclusive but not exhaustive

4.

$$\text{(a) } \frac{4}{9}, \quad \text{(b) } \frac{2}{9}, \quad \text{(c) } \frac{2/9}{4/9} = \frac{1}{2}\left(\text{or } \frac{2}{4} \text{ by directing counting}\right)$$

5.

$$p(\mathbf{A}) = \frac{3}{7}, \quad p(\mathbf{B}) = \frac{3}{7}, \quad p(\mathbf{A} \wedge \mathbf{B}) = \frac{1}{7}, \quad \frac{3}{7} \times \frac{3}{7} \neq \frac{1}{7} \text{ so not independent}$$

6.

$$p(\mathbf{A}) = \frac{1}{6}, \quad p(\mathbf{B}) = \frac{6}{36} = \frac{1}{6}, \quad p(\mathbf{A} \wedge \mathbf{B}) = \frac{1}{36}, \quad \frac{1}{6} \times \frac{1}{6} = \frac{1}{36} \text{ so independent}$$

Or $p(\mathbf{B}) = p(\mathbf{B}|\mathbf{A})$, so independent

7. p(current) = 0.452 p(deposit) = 0.336 p(current) $\times p$(deposit) = 0.151872
   But p(current $\wedge$ deposit) = 0.302, so not independent.

8. (a) $0.6^3 = 0.216$ (b) $3 \times 0.4 \times 0.6^2 = 0.432$

$$(0.4 \times 0.6 \times 0.6) + (0.6 \times 0.4 \times 0.6) + (0.6 \times 0.6 \times 0.4)$$

Multiplying justified by independence. Adding because mutually exclusive.

9.

$$(a) \ \frac{197}{360} \approx 0.547, \quad (b) \ \frac{165}{191} \approx 0.864, \quad (c) \ \frac{165}{197} \approx 0.838$$

10. For ease of computation consider a population of size 10 000:

scored as

		poor	~poor	
actual	poor	490	10	500
	~poor	1425	8075	9500
		1915	8085	10000

p(poor | scored as potentially poor)

$$= \frac{490}{1915} \approx 0.256$$

Loses a surprising (?) amount of good business.

11.

$$E(X) = 0 \times 0.5 + 1 \times 0.3 + 2 \times 0.2 = 0.7$$
$$E(3 + X) = 3 \times 0.5 + 4 \times 0.3 + 5 \times 0.2 = 3.7 = 3 + E(X)$$
$$E(3X) = 0 \times 0.5 + 3 \times 0.3 + 6 \times 0.2 = 2.1 = 3E(X)$$
$$E(X^2) = 0 \times 0.5 + 1 \times 0.3 + 4 \times 0.2 = 1.1 \ (\neq (E(X))^2!)$$
$$Var(X) = E(X^2) - (E(X))^2 = 0.61$$

12.

X	0	1	2	3	4	5	6	7	8	9	10
Freq.	146	97	73	34	23	10	6	3	4	2	2
Prob.	0.365	0.2425	0.1825	0.085	0.0575	0.025	0.015	0.0075	0.01	0.005	0.005
Cum.	0.365	0.6075	0.79	0.875	0.9325	0.9575	0.9725	0.98	0.99	0.995	1

$P(X = 0) = 0.365$; $P(X \leqslant 1) = 1 - 0.365 = 0.635$; $P(X > 5) = 1 - 0.9575 = 0.0425$
$P(X < 6) = 0.9575$
$E(X) = 1.535$; $Var(X) = 3.378775$

13. Expected return = $(0.35 \times 0.08) + (0.65 \times 0.11) = 0.0995 \approx 10\%$

Variance = $(0.35^2 \times 0.07^2) + (2 \times 0.35 \times 0.65 \times 0.7 \times 0.07 \times 0.10) + (0.65^2 \times 0.10^2)$
$= 0.007055$

Standard deviation = 0.084

14. (All answers given to two decimal places)

(a) 0.12, (b) 0.09, (c) 0.84, (d) 0.20, (e) 0 98, (f) 0.53,
(g) 0.91, (h) 0.98, (i) 0.43, (j) 0.91, (k) 0.27, (l) 0.91,
(m) 0.25

15.

(a) $P\left(z \geqslant \dfrac{14-5}{6} = 1.5\right) \approx 0.07$

(b) $P\left(z \leqslant \dfrac{9.5-5}{6} = 0.75\right) \approx 0.77$

(c) $P\left(z \geqslant \dfrac{3.5-5}{6} = -0.25\right) \approx 0.60$

(d) $P\left(\dfrac{2-5}{6} \leqslant z \leqslant \dfrac{12.5-5}{6}\right) = P(-0.5 \leqslant z \leqslant 1.25) \approx 0.59$

16.

(a) $P\left(z \geqslant \dfrac{15.495-15.28}{0.12} = 1.792\right) \approx 0.04$

(b) $P\left(z \leqslant \dfrac{15.005-15.28}{0.12} = -2.29\right) \approx 0.01$

(c) $P\left(\dfrac{15.095-15.28}{0.12} \leqslant z \leqslant \dfrac{15.405-15.28}{0.12}\right) = P(-1.54 \leqslant z \leqslant 1.04) \approx 0.79$

(d) $P\left(\dfrac{15.045-15.28}{0.12} \leqslant z \leqslant \dfrac{15.155-15.28}{0.12}\right) = P(-1.96 \leqslant z \leqslant -1.04) \approx 0.12$

17.

$$P(z) \leqslant \left(\dfrac{20-\mu}{\sigma}\right) = 0.20, \text{ so } \dfrac{20-\mu}{\sigma} = -0.84$$

$$P(z) \geqslant \left(\dfrac{75-\mu}{\sigma}\right) = 0.25, \text{ so } \dfrac{75-\mu}{\sigma} = 0.67$$

Solving simultaneously, $\mu \approx 50.6$, and $\sigma \approx 36$.
   If

$$X \sim N(50.6,(36)^2) \text{ then } P(X < 0) = P\left(z < \left(\dfrac{50.6}{36}\right) \approx 1.41\right) \approx 0.08$$

Instead of modelling prices by a normal distribution, which allows negative outcomes, the logarithms of returns are modelled by a normal distribution. This encapsulates the multiplicative structure of returns. As a consequence prices are lognormally distributed and negative values are not allowed.

18. (a) $P(2 \text{ heads}) = {}^5C_2(0.8)^2(0.2)^3 = 0.0512$
    (b) $P(4 \text{ heads}) = {}^5C_4(0.8)^4(0.2) = 0.4096$
    (c) $P(\geqslant 2 \text{ heads}) = 1 - P(0 \text{ heads}) - P(1 \text{ head}) = 1 - (0.2)^5 - 5(0.8)(0.2)^4 = 0.99328$

19.

$10.40 = 10 \times 1.01^n \times 0.99^{12-n}$
$\log(10.40) = \log 10 + n \log(1.01) + (12-n) \log 0.99 \Rightarrow n = 8$
$\text{Probability} = {}^{12}C_8(0.6)^8(0.4)^4 \approx 0.213$

20. Let X represent number of days below target in a trading period.
    Assume $X \sim \text{Bin}(10, 0.15)$.

$P(X > 3) = 1 - P(X = 0) - P(X = 1) - P(X = 2) - P(X = 3)$
$\qquad = 1 - {}^{10}C_0 0.15^0 0.85^{10} - {}^{10}C_1 0.15 0.85^9 - {}^{10}C_2 0.15^2 0.85^8 - {}^{10}C_3 0.15^3 0.85^7$
$\qquad \approx 0.05$

$100 \times 25 = 2500$ periods to consider, giving $2500 \times P(X > 3) \approx 125$ lost bonuses.
Assumption of independence likely to be invalid.

21. $X \sim \text{Poisson}(3.5)$

$$P(X = 0) = e^{-3.5} \approx 0.03$$

$$P(X \geqslant 1) = 1 - P(X = 0) \approx 0.97$$

$$P(X = 2) = \frac{3.5^2 e^{-3.5}}{2!} \approx 0.185$$

$$P(X = 4) = \frac{3.5^4 e^{-3.5}}{4!} \approx 0.189$$

$Y \sim \text{Poisson}(5 \times 3.5) = \text{Poisson}(17.5)$
$P(Y > 20) \approx 0.231$.
Using $Y \approx N(17.5, 17.5)$, we need

$$P\left(z \geqslant \left(\frac{20.5 - 17.5}{\sqrt{17.5}}\right) \approx 0.717\right) \approx 0.236$$

Appendix: the mean and variance of the lognormal distribution

X and Y are random variables. $Y = \ln(X)$, and $Y \sim N(\mu, \sigma^2)$, i.e. X is lognormally distributed. What is $E(X)$ and var(X)?

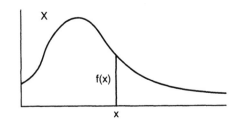

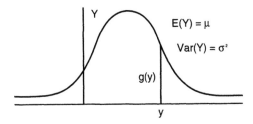

Figure 4.11

$y = \ln(x)$

Preliminary result

$$\int_x^\infty f(x)dx = \int_y^\infty g(y)dy \Rightarrow f(x) = g(y)\frac{dy}{dx} \qquad (*)$$

Expectation

$$E(X) = \int_0^\infty xf(x)\,dx$$

$$= \int_{-\infty}^\infty \exp(y)g(y)\,dy \;(\text{using} *)$$

$$= \int_{-\infty}^\infty \exp(y)\left(\frac{1}{\sigma\sqrt{2\Pi}}\exp\left(\frac{-(y-\mu)^2}{2\sigma^2}\right)\right)dy$$

$$= \frac{1}{\sigma\sqrt{2\Pi}}\int_{-\infty}^\infty \exp\left(y - \frac{(y-\mu)^2}{2\sigma^2}\right)dy$$

$$= \frac{1}{\sigma\sqrt{2\Pi}} \int_{-\infty}^{\infty} \exp\left(-\frac{1}{2\sigma^2}(y^2 + (-2\sigma^2 - 2\mu)y + \mu^2)\right) dy$$

$$= \frac{1}{\sigma\sqrt{2\Pi}} \int_{-\infty}^{\infty} \exp\left(-\frac{1}{2\sigma^2}(y - (\sigma^2 + \mu))^2 - \frac{1}{2\sigma^2}(-2\mu\sigma^2 - \sigma^4)\right) dy$$

$$= \frac{1}{\sigma\sqrt{2\Pi}} \int_{-\infty}^{\infty} \exp\left(-\frac{1}{2\sigma^2}(y - (\sigma^2 + \mu))^2\right) \exp\left(\mu + \frac{\sigma^2}{2}\right) dy$$

$$\exp\left(\mu + \frac{\sigma^2}{2}\right) \frac{1}{\sigma\sqrt{2\Pi}} \int_{-\infty}^{\infty} \exp\left(-\frac{1}{2\sigma^2}(y - (\sigma^2 + \mu))^2\right) dy$$

$$= \exp\left(\mu + \frac{\sigma^2}{2}\right)$$

Variance

$$VAR(X) = E(X^2) - (E(X))^2$$

$$E(X^2) = \int_0^{\infty} x^2 f(x)\, dx$$

$$= \int_{-\infty}^{\infty} \exp(2y)g(y)\, dy$$

$$= \int_{-\infty}^{\infty} \exp(2y) \frac{1}{\sigma\sqrt{2\Pi}} \exp\left(-\frac{(y-\mu)^2}{2\sigma^2}\right) dy$$

$$= \frac{1}{\sigma\sqrt{2\Pi}} \int_{-\infty}^{\infty} \exp\left(-\frac{1}{2\sigma^2}((y-\mu)^2 - 4\sigma^2 y)\right) dy$$

$$= \frac{1}{\sigma\sqrt{2\Pi}} \int_{-\infty}^{\infty} \exp\left(-\frac{1}{2\sigma^2}(y^2 - (2\mu + 4\sigma^2)y + \mu^2)\right) dy$$

$$= \frac{1}{\sigma\sqrt{2\Pi}} \int_{-\infty}^{\infty} \exp\left(-\frac{1}{2\sigma^2}(y - (\mu + 2\sigma^2))^2 - \frac{1}{2\sigma^2}(-(\mu + 2\sigma^2)^2 + \mu^2)\right) dy$$

$$= \frac{1}{\sigma\sqrt{2\Pi}} \int_{-\infty}^{\infty} \exp\left(-\frac{(y - (\mu + 2\sigma^2))^2}{2\sigma^2}\right) \exp\left(\frac{4\mu\sigma^2 + 4\sigma^4}{2\sigma^2}\right) dy$$

$$= \exp(2\mu + 2\sigma^2) \frac{1}{\sigma\sqrt{2\Pi}} \int_{-\infty}^{\infty} \exp\left(-\frac{(y-(\mu+2\sigma^2))^2}{2\sigma^2}\right) dy$$

$$= \exp(2\mu + 2\sigma^2)$$

$$\therefore \text{VAR}(X) = \exp(2\mu + 2\sigma^2) - \exp 2\left(\mu + \frac{\sigma^2}{2}\right) = \exp(2\mu + 2\sigma^2) - \exp(2\mu + \sigma^2)$$

$$= \exp(2\mu + \sigma^2)(\exp(\sigma^2) - 1)$$

5

Statistical inference: confidence intervals and hypothesis testing

Introduction

The data that we have to work with may constitute the whole body of data, in which case it is known as a **population** and the descriptive statistics are known as **population statistics** or **population parameters**. Alternatively, the data may represent only a sample of the total observations of that variable, in which case the descriptive statistics are known as **sample statistics**.

Statistical analysis can be broadly divided into two forms, **descriptive** and **inferential**. Descriptive statistics were covered in Chapter 2 where we calculated various types of descriptive statistics in order to summarize certain qualities of the data. However, no attempt was made to use those statistics to estimate or infer the characteristics of the population being analysed. In the case of inferential statistics, the information gained from the descriptive statistics of sample data is used to generalize to the characteristics of the whole population.

The topic of **inferential statistics** itself has two broad areas of application, **estimation** and **hypothesis testing**. Estimation applies where we have no prior (or *a priori*) knowledge of the magnitude of the population parameters. Under these circumstances we can create **confidence intervals** to **estimate** the true population parameter, and to give a measure of the accuracy of our estimation.

When we have *a priori* knowledge concerning the population parameter, this knowledge can be formulated into a hypothesis which can be tested. For example, we can **test the hypotheses** that the population parameter has a specified range. These two uses, estimation and hypothesis testing, constitute the body of this chapter.

It is appropriate at this stage to introduce notation which differentiates population parameters from their appropriate sample statistics as follows:

μ = population mean $\overline{X}$ = sample mean

σ = population standard deviation s = sample standard deviation

If we are fortunate enough to be able to work with the population data, then clearly, as we have taken all the possible data into account, the population statistic will be the **true** statistic. No inference or estimation will be required. However, it is unusual in finance or other social sciences to have data of the whole population of possible observations. We have to work with samples, and we do not know whether the sample statistic coincides with the true population statistic or is significantly different from it. We therefore have to develop techniques to determine the degree of confidence with which we can infer the true population statistics from the sample statistics.

Sampling theory

When working with samples of data we have to rely on **sampling theory** to give us the probability distribution pertaining to the particular sample statistic. This probability distribution is known as the **sampling distribution**.

To understand the intuition behind sampling distributions, consider selecting balls from a barrel for a lottery draw. When one ball has been removed and the number observed, that ball is replaced into a barrel and could be selected again. This form of sampling is known as **random sampling with replacement**, and is the form of sampling that is assumed in this chapter because it is particularly relevant as a model for sampling from large or infinite populations.

To apply this idea of sampling to statistics, consider drawing 50 samples, each of 20 balls, from the barrel and replacing them, and then calculating the sample statistic, say the arithmetic mean, for each sample. Each of the calculations of the sample mean, one from each sample, would be slightly different, but all would cluster around the true mean, some above and some below. As the samples are assumed to be selected randomly, the sample statistics are treated as random variables. Continuous sampling with replacement will enable a probability distribution of the sample statistic to be generated. This probability distribution is known as the **sampling distribution** of the sample statistic. It is our knowledge of the sampling distribution of each type of sample statistic which enables us to infer the value of the population parameter from the sample statistic.

We are concerned with the normal distribution and Student-t distribution for means, the χ^2 distribution for variances and the F-distribution for coefficients of determination.

As the latter is related to the goodness of fit of regression lines which are dealt with in the next chapter, we will defer coverage of the F-distribution until then.

Sampling distributions of sample statistics

Sampling distribution of the sample arithmetic mean

In Chapter 4 we learned that through the application of the central limit theorem, the means of additive processes (arithmetic means) will be normally distributed, irrespective of the distribution of the underlying variable, provided that the samples are large, i.e. where the sample size exceeds 30. If the underlying population is normally distributed, but the sample is smaller that 30, the distribution of sample means will be a Student-t distribution.

The expected value of the mean of the repeated calculations of the sample means is the mean of the population. The **standard deviation of the sample means** is known as the **standard error**, and is calculated as the population standard deviation divided by the square root of the sample size. Formally, this is referred to as the **standard error of the mean** and given as

$$SE = \frac{\sigma}{\sqrt{n}} \tag{5.1}$$

where

σ = the population standard deviation
n = the number in the sample from which the means are calculated.

Thus the standard error is given by the standard deviation of the original population divided by the square root of the sample size. However, the population standard deviation is not known, so the sample standard deviation, s, is taken to be the estimate. Thus the standard error is estimated by

$$SE = \sqrt{\frac{s^2}{n}} = \frac{s}{\sqrt{n}} \tag{5.2}$$

Thus with the help of the central limit theorem we can say that the mean of a large sample is approximately normally distributed with a mean equal to the population mean and a standard deviation equal to the standard error of the mean. Formally

$$\overline{X} \sim N(\mu, s^2/n)$$

where $\sim$ means "approximately distributed".

An explanation as to why n is used as the divisor is provided in Appendix 5.1.

For small sample sizes from normal distributions the uncertainty introduced by estimating σ by s (which is greater with a smaller sample size) is allowed for by using the t distribution.

Sampling distribution of the sample variance

The sampling distribution of the sample variance is a form of gamma distribution known as the chi-square distribution, indicated as χ^2. This distribution has a different shape for different degrees of freedom. It is necessary to standardize the sample variance in a manner analogous to the way in which the normal distribution was standardized in Chapter 4. The standardization takes the form

$$\chi^2_{n-1} = (n-1)s^2/\sigma^2 \tag{5.3}$$

The subscript $n-1$ refers to the degrees of freedom, which, in the case of χ^2, are the number of observations minus 1. With a small sample the shape of the probability distribution is skewed to the right, but as the sample gets larger, the distribution becomes more symmetric.

Thus, if the mean of our variable is normally distributed, then

$$(n-1)s^2/\sigma^2 \tag{5.4}$$

will have a χ^2 distribution with $n-1$ degrees of freedom. We do not know σ, but the result can be used in testing hypothesized values of σ or in constructing confidence intervals.

Estimation and confidence intervals

Armed with our sample statistics and with knowledge of the sampling distribution of those sample statistics, we are in a position to **estimate** the population parameters of the data analysed. When using the sample statistics to estimate the population parameters, the sample statistics are referred to as **estimators**. It is desirable that these estimators are **BLUE**. That is that they are the **B**est, **L**inear, **U**nbiased **E**stimators. However, it may be necessary to achieve a degree of bias in order to achieve a lower variance.

Best refers to the property of having the smallest variance of all the possible unbiased estimators.

Linear refers to the property of the estimator of being a linear function of the sample observations. For example

$$\hat{\theta} = c_1 X_1 + c_2 X_2 + \dots + c_n X_n$$

where c is constant.

The importance of this property may not be obvious, but suffice it to say that the mathematical properties of a linear estimator are much easier to analyse.

Unbiased refers to the property where the expected value of the estimator (the mean of the sampling distribution) is equal to the population parameter. Thus when taking many samples of the estimator, some will be higher than the population parameter, some will be lower, but the average will be the same. In contrast a biased estimator would be one where the average value would be above or below the population parameter.

Estimators that are both **unbiased** and have the **smallest variance** are referred to as **efficient** estimators.

If an estimator is biased and/or inefficient it is desirable that it possesses **asymptotic** properties.

Asymptotic unbiasedness refers to the property that any bias that exists in small samples, i.e. < 30, will get smaller as the sample increases and ultimately reduces to zero as the sample size tends to infinity.

Asymptotic efficiency refers to the property where the estimator is both consistent and has a smaller asymptotic variance than any other consistent estimator.

Consistency relates to the property whereby the variance of the estimator falls to zero as the sample size increases to infinity.

Confidence intervals

Calculating sample statistics as estimates of the population parameters gives what are known as **point estimates**. However, we know that these point estimates will be made with error, i.e. **estimation error**. We therefore need a mechanism that will provide insight into the degree of confidence that we can place on these point estimates. That brings us to **confidence intervals** or **interval estimates**.

We will illustrate the principle of confidence intervals by applying it to the mean. We will also adapt the process to determine the sample size that is required in order to achieve the degree of confidence required.

Confidence intervals for means (large samples)

Recall that our problem is that we do not know the population mean, only the sample mean. However, from the central limit theorem, we know that the sampling distribution of the means has a mean which itself equals the population mean, and a standard deviation (the standard error) equal to $\sigma/\sqrt{n}$, where σ is the population standard deviation.

Now another problem has arisen – we do not know the population standard deviation, only the sample standard deviation. However, we can utilize another part of sampling theory that states that the best estimator of σ is

$$s = \sqrt{\frac{(X - \bar{X})^2}{(n-1)}} \qquad (5.5)$$

In other words, provided that the sample standard deviation, s, is calculated using $n - 1$ as the divisor, it is an unbiased estimator of the population standard deviation. The analysis behind this point is the subject of the appendix to Chapter 2.

We know that for a normally distributed variable, 95% of the observations will be plus or minus 1.96 standard deviations of the mean. As the standard deviations of sampling distributions are standard errors, we can state that the sample mean will lie within an **interval** plus or minus 1.96 **standard errors** of the population mean 95% of the time. More formally, the **confidence interval** is given as

$$\mu \pm 1.96 \frac{s}{\sqrt{n}} \qquad (5.6)$$

where s is the sample standard deviation.

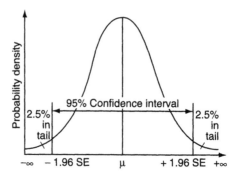

Figure 5.1

This can be illustrated with the aid of Fig. 5.1. This diagram shows that 95% (2.5% in each tail) of the time, the sample mean will lie within 1.96 standard errors of the population mean. We can transform this confidence interval into a 95% probability statement thus

$$p\left[\mu - 1.96\frac{s}{\sqrt{n}} < \overline{X} < \mu + 1.96\frac{s}{\sqrt{n}}\right] = 0.95 \qquad (5.7)$$

A little algebraic manipulation of these two inequalities leads to

$$p\left[\overline{X} - 1.96\frac{s}{\sqrt{n}} < \mu < \overline{X} + 1.96\frac{s}{\sqrt{n}}\right] = 0.95 \qquad (5.8)$$

To illustrate this with a concrete example, consider that we have data on 60 monthly observations of the returns to the FTSE 100 index. The sample mean monthly return is 1.125% with a standard deviation of 2.5%. What is the 95% confidence interval for this mean?

Firstly the standard error is calculated as

$$SE = \frac{2.5}{\sqrt{60}} = 0.3227$$

The **confidence interval** would be

$$\mu = 1.125 \pm 1.96 * 0.3227$$
i.e. $1.125 - 0.6325 \leqslant \mu \leqslant 1.125 + 0.6325$
i.e. $0.4925 \leqslant \mu \leqslant 1.7575$

This confidence interval can be illustrated with Fig. 5.2.
The **probability statement** would be

$$p\left[\overline{X} - (1.96 * 0.3227) \leqslant \mu \leqslant \overline{X} + (1.96 * 0.3227)\right] = 0.95$$

$$p\left[[1.125 - 0.6325] \leqslant \mu \leqslant [1.125 + 0.6325]\right] = 0.95$$

$$p[0.4925 \leqslant \mu \leqslant 1.7575] = 0.95$$

How does the analyst use this information? It is for the analyst to decide whether or not

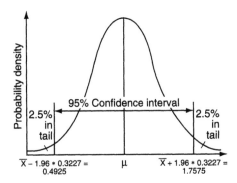

Figure 5.2

the band within which the population mean will be too wide for the confidence interval to be of any practical use. For example, the confidence interval given above spans 0.4925 to 1.7775, a range of 1.265, which is greater than the estimate of the mean itself.

Clearly, the analyst may not wish to have such a broad range and thus would like to reduce it. Given that the sample mean is fixed and the 1.96 is tied in with the 95% probability, the variable to try to influence is the standard error. This is a function of the sample standard deviation and the sample size. Thus one way to reduce the size of the standard error is to increase the size of the sample.

To illustrate the influence of the sample size consider the above example but with the sample increased to 120 and the sample standard deviation remaining unchanged at 2.5. The standard error would be

$$SE = \frac{2.5}{\sqrt{120}} = \frac{2.5}{10.95} = 0.2282$$

So the confidence interval would be

$$\mu = 1.125 \pm 1.96 * 0.2282$$
$$\text{i.e. } 1.125 - 0.4473 \leqslant \mu \leqslant 1.125 + 0.4473$$
$$\text{i.e. } 0.6777 \leqslant \mu \leqslant 1.5723$$

a range of only 0.8946.

What about small samples?

The central limit theorem can be used to justify the assumption that the mean of a sample is normally distributed if the size of the sample is over 30. With small samples we need to be able to assume that we are sampling from a normal distribution in order to have our sample mean normally distributed. Furthermore with only a small sample size, our estimate of the population variance will not be reliable. The *t*-distribution allows for this extra degree of variability.

Like the normal distribution, the *t*-distribution is symmetrical, but is slightly flatter. The actual shape depends upon the degrees of freedom, given as $n - 1$. As the sample gets larger the *t*-distribution becomes more like the normal distribution.

The confidence interval for a small sample, two-tailed test, thus becomes

$$\left[\overline{X} - t_{n-1,\alpha/2} \frac{S_{n-1}}{\sqrt{n}} < \mu < \overline{X} + t_{n-1,\alpha/2} \frac{S_{n-1}}{\sqrt{n}} \right]$$

(5.9)

where

s_{n-1} = the sample standard deviation

t_{n-1} = the value given in the t-statistic tables for a sample size n and $n - 1$ degrees of freedom

α = the level of significance.

The probability statement pertaining to this confidence interval is

$$P\left(\overline{X} - t_{n-1,\alpha/2} \frac{S_{n-1}}{\sqrt{n}} < \mu < \overline{X} + t_{n-1,\alpha/2} \frac{S_{n-1}}{\sqrt{n}} \right) = 1 - \alpha$$

(5.10)

To illustrate the use of the t-statistic, consider testing the sample mean of the quarterly returns of a particular group of fund managers. From 20 observations, and therefore $20 - 1 = 19$ degrees of freedom, the sample mean is calculated as 4.5%. The sample standard deviation is 5%. At the 95% level of confidence, the confidence interval is

$$\overline{X} - 2.093 * 5\sqrt{(19)} \leqslant \mu < \overline{X} + 2.093 * 5\sqrt{(19)}$$

$$4.5 - (2.093 * 1.147) \leqslant \mu < 4.5 + (2.093 * 1.147)$$

$$4.5 - 2.401 \leqslant \mu \leqslant 4.5 + 2.401$$

$$2.099 \leqslant \mu \leqslant 6.901$$

The probability statement is

$$P(4.5 - 2.093 * 1.147 \leqslant \mu \leqslant 4.5 + 2.093 * 1.147) = 95\%$$

$$p[4.5 - 2.401 \leqslant \mu \leqslant 4.5 + 2.401] = 95\%$$

$$p[2.099 \leqslant \mu \leqslant 6.901] = 95\%$$

Sample size

We have already seen that the width of the confidence interval is influenced by the size of the sample. So it is often appropriate to determine the size of the sample that will give an estimate of the population parameter with the required degree of confidence. The formula is

$$n = \left(\frac{zs}{e} \right)^2$$

(5.11)

where

    n = the required sample size
    z = the critical value in the distribution tables for the required level of confidence
    e = the desired half-width of the confidence interval (the half-width is the difference between μ and the tail)
    s = the estimate of the population standard deviation.

Confidence intervals for variances

We noted earlier that the sampling distribution of the variance is, after appropriate standardization, a χ^2 distribution. To develop confidence intervals for variances, we are not concerned about the sampling distribution of the variance itself, but of the sampling distribution of the standardized variable

$$\frac{(n-1)s^2}{\sigma^2} \sim \chi^2_{n-1} \tag{5.12}$$

To produce a 95% confidence interval for the point estimate of our variance, we need to determine the values of χ^2 that account for 2.5% in each tail of the distribution (see Fig. 5.3). Thus we need a value of χ^2 that assumes 97.5% of the values are to the right and another value which assumes 2.5% are to the right. If we refer to the degree of confidence as $1 - \alpha$, we need the values of $\chi^2_{1-\alpha/2}$ and $\chi^2_{\alpha/2}$. If we are working to 95% confidence, the value of α will be 0.05, and the relevant χ-squares will be $\chi^2_{n-1,0.975}$ and $\chi^2_{n-1,0.025}$.

The confidence interval is given as

$$\left[\frac{(n-1)s^2}{\chi^2_{n-1,1-\alpha/2}} \leqslant \sigma^2 \leqslant \frac{(n-1)s^2}{\chi^2_{n-1,\alpha/2}} \right] \tag{5.13}$$

and the probability statement is

$$p\left[\frac{(n-1)s^2}{\chi^2_{n-1,1-\alpha/2}} \leqslant \sigma^2 \leqslant \frac{(n-1)s^2}{\chi^2_{n-1,\alpha/2}} \right] = 1-\alpha \tag{5.14}$$

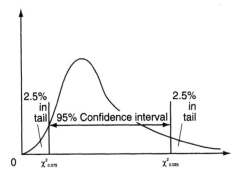

Figure 5.3

To illustrate how this works, consider that from a sample of 30 monthly observations the variance of the FTSE 100 index is 0.0225. With $n - 1 = 29$ degrees of freedom the critical values of the χ^2 distribution leaving 2.5% in each tail are 45.72 and 16.05, so

$$P\left(\frac{29*0.0225}{45.72} \leqslant \sigma^2 \leqslant \frac{29*0.0225}{16.05}\right) = 0.95$$

$$P(0.0143 \leqslant \sigma^2 \leqslant 0.04066) = 0.95$$

The confidence interval for the standard deviation is given by the square root of the confidence interval for the variance, for example

$$P(0.1195 \leqslant \sigma \leqslant 0.2016) = 0.95$$

Hypothesis testing

Hypothesis testing has two broad approaches, the **classical approach** which is the older of the two, and the **p-value** approach which is gaining popularity with the advent of more sophisticated software packages. We will study the classical approach first and then go on to study the p-value approach. However, before we proceed we must explain some terms.

What we have seen so far is that, given our knowledge of the sampling distribution of the various descriptive statistics, the size of the sample and the sample statistics themselves, we are able to create confidence intervals around our point estimates. However, we often have some *a priori* suspicion regarding the value of the population parameter.

Armed with this *a priori* knowledge we are able to test the hypothesis that our suspected value is in fact the population parameter. A **statistical hypothesis** is an **assumption** about the value of a population parameter of the probability distribution under consideration. When testing a hypothesis, two hypotheses are established, the **null hypothesis** and the **alternative hypothesis**. In effect we set up two competing hypotheses and test which one applies.

The **null hypothesis**, usually designated H_0, is the assumption which will be accepted pending evidence against it from a statistical test procedure. The **alternative hypothesis**, usually designated H_1, is that hypothesis which will be accepted if the statistical test leads to a rejection of the null hypothesis.

The exact formulation of the hypothesis depends upon what we are trying to establish. For example, imagine that we simply wish to know whether or not a population parameter, say the mean, μ, has a value of μ_0. The hypotheses would then be formulated as

$$H_0: \mu = \mu_0$$
$$H_1: \mu \neq \mu_0$$

(5.15)

However, if we wished to know whether or not a population parameter is greater than a given figure, μ_0, the hypothesis (in relation to means) would be

$$H_0: \mu = \mu_0$$
$$H_1: \mu > \mu_0$$

(5.16)

If we wished to know whether the population parameter was less than μ_0, the hypotheses would be

$$H_0: \mu = \mu_0$$

$$H_1: \mu < \mu_0$$

(5.17)

In order to test our hypothesis we have to carry out a statistical test. A **statistical test** consists of using a **standardized test statistic**, computed from a sample, to decide whether to reject a hypothesis about the population parameter in question.

The standardized test statistic

We learned in Chapter 2 that in order to compare one normally distributed variable with another, the variables had to be standardized so that they had a mean of zero and a standard deviation of one. The probability distribution of such a standardized variable is known as the standard normal distribution. In hypothesis testing we have to standardize the test statistic so that meaningful comparison can be made with the standard normal or t-distributions in the case of means, and χ^2 distributions in the case of variances.

In the case of the test statistic for means, recall that if X is normally distributed, i.e. $X \sim N(\mu, \sigma^2)$, then $\overline{X}$ is approximately distributed $\overline{X} \sim N(\mu, s^2/n)$. Because the shape of the normal distribution depends on the magnitude of the mean and the standard deviation, we must standardize the variable before comparisons can be made. In the case of hypothesis testing of the mean the standardization takes the form

$$\frac{\overline{X} - \mu_0}{s/\sqrt{n}}$$

(5.18)

This is the **standardized test statistic.** If $\overline{X}$ is μ_0, i.e. if H_0 is true, the test statistic has a standardized normal distribution if the sample is large, but has a standardized t-distribution with $n - 1$ degrees of freedom for small samples. Both distributions have a mean of zero and a standard error (because they relate to sample statistics) of one. Having standardized the test statistic we can compare its value directly with the values given in the appropriate standardized distributions for given probabilities.

If the test statistic that results from this standardization process is located in the critical region of the distribution, then it provides evidence that $\overline{X}$ does not have a mean equal to μ_0, so our assumption is wrong and H_0 is false. The point here is that if the statistic is in the critical region, this evidence is unlikely to occur given our assumptions. However, we have seen the evidence, it has occurred, so we must question our assumption.

The hypothesis test may be a **one-tailed test** or a **two-tailed test. One-tailed tests** are applied when we are particularly concerned that the population parameter is strictly larger (right-tailed test) or strictly smaller (left-tailed test) than the hypothesized value. The hypotheses given in equations (5.16) and (5.17) above would be tested by using one-tailed tests. Two-tailed tests are applied when we are concerned that the real value of the parameter is simply different to the hypothesized value. For example, the hypotheses given in equation (5.15) would be tested using a two-tailed test.

The **critical region** consists of those values of the sample statistic which lead to the rejection of the null hypothesis. In terms of Fig. 5.4, the critical region is that area in the tails of the distribution.

The **significance level** of a test is the probability that the **test statistic** lies in the critical region when H_0 is true – commonly 5% or 1%. If a test at the 5% level leads to the rejection of H_0 then the value of the test statistic is said to be "significant". If H_0 is rejected at the 1% level the term "highly significant" is used.

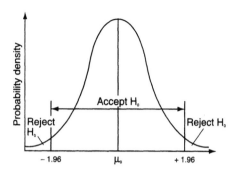

Figure 5.4

The concept of the critical region and the significance level can be illustrated by means of Fig. 5.4 which relates to the confidence interval of a mean. The critical regions are the areas in the tails to the left of -1.96 and to the right of $+1.96$. The confidence interval is the area between those two points. If $\overline{X}$ is positively or negatively more than 1.96 standard errors from μ_0, it is so large (or so small) compared to the hypothesized μ_0 that there is only a small probability (reflected by the area in the tails) that $\overline{X}$ represents μ_0.

A **decision rule** for a statistical test is a model giving values of a sample statistic that will lead to acceptance or rejection of the null hypothesis.

However, before we look at hypothesis testing in detail we must consider two very important types of error found in hypothesis testing.

Type I and type II errors

When testing hypotheses there is a possibility that the null hypothesis will be rejected when in fact it should have been accepted. This is referred to as a **type I error**. The probability of making a type I error is the **significance level** of the test. Thus when we choose a 5% level of significance for our test, we are accepting a 5% chance that we will reject the null hypothesis when in fact we should have accepted it.

The second type of error is that the null hypothesis is accepted when in fact it should have been rejected. This is referred to as a type II error.

For an intuitive explanation of the relationship between a type I error and a type II error consider the following analogy relating to the jury system depicted graphically in Fig. 5.5.

		The actuality	
		H_0 true Defendant innocent	H_0 false Defendant guilty
The decsion	H_0 accepted Defendant freed	✔	Type II error
	H_0 rejected Defendant punished	Type I error	✔

Figure 5.5

The defendant can either be innocent or guilty, The jury can find the defendant innocent or guilty. The jury system is actually testing the null hypothesis that the defendant is not guilty. If the jury finds the defendant guilty when he or she is indeed guilty that is good – there is no error. Likewise, if the jury finds the defendant innocent when he or she is indeed innocent all well and good. But what if the defendant is innocent but the jury finds the defendant guilty, i.e. they reject the null hypothesis when actually it is correct? This is a major error – it is a **type I error**. It is the type of error that one most wants to avoid. Accordingly we set the significance level of the test to make the probability of this small (commonly 5% or 1%).

Having set the significance level we have little influence over the probability of a type II error – finding a guilty person to be innocent.

Hypothesis tests of the population mean

Two-tailed test of the mean

If we wished to simply test whether the population mean equals the sample mean, we set up the hypotheses as

$$H_0: \mu = \mu_0$$
$$H_1: \mu \neq \mu_0$$

(5.19)

and then the standardized test statistic is used in the following procedure to test the above hypothesis:

1. Decide on the level of significance for the test. This will usually be a 10, 5 or 1% level, and we need to establish the test so that 5, 2.5 or 0.5% is in each tail of the distribution.
2. Set the value of μ_0 in the null hypothesis.
3. Identify the appropriate critical value of z (or t, if the test is for a small sample) from the tables that reflect the percentages in the tails according to the level of significance chosen.
4. Apply the following **decision rule**:

$$\text{Accept } H_0 \text{ if } -z \leqslant \frac{\bar{X} - \mu_0}{\sqrt{s^2/n}} \leqslant z$$

(5.20)

Reject H_0 otherwise

To give some intuitive understanding of this, recall that for normally distributed variables, 95% of all observations fall within plus or minus 1.96 standard deviations of the mean. In an analogous manner, if $\bar{X}$ is statistically equivalent to μ_0, with 95% confidence, the standardized test statistic will lie plus or minus 1.96 standard errors of zero. This number 1.96 is the critical value for a test at the 5% level of significance. If the standardized test statistic had a value in excess of +1.96 or less than −1.96, there is evidence at the 5% level that $\bar{X}$ is not the same as μ_0.

To illustrate this consider a test of whether or not the mean of a portfolio manager's monthly returns of 2.4% is statistically significantly different from the industry average of 2.3%. We first choose critical values of the test statistics according to the level of significance that we require. For a two-tailed test these are plus and minus 1.64, 1.96 and 2.58, respectively assuming a normal test. With $\overline{X}$ at 24 and μ_0 at 23, from 36 observations with a standard deviation of 1.7% the value of z is

$$\frac{2.4 - 2.3}{1.7/\sqrt{36}} = 0.3529$$

Note: as the sample size is greater than 30 we can assume normality.

As 0.3529 is between the negative and positive values of all the critical values we conclude that $\overline{X}$ of 2.4% is no different to μ_0 of 2.3%, and thus the 0.1% difference is due purely to chance.

Classical one-tailed test

Right-tailed tests

If we wished to test that the population mean, μ, is **more than** a pre-specified figure, μ_0, we have to carry out a one-tailed test because we are only concerned with the population mean being more than a given value. We treat the null hypothesis that μ is equal to μ_0, and the alternative hypothesis that μ is greater than μ_0. The null and alternative hypotheses would be

$$H_0: \mu = \mu_0$$
$$H_1: \mu > \mu_0$$

(5.21)

We would carry out a **right-tailed test**, a test of the probability of μ being in the right tail of the distribution, because if μ is in the right tail it will be greater than our hypothesized value of μ_0 and we would reject the null hypothesis.

The standardized test statistic is then calculated in the same manner as for the two-tailed test, except that the level of significance relates only to the right tail of the distribution. The decision rule is:

$$\text{Accept } H_0 \text{ if } \frac{\overline{X} - \mu_0}{s/\sqrt{n}} \leq z$$

(5.22)

$$\text{Reject } H_0 \text{ if } \frac{\overline{X} - \mu_0}{s/\sqrt{n}} > z$$

To understand the intuition behind this rule consider the interpretation of the test statistic being above z. This means that $\overline{X}$ is greater than μ_0 by an amount that puts it into the critical or reject region. $\overline{X}$ is so large that the probability of it being the population mean is smaller than the level of significance that we have set for the test. We therefore reject the null hypothesis. If the test statistic were less than z, it would mean that $\overline{X}$ was not so large as to be in the reject region. We would not, therefore, reject the null hypothesis.

To illustrate this assume that we wish to test that the mean monthly return on the FTSE 100 index for a given period is more than 1.2%. From 60 observations we calculate the mean as 1.25% and the standard deviation as 2.5%. The test statistic is

$$\frac{1.25-1.20}{2.5/\sqrt{60}} = 0.1549$$

Note: again the sample size is large, so we can assume normality of the sample mean.

Looking at the normal distribution tables we see that z for 10, 5 and 1% one-tailed levels of significance is 1.28, 1.64 and 2.33, respectively. Thus the test statistic of 0.1549 is not significant.

Left-tailed test

If we wished to test the hypothesis that μ is less than μ_0, we would carry out a left-tailed test to test for the probability of μ being in the left tail of the distribution. The null and alternative hypotheses would be

$$H_0: \mu = \mu_0$$
$$H_1: \mu < \mu_0$$

(5.23)

and the criteria for acceptance and rejection are

$$\text{Accept } H_0 \text{ if } \frac{\overline{X}-\mu_0}{s/\sqrt{n}} \geqslant z$$

$$\text{Reject } H_0 \text{ if } \frac{\overline{X}-\mu_0}{s/\sqrt{n}} < z$$

(5.24)

To illustrate this assume that we wish to test that the mean monthly return on the S&P 500 index is less than 1.30%. Assume also that the mean return from 75 observations is 1.18%, with a standard deviation of 2.2%. The test statistic would be calculated as

$$\frac{1.18-1.30}{2.2/\sqrt{75}} = -0.4724$$

Again choosing the 10, 5 or 1 levels of significance, with critical values of -1.28, -1.64 and -2.33, respectively, we see that test statistic is not significant, and we conclude that there is no evidence to reject the null hypothesis.

Hypothesis testing of the variance

The standardized test statistic for the population variance is

$$\frac{(n-1)s^2}{\sigma_0^2}$$

(5.25)

where σ_0^2 is the hypothesized variance.

We noted earlier that this standardized test statistic has a χ^2 distribution. The decision rules for left-tailed, right-tailed and two-tailed tests with χ^2 distributions are given below.

For the left-tailed test

The null and alternative hypotheses are

$$H_0: \sigma_0^2 = \sigma^2$$
$$H_1: \sigma_0^2 < \sigma^2 \tag{5.26}$$

The decision rule is

$$\text{Accept } H_0 \text{ if } \frac{(n-1)s^2}{\sigma_0^2} \geq \chi^2_{(1-\alpha)}$$

$$\tag{5.27}$$

$$\text{Reject } H_0 \text{ if } \frac{(n-1)s^2}{\sigma_0^2} < \chi^2_{(1-\alpha)}$$

where α is the chosen level of significance.

To illustrate this, assume that we wish to test that the variance of share B is below 25. The sample variance is 23 and the number of observations is 40, and therefore the degrees of freedom are $40 - 1 = 39$. As the χ^2 tables give the probabilities in the left tail, a left-tail test at the 5% level of significance will have 95% in the right tail. The critical value in such a situation with 39 degrees of freedom is approximately 26.5. The test statistic is

$$\frac{39 * 23}{25} = 35.88$$

As the test statistic of 35.88 is greater than the critical value of 26.5, we accept the null hypothesis that in this case there is insufficient evidence to conclude that the variance of share B is less than 25.

For the right-tailed test

$$\text{Accept } H_0 \text{ if } \frac{(n-1)s^2}{\sigma_0^2} \leq \chi^2_{\alpha}$$

$$\tag{5.28}$$

$$\text{Reject } H_0 \text{ if } \frac{(n-1)s^2}{\sigma_0^2} > \chi^2_{\alpha}$$

Assume that we wish to test that the variance of the returns on bond A is above 8%. The sample variance has been calculated as 9% from a sample of 35 observations. The test statistic is

$$\frac{34 * 9}{8} = 38.25$$

At the 95% level of significance, the right tail will contain 5%. Thus the critical value with 34 degrees of freedom is approximately 48.6. Thus we accept the null hypothesis, and conclude that the variance of returns of bond A is equal to 8%.

For the two-tailed test

$$\text{Accept } H_0 \text{ if } \chi^2_{(1-(\alpha/2))} \leqslant \frac{(n-1)s^2}{\sigma_0^2} \leqslant \chi^2_{(\alpha/2)}$$

(5.29)

Reject otherwise

If we wanted to test that the variance of bond B is equal to seven we must find the critical values for 0.975 and 0.025. With 34 degrees of freedom these are approximately 19.8 and 51.9, respectively. As the test statistic of

$$\frac{34*9}{7} = 43.71$$

lies between these critical values, we accept the null hypothesis that the variance is seven.

The reader should note from the above examples that there are many null hypotheses that cannot be rejected given a particular piece of evidence. In fact the set of hypothesized values which cannot be rejected is the corresponding confidence interval.

The p-value method of hypothesis testing

The p-value is the value which, if the null hypothesis is correct, represents the probability of getting a value for the standardized test statistic that is more extreme than the one observed.

In the case of a one-tailed test, this is equal to the area in the tail to the left (left-tailed test) or to the right (right-tailed test) of the value of the test statistic. In the case of a two-tailed test it is equal to double the area in the tail to the right or left of the test statistic.

Under the p-value method, the decision rule is the same whether carrying out left-tailed tests, right-tailed tests or two-tailed tests. Assuming the degree of significance of the test is designated α, the decision rule is

$$\text{Accept } H_0 \text{ if } p\text{-value} \geqslant \alpha$$

(5.30)

Reject H_0 otherwise

It should be noted that in the case of a two-tailed test, the p-value as displayed by a computer package may have to be doubled in order to find the probability of getting a test statistic more extreme than the one observed.

Calculation of the p-value

In order to calculate the p-value, we first calculate the standardized test statistic and then, with knowledge of the degrees of freedom, we find the probabilities (areas in the tail) associated with t-statistics or z-statistics which straddle the observed statistic. It is then necessary to interpolate between the probabilities to derive the p-value.

We will now illustrate this using tables for the t-statistic (given the sample size in our illustration we could assume normality, but that assumption, whilst a simplifying assumption, is not necessary). Imagine that we wish to find an investment which gives a consistent return of at least 13.2%. We wish to test whether or not a sample mean provides evidence that the true return is greater than the hypothesized mean return. Assume that the mean annualized monthly return of a given bond index is 14.4% and the sample standard deviation of those returns is 2.915%, there were 40 observations and the returns are normally distributed. The hypotheses to test are:

$$H_0: \mu = 13.2$$
$$H_1: \mu > 13.2$$

The test statistic is

$$\frac{(\bar{X} - \mu)}{\frac{s}{\sqrt{n}}} = \frac{14.4 - 13.2}{\frac{2.915}{\sqrt{40}}} = \frac{1.2}{0.461} = 2.604$$

With 39 (40 − 1) degrees of freedom, a t-value of 2.423 leaves 1% in the tail, whereas a figure of 2.704 leaves 0.5% in the tail. Thus the probability associated with a t-value of 2.604 can be approximated by interpolation as follows.

Firstly we must interpolate the position of the t-value 2.604 in the interval between 0.01 and 0.005, thus

$$\frac{2.604 - 2.423}{2.704 - 2.423} = 0.644$$

so it is just over half-way between 2.704 and 2.423.

To calculate the p-value we must multiply the interpolated position figure 0.644 by the difference between the two significance levels, and subtract the result from the larger significance level. The p-value is thus

$$0.01 - (0.644 * (0.01 - 0.005)) \approx 0.01 - 0.0032 \approx 0.0068$$

Thus the probability of obtaining a test statistic with a value in excess of 2.604 is 0.0068. The decision rule is to accept H_0 if p is greater than α. As $0.0068 < 0.05$ we reject the null hypothesis that $\mu = \mu_0$.

The p-test for variances

Recall that in the left-tailed test of the variance of the returns of share B the test statistic was 35.88, with 39 degrees of freedom. Tables indicate that for 39 degrees of freedom, 33.93 will leave 30%, i.e. 100 − 70, in the left tail whereas 36.16 will leave 40% in the left tail.

To interpolate the probability associated with a test statistic of 35.88 we start as follows:

$$\frac{35.88 - 33.93}{36.16 - 33.93} = \frac{1.95}{2.23} = 0.8744$$

Next we multiply 0.874 by the range of probability levels spanning the test statistic, 10 in this example, and add to the lower value of the range, thus

$$30 + (0.8744 * 10) = 38.74$$

Thus our p-value is 0.3874. As this is greater than the 5% (0.05) probability level set at the outset of the test, we accept the null hypothesis as before.

Testing goodness of fit

We can use the χ^2 distribution to determine how well a sample or a population is approximated by a given probability distribution, for example by a normal, a binomial or a Poisson distribution.

The general principle is to create an expected notional distribution according to the parameters of the data, e.g. the mean and standard deviation, in the case of a normal distribution. This expected distribution is then divided into equiproportional intervals but with the restriction that no interval should have less than five observations. Next we determine the number of actual data observations that fall into each equiproportional interval. We then use a χ^2 test, to compare the sum of the squared differences between the expected observations and actual observations, each difference being divided by the expected observation.

To illustrate this technique, consider a data set of monthly returns to the FTSE 100 index has a mean return of 0.6905% per month and a variance of 4.7080. There are 50 observations.

We begin by determining what value of X will leave 0.1 in the tail.

We know that $X = zs + \mu$, where s is the sample standard deviation. To construct equally likely intervals (convenient but not necessary) we thus need to use the z tables to determine the value of z that relates to 10% remaining in the left tail, i.e. 1.282, the value of z that gives 20% in the left tail, i.e. 0.842, and so on. The complete list of values for 10% through to 90% in 10% intervals is given below:

% in left tail	z-value
10	−1.282
20	−0.842
30	−0.524
40	−0.253
50	0
60	0.253
70	0.524
80	0.842
90	1.282

We calculate the value of X for the first interval, i.e. with 10% in the left tail, as

$$X = -1.282 * \sqrt{4.708} + 0.6905$$
$$= -1.282 * 2.1698 + 0.6905$$
$$= -2.7817 + 0.6905 = -2.0912$$

This is repeated until the values of X for each interval are known. The values for our example with $s = 4.7080$ and $\mu = 0.6905$ are

−2.0912, −1.1365, −0.4465, 0.1415, 0.6905, 1.239, 1.827, 2.517, 3.472

Next we create a frequency distribution from our data, noting the number of actual observations in each interval. This is shown in Table 5.1.

As the intervals were equiprobability intervals, we would expect an equal number of observations in each interval. We can therefore determine the deviation of the actual observations from the expected observations. As some will be negative and some will be positive, we must square the deviations in the same way as we did for the variance. We then divide each by the expected number of observations in a single interval. Then we sum the result for each interval, and this result should have a χ^2 distribution with $k - r - 1$ degrees of freedom, where k refers to the number of intervals in the analysis and r refers to the number of parameters calculated from the original sample, i.e. the mean and standard deviation in the case of a normal distribution.

Table 5.1

Returns	Frequency (O)	Expected frequency (E)	$(O - E)^2$	$(O - E)^2/E$
<−2.0912	14	5	81	16.2
−2.0912 to −1.1365	3	5	4	0.8
−1.1365 to −0.4465	2	5	9	1.8
−0.4465 to 0.1415	4	5	1	0.2
0.1415 to 0.6905	2	5	9	1.8
0.6905 to 1.239	2	5	9	1.8
1.239 to 1.827	4	5	1	0.2
1.827 to 2.517	3	5	4	0.8
2.517 to 3.472	4	5	1	0.2
3.472 and over	12	5	49	9.8
			Sum $(O - E)^2/E$	33.6

The test statistic for this hypothesis test is

$$\sum \left[\frac{(O_i - E_i)^2}{E_i} \right] \tag{5.31}$$

where

O_i = the observed frequency in interval i
E_i = the expected frequency in interval i.

Thus to test the hypothesis that the sample comes from a normally distributed population:

$$H_0 = \text{population is normal}$$

$$H_1 = \text{population is not normal}$$

The decision rule is

$$\text{Accept } H_0 \text{ if } \sum \left[\frac{(O_i - E_i)^2}{E_i} \right] \leq X^2_{k-r-1},\ (\alpha)$$

Reject H_0 otherwise $\tag{5.32}$

The test statistic relating to the example in Table 5.1 is given at the foot of the final column of that table. The value is 33.6. However, at the 5% significance level, i.e. 5% in the right tail, the χ^2 value for seven degrees of freedom $(10 - 2 - 1)$ is 14.06713. Thus, as the test statistic is greater than the critical value we reject the null hypothesis in this case.

Exercises

1. Distinguish between sample statistics and population statistics.

2. What do you understand by the terms "sampling distribution of the sample mean" and "sampling distribution of the sample variance"? With respect to the returns of a financial index, with a mean return of 10%, a standard deviation of 16% and 60 observations, calculate the standard error of the mean.

3. Explain the characteristics and importance of BLUE estimators.

4. Consider the statement: "A 95% confidence interval for the population mean lies between 0.046 and 0.152." Explain what this means.

5. Using the data in question 2 calculate the 95% confidence interval for the sample mean and for the sample standard deviation.
   Now repeat this process but construct a 90% confidence interval. Compare the widths of your confidence intervals. Note that the lower the degree of confidence, the greater the precision and vice versa.

6. A sample of 25 weekly observations of the FTSE 100 index returns has a mean of 0.005 and a standard deviation of 0.02. Assuming that weekly returns are normally distributed, calculate a 95% confidence interval for the mean weekly return.

7. Using the data in question 6 how large will the sample size have to be to estimate the mean weekly return with 95% confidence to within a maximum error bound of 0.004?

8. Using the data in question 6 calculate the 95% confidence interval for the standard deviation of weekly returns.

9. What do you understand by the following: null hypothesis; alternative hypothesis; standardized test statistic; critical region; significance level; decision rule; one-tailed test; two-tailed test; p-value; type I errors; type II errors.

10. Explain the relationship between a confidence interval and the corresponding hypothesis test.

11. ABC Fund Managers Plc claim that the monthly returns to their high-income tracker fund exceed the index return by 0.3% or 0.003. Over a one-year period the mean return to the index is 0.005 and the mean return to the fund is 0.0065, with a standard deviation of 0.019. Carry out a one-tailed test to investigate their claim.

12. Given the data in question 11 carry out a one-tailed test to test the hypothesis that the returns to the portfolio actually exceed the returns to the index.

13. A portfolio manager is concerned not to invest in securities with a variance of annualized returns greater than 0.04. A sample of 52 observations of the returns to asset "A" indicates a variance of 0.045. Test the hypothesis that the returns to asset "A" have a variance less than or equal to 0.04.

14. Explain what you understand by the term p-value.

15. Tom and Dick have to take turns making tea for the senior female staff in a firm of city stockbrokers. They toss a coin to decide whose turn it is, and they keep a special coin for this. Tom lost the toss 116 times in the last 200. Should he insist on using a different coin in future?

Answers to selected questions

2. $SE = 0.16/\sqrt{(60)} \approx 0.02$

5. 95% Mean

$$0.10 \pm 1.96 \frac{0.16}{\sqrt{60}} \approx 0.10 \pm 0.04$$

$$= 0.06 \text{ to } 0.14 = 6\% \text{ to } 14\%$$

95% Standard deviation

$$\frac{59*0.16^2}{\chi^2_{59(0.975)}} < \sigma^2 < \frac{59*0.16^2}{\chi^2_{59(0.925)}}$$

$$\frac{59*0.16^2}{83.3} < \sigma^2 < \frac{59*0.16^2}{40.5}$$

$$0.0181 < \sigma^2 < 0.0373$$

$$0.135 < \sigma < 0.193$$

90% Mean

$$0.10 \pm 1.64 \frac{0.16}{\sqrt{60}} \approx 0.10 \pm 0.034$$

$$= 0.066 \text{ to } 0.134 = 6.6\% \text{ to } 13.4\%$$

90% Standard deviation

$$\frac{59*0.16^2}{\chi^2_{59(0.95)}} < \sigma^2 < \frac{59*0.16^2}{\chi^2_{59(0.05)}}$$

$$\frac{59*0.16^2}{79.1} < \sigma^2 < \frac{59*0.16^2}{43.2}$$

$$0.0191 < \sigma^2 < 0.0350$$

$$0.138 < \sigma < 0.187$$

6.

$$0.005 \pm t_{24(0.975)} * \frac{0.02}{\sqrt{25}} = 0.005 \pm 2.064 * \frac{0.02}{\sqrt{25}} = -0.0033 \text{ to } +0.0133$$

7. Require

$$1.96 * \frac{0.02}{\sqrt{n}} = 0.004$$

$$\text{thus } n = \left(\frac{1.96 * 0.002}{0.004}\right)^2 \approx 96$$

8. 95% Standard deviation

$$\frac{24 * 0.02^2}{\chi^2_{24(0.975)}} < \sigma^2 < \frac{24 * 0.02^2}{\chi^2_{24(0.025)}}$$

$$\frac{24 * 0.02^2}{39.3641} < \sigma^2 < \frac{24 * 0.02^2}{12.4012}$$

$$0.000244 < \sigma^2 < 0.000774$$

$$0.0156 < \sigma < 0.0278$$

10. A confidence interval is the set of all possible parameter values which would not be rejected as hypothesized values by the corresponding test.

11.

$$t = \frac{0.0065 - 0.008}{0.019 / \sqrt{12}} = -0.2735$$

$$t_{11(0.05)} = 1.796$$

so we reject the claim.

12.

$$t = \frac{0.0065 - 0.005}{0.019 / \sqrt{12}} = 0.2735$$

$$t_{11(0.05)} = 1.796$$

so we reject the hypothesis that they earn only the index return.

13. Test statistic

$$\frac{51*0.045^2}{0.04^2} = 64.55$$

$$\chi^2_{51(0.05)} = 67.5$$

Therefore, insufficient evidence to reject the hypothesis that the variance does not exceed 0.04.

14. The probability under H_0 of observing a value equal to or more extreme than the test statistic.

15.

$$\chi^2 = \frac{(-16)^2}{100} + \frac{16^2}{100} = 5.12$$

$$\chi^2_{1(0.05)} = 3.84$$

Therefore reject the hypothesis that the coin is fair.
NB. When v (degrees of freedom) $= 1$, a correction should be made. The modulus of "observed–expected" should be reduced by 0.5. This gives

$$\chi^2 = \frac{(-15.5)^2}{100} + \frac{15.5^2}{100} = 4.81$$

which is still significant.

Further reading

Bowers, D. (1991) *Statistics for Economics and Business*. Macmillan, London.
Curwin, J. and Slater, R. (1996) *Quantitative Methods for Business Decisions*, 4th edn. Chapman & Hall, London.
Silver, M. (1992) *Business Statistics*. McGraw-Hill, London.

Appendix 5.1: standard error of the mean

To explain why the standard error of the mean is given by $\sigma/\sqrt{(n)}$, consider the random variable X defined by:

Possibility	−1	1
Probability	0.5	0.5

(We have seen this before in Chapter 4.)

We could simulate the process described by X by repeatedly tossing a coin and recording one for a head and minus one for a tail. Let us perform this experiment 16 times, calculating what we might expect to see, and what actually happens.

Expectations

$E(X) = 0.5 \times (-1) + 0.5 \times 1 = 0$

$\begin{aligned} Var(X) &= 0.5 \times (-1)^2 + 0.5 \times 1^2 \\ &= 1 \\ &= \sigma^2 \end{aligned}$

Results

1, −1, −1, 1, 1, 1, −1, −1, −1, −1, −1, 1, 1, −1, 1, −1

$\overline{X} = -0.125$

$s^2 = 1.05$ (Note that dividing by 15 instead of 16 tends to compensate for using −0.125 instead of 0 as the mean.)

We now question what would have happened if instead of producing individual observations, we had produced pairs of observations, in each case taking the mean of the pair. In Chapter 4 we showed how to add random variables and how to multiply by constants. Our averaging process is given by adding two identical random variables and by multiplying that sum by 0.5. Applying the methods of Chapter 4 this gives us a distribution of

Possibility	−1	0	1
Probability	0.25	0.5	0.25

We can use the same data as before to compare what actually happens with what we expect:

Expectations

$E\left(\dfrac{X+X}{2}\right) = 0.25 \times (-1) +$

$0.5 \times 0 + 0.25 \times 1 = 0$

$Var\left(\dfrac{X+X}{2}\right) = 0.25 \times (-1)^2 +$

$0.25 \times 1^2 = 0.5 = \dfrac{\sigma^2}{2}$

Results

Pairing our previous results:
(1,−1) (−1,1) (1,1) (−1,−1) (−1,−1) (−1,1) (1,−1) (1,−1)
This gives means of:
0, 0, 1, −1, −1, 0, 0, 0
The mean of these is −0.125, and the variance is 0.411.
Note the reduction in variance achieved by taking means of samples of size 2. We expect to see a 50% reduction and have noted a reduction of approximately 60%. A larger sample would confirm the 50% figure.

We could repeat this for means of samples of size 3, 4 and so on, though we would need a larger data sample to get a reliable picture of the results. We would expect the mean to stay unchanged but the variance to be divided by the size of the sample.

Appendix 5.2: FTSE 100 index data for goodness of fit tests

FTSE100	Returns
2407.5	
2289.2	
2160.1	−5.8048
2311.1	6.7569
2422.7	4.7159
2345.8	−3.2256
2238.4	−4.6865
2221.6	−0.7534
2117.9	−4.7803
2371.4	11.3055
2372	0.0253
2339	−1.4010
2166.6	−7.6564
2030.8	−6.4729
2028	−0.1380
2162.7	6.4307
2143.5	−0.8917
2165.7	1.0304
2386.9	9.7252
2456.5	2.8742
2508.4	2.0908
2515.8	0.2946
2443.6	−2.9118
2591.7	5.8842
2679.6	3.3353
2645.6	−1.2770
2549.5	−3.7001
2414.9	−5.4239
2493.1	3.1869
2560.2	2.6558
2554.3	−0.2307
2408.6	−5.8733
2659.8	9.9205
2697.6	1.4112
2493.9	−7.8515
2420.2	−2.9998
2298.4	−5.1637
2572.3	11.2587
2687.8	4.3923
2792	3.8035
2846.5	1.9332
2851.6	0.1790
2882.6	1.0812
2878.4	−0.1458
2813.1	−2.2948
2849.2	1.2751
2888.8	1.3803
2941.7	1.8146
3085	4.7564
3039.3	−1.4924
3164.4	4.0336
3233.2	2.1509
Mean	0.6905
Standard deviation	4.7080

6

Regression analysis

Introduction

Regression analysis constructs and tests a mathematical model of the relationship between one dependent (i.e. endogenous), and one or more independent (i.e. exogenous) variables. The dependent variable is usually depicted as the Y variable, and the independent variable(s), also referred to as regressors, are depicted as the X variable(s).

The direction of causality between the variables is determined from *a priori* reasoning and embodied in the model by way of the hypothesis. The regression analysis tests the statistical strength of the model as hypothesized.

As an example, imagine that it has been hypothesized that the level of the FTSE stock exchange index (FTSE 100) is linearly dependent on the level of the S&P 500 stock exchange index (S&P 500). Thus when the S&P 500 rises, the FTSE 100 also rises, and when the S&P 500 falls the FTSE 100 falls. We can test this hypothesis using simple linear regression involving only two variables.

An alternative hypothesis may be that the FTSE 100 is influenced, not just by one factor but by several. For example, the current level of the FTSE 100 may be influenced by the S&P 500, the level of the UK bond market and by the $/£ exchange rate. This hypothesis can be tested by multiple regression.

A distinction is made between **cross-section regression** and **time-series regression**. Cross-section regression tests the relationship between variables at a particular point in time. As an example of cross-section regression, we may wish to measure the relationship between company size and the returns to investing in shares. To do this we could collect data on share returns for a single period, say one year, and data on company size at the beginning of the same period for many companies. The returns data would represent the dependent variable, while the size data would be the independent variable. Thus the regression analysis tells us the average relationship between the variable at a single point in time.

With time-series regression, the data for each variable is collected over successive time periods. The example of the stock exchange indices given earlier is a time-series regression because the data of the FTSE 100 and the S&P 500 would be collected over time. The regression analysis would give an average relationship over the time period covered by the data.

Irrespective of whether the analysis is cross-sectional or time-series the basic principles of applying regression analysis are the same.

Simple linear regression

In this section we will apply regression analysis to a simple linear relationship between the dependent (Y) variable and one independent (X) variable.

By linear we mean that the variable Y is assumed to be influenced by variable X in the following manner:

$$Y = \alpha + \beta X + e \tag{6.1}$$

where

α = the constant; that is even if X had zero value, Y would have some positive or negative value. Whether it is rational for Y to have a value even if X has zero value will depend upon the hypothesis to which the regression analysis is being applied

β = the regression coefficient. This represents the slope of the line along which the scatter of data observations lies. It can be interpreted as indicating the percentage change in variable Y that is caused by a one unit change in the value of X. Thus if Y is the FTSE 100 and X is the S&P 500, β will indicate by how many points the FTSE 100 will change for a one point change in the S&P 500. If the sign of β

is positive, the variables will be positively correlated. If the sign is negative the variables are negatively correlated

e = the error or disturbance term, also referred to as the residual. It reflects the fact that usually, at least, the movement in Y will be imperfectly described by movements in X alone. There will be other factors not captured by the model. However, if the underlying hypothesis is realistic these other variables should be relatively unimportant.

Referring again to the relationship between the FTSE 100 and the S&P 500 the FTSE 100 is the dependent variable, Y, because we have hypothesized that its movements are influenced by, i.e. dependent on, the S&P 500 which represents X. In our hypothesis we assume that the many other influences, trivial and unrelated, are represented by e.

If the economic arguments are strong enough we could develop the hypothesis that the level of the S&P 500 is influenced by the FTSE 100. If we believed that, the S&P 500 would become the Y variable and the FTSE 100 would become the X variable.

Plotting the data from which Table 6.1 is a sample on to a scatter diagram (see Fig. 6.1) below we do indeed see that high(low) values of the S&P 500 are associated with high(low) values of the FTSE 100. Thus the two sets of data seem to rise or fall together.

Table 6.1

FTSE 100 (Y)	$Y - \bar{Y}$	S&P 500 (X)	$X - \bar{X}$	$(X - \bar{X})^2$	$(X - \bar{X})(Y - \bar{Y})$
2851.6	320.8596	442.52	51.10135	2611.348	16396.358
2882.6	351.8596	442.01	50.59135	2559.484	17801.052
2878.4	347.6596	450.3	58.88135	3467.013	20470.666
2813.1	282.3596	442.46	51.04135	2605.219	14412.015
2849.2	318.4596	453.83	62.41135	3895.176	19875.493
2888.8	358.0596	449.02	57.60135	3317.915	20624.716
2941.7	410.9596	450.15	58.73135	3449.371	24136.211
3085	554.2596	463.15	71.73135	5145.386	39757.788
3039.3	508.5596	461.28	69.86135	4880.608	35528.659
3164.4	633.6596	469.1	77.68135	6034.392	49223.532
3233.2	702.4596	461.89	70.47135	4966.211	49503.275
$\bar{Y} = 2530.740$		$\bar{X} = 391.419$		$\Sigma(X - \bar{X})^2 =$ 108046.7	$\Sigma(X - \bar{X})(Y - \bar{Y}) =$ 644387.4688

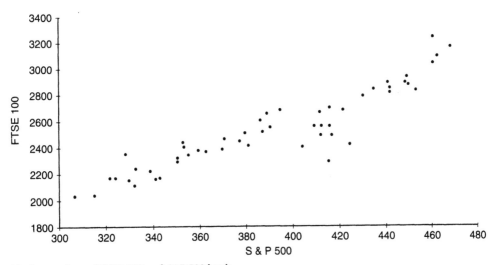

Figure 6.1 Scatter chart of FTSE 100 and S&P 500 levels.

However, the data tell us nothing about the causality. Our understanding of causality comes from our *a priori* hypothesizing. As we noted in the previous paragraph, the direction of cause and effect, i.e. which is the dependent variable and which is the independent variable, is determined by our hypothesis.

To illustrate this, consider again our hypothesis that the level of the FTSE 100 is influenced by the level of the S&P 500. The data would support such an idea but does your understanding of financial economics support it? The S&P 500 may influence the FTSE 100 because of the sheer size of the US economy and the international flow of funds. However, an alternative suggestion is that as both markets are open to international investors, they may both be influenced by a third factor, maybe the expectations of Japanese or European investors.

Clearly the hypothesis has to be developed independently of the regression model in order that the regression analysis can validly support or not support the hypothesis. Regression analysis cannot "prove" a hypothesis, it can only support it statistically or reject it.

Referring back to our scatter diagram, a number of straight lines may be drawn through the data, all satisfying equation (6.1), indeed it is not possible to produce one straight line that will pass through all the data observations. Consequently we have to choose one line, the one that best fits the data.

Ordinary least-squares regression

To be able to test statistically the relationship between the dependent and the independent variables it is necessary to derive the values of α, β and e in equation (6.1). The estimation method should be such that it is the **Best, Linear, Unbiased Estimator (BLUE)**.

Best refers to the desire for the estimators of the parameters to be the most efficient. That is that the variances around the estimates of the parameters are the smallest possible. This is achieved by choosing α and β that minimize the sum of the e_i^2 values.

Linear simply reiterates that the relationship is linear.

Unbiased requires that the expected values of the regression coefficients are the true coefficients.

If the data complies with the assumptions stated below, the method most commonly used and known as ordinary least-squares regression gives the best linear unbiased estimators. It derives its name because in calculating the straight line **that best fits the data,** it does so by calculating the line that minimizes the sum of the squares of the error terms between the value of Y predicted by the straight line, $\hat{Y}$, and the actual observations. This is shown in Fig. 6.2 below.

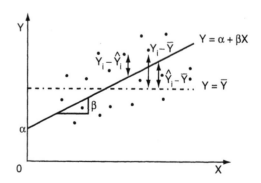

Figure 6.2

The statistical assumptions of ordinary least-squares regression

To be able **validly to apply** the **ordinary least-squares regression model (OLS)** to the data and be able to test the relationship between the variables, the data must comply with the assumptions underlying the regression model. These are as follows:

1. The mathematical form of the relationship between the **true dependent variable** Y, and the **independent variable X,** is

$$Y = \alpha + \beta X + e \qquad (6.2)$$

At this point it is useful to differentiate between deterministic models and probabilistic models. Deterministic models are those where once the value of X is known, the value of Y is also known exactly. For example, the income of a sales person may be determined by a fixed salary of "α" and a bonus of "β%" of sales. Once we know the value of sales we can determine Y exactly as $Y = \alpha + \beta X$. We do not need to apply regression analysis to such models. We simply have to observe two Xs and their associated Ys and draw a line between the points.

More usually we must treat Y as a random variable. Once the value of X is known, we have a probability distribution for Y. Thus our model is probabilistic, and is given by

$$Y = \alpha + \beta X + e$$

From the data we estimate α and β, the estimated values being denoted by $\hat{\alpha}$ and $\hat{\beta}$, respectively. We can then compute the expected value of Y, $\hat{Y}$, from

$$\hat{Y} = \hat{\alpha} + \hat{\beta} X \qquad (6.3)$$

Therefore, for each data value of X_i there is an actual value for Y_i but applying equation (6.3) there is also an estimated value, $\hat{Y}_i$. The difference between Y_i and $\hat{Y}_i$ is the estimation error e_i. OLS regression produces a straight line through the data which minimizes the sum of the squares of the e_is, i.e. minimizes $\Sigma e_i^2 = \Sigma(Y_i - \hat{Y}_i)^2$.

2. The error term, e_i, is **normally distributed,** with a **mean of zero** and a **constant variance, σ^2**. This is often stated as $e_i \sim N(0,\sigma^2)$.
3. The successive error terms are independent of each other, i.e. the covariance between pairs of error terms is zero (cov $e_i e_j = 0$).
4. The independent variable is non-stochastic.

The first assumption simply reiterates the point that the we are only dealing with linear relationships with this model, and that the values of the dependent variable, Y, are determined by only one significant factor, the independent variable, X.

The second assumption implies that although there is one major factor (X) that determines the value of Y, there are also many minor factors, some of which will have a positive influence and others that will have a negative influence. There will be many negative influences and many positive influences, and the error term will be normally distributed. The assumption of a constant variance to the error term means that however large or small the value of the independent variable X the spread of the e values is constant. The error term is said to be **homoscedastic.** If the spread of the error terms is not constant, the errors are said to be **heteroscedastic.**

The third assumption, that the e values are independent of each other, simply means that the minor factor or factors that caused one value of Y to show error does not automatically cause all the observations of Y to show error. When the e values are independent the data are said to be **non-autocorrelated**. If the e values are not independent the data are said to be **autocorrelated** or exhibit **autocorrelation**. The term "serial correlation" is sometimes used to refer to autocorrelation.

As Y is related to e in a linear form, Y itself is a random variable. For any values of X, Y will be normally distributed and therefore the statistical distribution of Y_i can be fully described by its mean and its variance, i.e.

$$E(Y_i) = E(\alpha + \beta_1 X_i + e_i) \tag{6.4}$$

Because α and β_1 are constants and X_i is non-stochastic, this becomes

$$E(Y_i) = \alpha + \beta_1 X_i + E(e_i) \tag{6.5}$$

However, the expected value of e_i is zero by assumption, therefore equation (6.5) becomes

$$E(Y_i) = \alpha + \beta_1 X_i \tag{6.6}$$

As the expected value of e_i is zero, the variance of Y_i, which is also the variance of e_i, is the mean value of e_i^2, i.e. $\Sigma(e_i - 0)^2/n = \Sigma e_i^2/n = E(e_i^2) = \sigma^2$.

Thus Y_i is distributed $N(\alpha + \beta_1 X_i, \sigma^2)$.

This is demonstrated in Fig. 6.3. For each value of X there is an expected value of Y_i which is normally distributed and thus to which we can attach probabilities. Hence the probabilistic model.

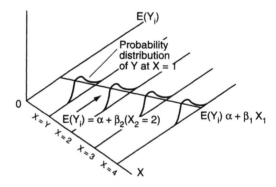

Figure 6.3

Fitting the regression line

The values of α and β that minimize the sums of the squared deviation of Y from $\hat{Y}$ are derived as follows:

$$\beta = \frac{\text{cov}_{XY}}{\text{var}_X} = \frac{\Sigma[(X - \bar{X})(Y - \bar{Y})]}{\Sigma(X - \bar{X})^2} \tag{6.7}$$

$$\alpha = (\bar{Y} - \beta\bar{X}) \tag{6.8}$$

Note that both the covariance and the variance normally have a denominator of $n - 1$. However, they cancel out in the equation above.

The error terms, also referred to as the residuals, are calculated as

$$e_i = (Y_i - \hat{Y}_i) \tag{6.9}$$

where Y_i is the observed value of Y, and $\hat{Y}_i$ is the estimated value of Y.

These results can be derived from calculus as follows: recall that the ordinary least-squares regression minimizes the sum of squares of the deviations of each observed value of Y from the estimated value of Y.

The sum of squares is

$$SS = \Sigma(Y - \hat{Y})^2 = \Sigma(Y - \alpha - \beta X)^2 \tag{6.10}$$

where α and β are the parameters which are to be determined.

$$\frac{\partial SS}{\partial \alpha} = \Sigma(-2(Y - \alpha - \beta X)) \tag{6.11}$$

and

$$\frac{\partial SS}{\partial \beta} = \Sigma(-2X(Y - \alpha - \beta X)) \tag{6.12}$$

The sum of squares is minimized when these partial derivatives are both zero, i.e. when

$$\Sigma(Y - \alpha - \beta X) = 0$$

$$\Sigma X(Y - \alpha - \beta X) = 0$$

This is achieved when

$$\Sigma Y = n\alpha + \beta \Sigma X$$

$$\Sigma XY = \alpha \Sigma X + \beta \Sigma X^2 \tag{6.13}$$

We now have to solve a simultaneous equation problem. We begin by multiplying the first equation by ΣX and the second by n, giving

$$\Sigma X \Sigma Y = n\alpha \Sigma X + \beta (\Sigma X)^2$$

$$n\Sigma XY = n\alpha \Sigma X + n\beta \Sigma X^2 \tag{6.14}$$

Subtracting the first equation from the second gives

$$n\Sigma XY - \Sigma X \Sigma Y = n\beta \Sigma X^2 - \beta(\Sigma X)^2 = \beta(n\Sigma X^2 - (\Sigma X)^2) \tag{6.15}$$

Therefore

$$\beta = \frac{n\Sigma XY - \Sigma X \Sigma Y}{n\Sigma X^2 - (\Sigma X)^2} \tag{6.16}$$

By replacing ΣX and ΣY by n times their means we get

$$\beta = \frac{n\Sigma XY - n\bar{X}n\bar{Y}}{n\Sigma X^2 - (n\bar{X})^2} \qquad (6.17)$$

Dividing top and bottom by n gives

$$\beta = \frac{\Sigma XY - n\overline{XY}}{\Sigma X^2 - n\bar{X}^2} \qquad (6.18)$$

This is exactly the same as

$$\beta = \frac{\Sigma[(X - \bar{X})(Y - \bar{Y})]}{\Sigma(X - \bar{X})^2} \qquad (6.19)$$

To arrive at α we divide the top equation (6.13) by n thus

$$\beta = \frac{\Sigma Y}{n} = \frac{n\alpha + \beta\Sigma X}{n}$$

$$\bar{Y} = \alpha + \beta\bar{X} \qquad (6.20)$$

$$\text{so } \alpha = \bar{Y} - \beta\bar{X}$$

Using the data relating to the FTSE 100 and S&P 500 indices in Table 6.1, you will see that

$$\beta = \frac{\text{cov}_{XY}}{\text{var}_X} = \frac{\Sigma[(X_i - \bar{X})(Y_i - \bar{Y})]}{\Sigma(X_i - \bar{X})^2} = \frac{644387.5}{108046.7} = 5.9640$$

and

$$\alpha = \bar{Y} - \beta\bar{X} = 2530.74 - (5.9640 * 391.4187) = 196.3298$$

Interpretation of the regression equation

These regression results are presented as follows:

$$\hat{Y} = +196.3298 + 5.9640X$$

This is interpreted as saying the estimated value of the dependent variable Y (FTSE 100) equals a constant amount $+196.3298$ plus 5.9640 for every unit of the independent variable X (S&P 500). The constant, or intercept, represents the value of the dependent variable when the independent variable has a value of zero. Schematically this is represented by the vertical distance between the origin and the point where the regression line intersects the vertical axis in Fig. 6.2.

Testing the model

We noted earlier that probabilistic models only provide estimates of the regression coefficients. It is important therefore to test how representative these estimates are of the true coefficients. This is achieved by testing the statistical significance of the regression coefficients and the closeness of fit of the data to the estimated regression line.

Significance tests of coefficients

As Fig. 6.3 shows, in the case of probabilistic models, calculating the regression coefficients using equations (6.7) and (6.8) gives single estimates of Y, e.g. $E(Y_i)$. The estimated regression coefficients are also assumed to come from a normal distribution. We need to know the statistical significance of these coefficients. We do this by testing that the regression coefficients are significantly different from zero.

Referring back to Fig. 6.3 again, as each estimate of the coefficients is assumed to come from a normal distribution, to test for significance we test to see if the estimated value falls in the tail of the distribution thereby being due to chance, or falls in the body of the distribution.

Thus the statistical significance of the coefficients is measured by the degree of dispersion around the estimated value. As the errors or residuals are assumed to be normally distributed, the standard deviation of the errors is used to measure that dispersion. These standard deviations are known as **standard errors of the coefficients**. We use the **t-statistics** to indicate the degree of significance of the coefficients. To derive these measures we first need to know:

- the sampling distribution of those coefficients
- estimates of their variances and thus standard deviations.

We can then either test hypotheses concerning the coefficients or construct confidence intervals for them.

Sampling distribution

The sampling distribution of the intercept is

$$\hat{\alpha} \sim N\left(\alpha, \frac{\sigma^2 \Sigma X^2}{n\Sigma(X - \bar{X})^2}\right) \tag{6.21}$$

where σ^2 is the variance of the e_is.

The sampling distribution of β is

$$\hat{\beta} \sim N\left(\beta, \frac{\sigma^2}{\Sigma(X - \bar{X})^2}\right) \tag{6.22}$$

The estimated variances and standard deviations

We noted earlier that the mean or expected value of e_i is zero, thus $\Sigma(e_i - 0)^2$ is simply Σe_i^2. To get s^2, the unbiased estimate of σ^2, we must divide by $n - 2$ since we are estimating two parameters from the data α and β. Thus the estimated variance of e_i, s^2, is given as

$$s^2 = \frac{\Sigma \hat{e}_i^2}{n-2} \tag{6.23}$$

Recall that

$$\hat{Y}_i = \hat{\alpha} + \hat{\beta} X_i + \hat{e} \tag{6.24}$$

Thus

$$\hat{e}_i = \hat{Y}_i - \hat{\alpha} - \hat{\beta} X_i \tag{6.25}$$

and therefore

$$Y_i - \hat{Y}_i = \hat{e}_i \tag{6.26}$$

We will use these in quoting the variance of the regression coefficients.

The variance of the intercept, $\hat{\alpha}$, is estimated by

$$\text{var}\,\hat{\alpha} = \frac{\dfrac{\Sigma (Y_i - \hat{Y}_i)^2}{n-2} \Sigma X_i^2}{n\Sigma (X_i - \bar{X})^2} \tag{6.27}$$

The variance of the slope coefficient, $\hat{\beta}$, is estimated as

$$\text{var}\,\hat{\beta} = \frac{\left[\dfrac{\Sigma (Y_i - \hat{Y}_i)^2}{n-2} \right]}{\Sigma (X_i - \bar{X})^2} \tag{6.28}$$

The standard errors

The standard errors of the coefficients are simply their standard deviations, that is the square root of the variances of the coefficients.

The standard error of the intercept is calculated as

$$\text{SE of }\hat{\alpha} = \sqrt{\frac{\dfrac{\Sigma (Y_i - \hat{Y}_i)^2}{n-2} \Sigma X_i^2}{n\Sigma (X_i - \bar{X})^2}} \tag{6.29}$$

The **standard error** of the estimate of the slope coefficient, β, is calculated as

$$\text{SE of }\hat{\beta} = \sqrt{\frac{\left(\dfrac{\Sigma (Y_i - \hat{Y}_i)^2}{n-2} \right)}{\Sigma (X_i - \bar{X})^2}} \tag{6.30}$$

We learned in Chapter 5 that for data that have a normal or Gaussian distribution, the difference between a variable and its mean divided by the estimate of its standard deviation has a t-distribution. Thus the estimates divided by the standard errors as calculated above have t-distributions with $n-2$ degrees of freedom. With this information we can derive confidence intervals around the point estimates of the coefficients.

We will denote the level of confidence as $1-c$, c being the probability of the variable being in the tail of the t-distribution. The probability statements are

$$P\left(-t_{n-2,c/2} \leq \frac{\hat{\alpha}-\alpha}{\text{SE}_a} \leq t_{n-2,c/2}\right) = 1-c$$

$$P\left(-t_{n-2,c/2} \leq \frac{\hat{\beta}-\beta}{\text{SE}_\beta} \leq t_{n-2,c/2}\right) = 1-c$$

(6.31)

From these equations we can derive the confidence interval estimates of the coefficients

$$P(\hat{\alpha}-t_{n-2,c/2}\text{SE}_\alpha \leq \alpha \leq \hat{\alpha}+t_{n-2,c/2}\text{SE}_\alpha) = 1-c$$

$$P(\hat{\beta}-t_{n-2,c/2}\text{SE}_\beta \leq \beta \leq \hat{\beta}+t_{n-2,c/2}\text{SE}_\beta) = 1-c$$

(6.32)

Thus we have $1-c$ probability that the true value of the coefficients falls within the range specified. If that range includes zero, the coefficients are not statistically significantly different from zero.

In practice all of the above calculations and others in this chapter are produced automatically in the standard computer packages for applying regression analysis.

Hypothesis testing

Earlier we noted that the regression equation is frequently created to test a hypothesis. This is achieved by setting up the null hypothesis, H_0, that the coefficients are not statistically significantly different from zero. In the case of coefficients α and β, the hypotheses would be as follows:

$$H_0: \alpha = 0$$
$$H_1: \alpha \neq 0$$
$$H_0: \beta = 0$$
$$H_1: \beta \neq 0$$

To test these hypotheses we need to calculate the t-statistic for the appropriate regression coefficient. The **t-statistics** are calculated by dividing the regression coefficients by their standard errors

$$t = \frac{\beta}{\text{SE of } \beta}$$

(6.33)

which is t-distributed, subject to the assumptions of normal regression analysis.

It is usual to test for statistical significance at the 95 or 99% level of confidence. That means that there is 95 or 99% probability that the values of α and β are not due to chance. The probability distribution of the t-statistic is a t-distribution with $n-2$ degrees of freedom. Degrees of freedom in this context refers to the number of pairs of data points used in the regression. The regression coefficients are significant if the t statistic is greater than the value given in the t distribution tables.

To illustrate tests of significance using tables, note the value of β is 5.9640, and the standard error of β is 0.3476. The t-statistic is therefore:

$$t = \frac{5.9640}{0.3476} = 17.1577$$

The test statistic for α is

$$\frac{\alpha}{SE_\alpha} = \frac{196.3298}{136.991} = 1.4332$$

and 95% confidence intervals are:

for α: $196.3298 \pm 2 \times 136.991$, i.e. $-77.65 \leqslant \alpha \leqslant 470.31$
for β: $5.9640 \pm 2 \times 0.3476$, i.e. $5.27 \leqslant \beta \leqslant 6.66$

A one-tailed test or a two-tailed test?

We have to decide whether the significance test will be a "one-tailed test" or a "two-tailed test". This decision should be made before the regression results are known. The choice is determined by the theory underlying the model between X and Y which the regression is testing.

Recall from Chapter 5 that a "one-tailed test" tests the significance of the regression coefficient where the expected relationship between Y and X can either be positive or negative but not both. A "two-tailed test" tests the significance when the relationship between Y and X could be either positive or negative.

To illustrate this decision consider the relationship between the two stock exchange indices. If this model is based upon a strong belief that the US economy dominates the world economy and investors' expectations, the relationship between the two variables could only be positive and thus a one-tailed test would be appropriate.

However, consider an alternative view, that the relationship between the two markets depends on the relative behaviour of the two economies. Unless there is a very high correlation between the levels of activity in the two economies, they may diverge and the stock markets diverge also. Thus, they may be positively or negatively related. Consequently, a two-tailed test would be required in any test between the two markets.

To make a "one-tailed test" test that the regression coefficient is significantly different from zero at the 95 or 99% level of confidence we must recall that the number of pairs of observations is 52 and therefore there are 50 degrees of freedom. Referring to the t-distribution tables, we first move down the column headed degrees of freedom to 50. Next we move across to the column representing the level of confidence which we wish to test for. The column headed 0.05 represents the 95% level of confidence, whilst the column headed 0.01 represents the 99% level of confidence.

We will illustrate a one-tailed test for the 99% level of confidence. With 50 degrees of freedom the critical value of the t-statistic is 2.4. If the t-statistic for any coefficient is greater than 2.4 that regression coefficient is said to be significant at the 99% level of confidence.

In our example above, the t-statistic for the α coefficient is 1.4332 and for the β 17.15537, which is much larger than the critical level of 2.4. Thus, the α is not significantly different from zero but the β coefficient is significant at the 99% level of confidence.

To conduct a "two-tailed test" at the 95% level of confidence we would look under the column headed 0.025, i.e. half of 0.05, and find the critical value to be 2.0. To test at 99% confidence we would look under the column headed 0.005, and the critical value for this example would be 2.7.

Goodness of fit: coefficient of determination, R^2

The regression model indicates that variation in Y can be explained by variation in the independent variable X, and by the error term e. We wish to know how much of the variation in Y is caused by X and how much is caused by e. In other words we want to know how well the estimated regression equation fitted the data, i.e. how small was the dispersion of data around the regression line. This is shown schematically below. Figure 6.4(a) shows the data to be widely dispersed about the regression line. Thus the errors are large. Figure 6.4(b) shows data tightly clustered about the regression line. Thus these errors are small and the relationship tested by that regression line seems to reflect the true relationship to a high degree.

To evaluate the goodness of fit of the regression line, we need to calculate the **total sum of squares**, the **sum of squares due to regression** and the **sum of squares due to error** in order to calculate the **coefficient of determination** or R^2.

The **total sum of squares** (SST) is the sum of the squared differences between observed values of the dependent variable, Y_i, and the mean of the observed values of the dependent variable $\overline{Y}$

$$SST = \Sigma(Y_i - \overline{Y})^2 \tag{6.34}$$

The **sum of squares due to regression** is the sum of the squared differences between the predicted values of the dependent variable, $\hat{Y}$, and the mean of the observed values of the dependent variable, $\overline{Y}$

$$SSR = \Sigma(\hat{Y}_i - \overline{Y})^2 \tag{6.35}$$

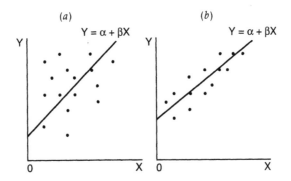

Figure 6.4

The sum of **squares due to error** is the sum of the squared differences between $\hat{Y}$ and Y_i

$$\text{SSE} = \Sigma(Y_i - \hat{Y}_i)^2 \tag{6.36}$$

The total sum of squares is the sum of the sum of squares due to regression plus the sum of squares due to error. These terms are shown diagrammatically in Fig. 6.2 earlier in this chapter.

The ratio of the **sum of squares due to regression to the total sum of squares** gives the proportion of the variation in Y explained by the variation in X and is referred to as R^2 or the **coefficient of determination**

$$R^2 = \frac{\text{SSR}}{\text{SST}} = \frac{\Sigma(\hat{Y}_i - \bar{Y})^2}{\Sigma(Y_i - \bar{Y})^2} \tag{6.37}$$

In the case of a simple two-variable regression R^2 is equivalent to the square of the correlation coefficient R.

The **coefficient of determination** will have a value ranging from zero where X has no influence upon Y to one where all the variation in Y is explained by variation in X. The R^2 for the regression of the FTSE 100 and S&P 500 data given in Table 6.1 is 0.8548. The normal interpretation of the coefficient of determination is that the number, say $R^2 = 0.8548$, is multiplied by 100 and expressed as a percentage of the variation in Y that is explained by the variation in X. Thus in this example 85.48% of the variation in Y (the FTSE 100) index is explained by the variation in X (the S&P 500 index). Whether or not you believe that the US stock market has a strong causal influence of the UK market, will depend how thoroughly you have examined competing theories. Remember this regression model is only a mathematical statement of the one hypothesis that has been tested.

R^2 is itself a random variable, because both X and Y are random variables. The test statistic has an F distribution. The F distribution is different from the other distributions in that it has two sets of degrees of freedom, one (often designated v_1) in the numerator of the test statistic and a second (designated v_2) in the denominator. In the test statistic for R^2 there is one degree of freedom in the numerator and $(n - 2)$ degrees of freedom in the denominator. We calculate the test statistic for the above R^2 as follows:

$$\frac{R^2}{\left(\dfrac{1-R^2}{n-2}\right)} = \frac{0.8548}{\left(\dfrac{1-0.8548}{50}\right)} = 294.4$$

Referring to the F tables we see that the 5% critical value for $v_1 = 1$ and $v_2 = 50$ is approximately four. As the value of the test statistic is greater than four, we reject the null hypothesis that $R^2 = 0$.

However, the test is not particularly useful. It only tells us that there *is* correlation between Y and X. When interpreting the coefficient of determination, care is necessary. For example, as an indicator of goodness of fit, the R^2 is often used to compare regression equations. However, it is only valid to use the R^2 in this way if the dependent variables in each equation being compared are identical. It is also inappropriate to use R^2 to compare the explanatory power of regression models that have differing numbers of explanatory variables. Thus using R^2 to compare the goodness of fit between a simple regression model and a multivariate model (i.e. one with several independent variables) would not be valid.

Using regression for prediction

The prediction interval

The results of applying the OLS model can be used for prediction. For example, suppose that we wish to predict the level of the FTSE 100 if the S&P 500 index rose to 550 on a particular day. The predicted value would be

$$\hat{Y} = + 196.3298 + (5.963972 * 550) = 3476.51 \approx 3477$$

When we use the regression model to predict the value of Y (level of the FTSE 100) given a value of X (level of the S&P 500) we will want to know the degree of confidence to place on that estimated value. For this purpose we calculate the **standard error of estimate** and then the **prediction interval**.

The standard error of estimate, also known as the standard error of the regression, is given as follows (see equation (6.23)):

$$s = \sqrt{\frac{\Sigma(Y_i - \hat{Y}_i)^2}{(n-2)}} \tag{6.38}$$

It is in fact the standard deviation of all the e_is.

The prediction interval would be calculated as

$$\hat{Y} \pm t_{99} * s \sqrt{1 + \frac{1}{n} + \frac{(X^* - \overline{X})^2}{\Sigma(X_i - \overline{X})^2}} \tag{6.39}$$

where the subscript 99 relating to the t indicates the level of confidence and X^* is the value of X used in the prediction, i.e. 550 in the above example.

The standard error of estimate relating to the FTSE/S&P regression is $s = 114.27$. The prediction interval is

$$3476 \pm 2.500 * 114.27 * \sqrt{1 + 0.0192 + \frac{(550 - 391.42)^2}{108046.7}}$$

$$= 3476 \pm 2.500 * 114.27 * 1.1189$$

$$= 3476 \pm 319.65$$

Thus we can consider with 99% confidence that if the S&P index rises to 550, the FTSE 100 index will rise to 3476 ± 320, i.e. between 3156 and 3796.

Spurious regression

The above example of regression analysis used levels data relating to stock exchange indices. However, using data relating to the levels or prices may give rise to what is known as **spurious regression**. Spurious regression may result from analysing data where the magnitude of the observations of each variable tends to increase (or decrease) over time. This tendency creates a degree of correlation that overstates any underlying causal relationship. Data relating variables expressed in monetary terms or financial aggregates tend to grow in magnitude over time and are therefore particularly susceptible to this problem.

An examination of the relationship between the changes in the level of one variable and the changes in the level of the other may tell a different story. For this reason, when applying regression analysis to time series of financial variables, the data used are often differenced (i.e. changes in levels are computed) or alternatively the changes may be converted to percentage returns [e.g. $(P_1 - P_0)/P_0$ or $\ln P_1/P_0$]. In essence we are looking to convert our data into stationary data, since there are techniques available for analysing such data. (See Chapter 7 for a development of this.) We have analysed levels data in this section because of the intuitive, if maybe spurious, link between the two stock markets. We rectify that situation in the next section.

Multiple regression

Rarely is the behaviour of the dependent variable explained by just one independent variable. Usually several independent variables, used in combination, offer the best explanation. A regression model incorporating several independent variables is known as **multiple regression**.

The true relationship between the dependent variable Y and the various independent variables, the X_is, is given by

$$Y = \alpha + \beta_1 X_1 + \beta_2 X_2 + \ldots + \beta_n X_n + e \tag{6.40}$$

However, just as in the case of the simple linear regression, we do not know the true relationship and have to estimate

$$\hat{Y} = \hat{\alpha} + \hat{\beta}_1 X_1 + \hat{\beta}_2 X_2 + \ldots + \hat{\beta}_n X_n \tag{6.41}$$

The $\hat{\beta}_i$s represent the partial derivatives of Y with respect to the appropriate X_i, for example

$$\hat{\beta}_1 = \frac{\partial \hat{Y}}{\partial X_1}, \qquad \hat{\beta}_2 = \frac{\partial \hat{Y}}{\partial X_2}, \qquad \hat{\beta}_n = \frac{\partial \hat{Y}}{\partial X_n} \tag{6.42}$$

assuming that all the other X_is are held constant.

Recall that in the case of the simple regression the constant represented the value of the dependent variable when the independent variable had a value of zero. However, in multiple regression, interpretation of the constant is more complex. In some models a constant term is expected *a priori*, in other cases a significant constant might represent the mean effect on Y of any independent variables not included in the model. Thus a significant constant may suggest that some important explanatory variable has been omitted from the model being tested.

To illustrate multiple regression, consider the hypothesis that returns to the FTSE 100 index are caused by returns to the UK government bond market, known as the Gilts market (variable X_1), by returns to the S&P 500 index (variable X_2) and by returns to the US\$/£ exchange rate (X_3).

There are many computer packages that solve such a multiple regression problem. The typical output would be similar to the following:

$$Y = -0.215 + 0.209 X_1 + 0.934 X_2 - 0.302 X_3$$
$$(-0.39) \quad (1.02) \quad (6.42) \quad (-2.54)$$

$$R^2 = 0.52 \qquad DF = 47$$

$$\text{Adjusted } R^2 = 0.49$$

$$DW = 2.3$$

$$F = 26.0$$

t statistics in parentheses.

Note: the data relating to this problem is given in Appendix 6.2 and consists of 51 sets of observations.

This may be interpreted as follows: if X_2 and X_3 are held constant a one-unit change in X_1 will cause a 0.209 change in Y. Similarly, if X_1 and X_3 are held constant, a one-unit change in X_2 will cause Y to change by 0.934 units.

The figures in parentheses may represent standard errors of the coefficients or t-statistics; there should be an indication of which on the print-out. DF refers to the degrees of freedom (see below) and DW refers to the Durbin–Watson statistic which relates to autocorrelation and is discussed later in this chapter.

The assumptions regarding multivariate ordinary least-squares are the same as for the univariate model. However, the multivariate model has the additional assumption that the independent variables are independent of each other, i.e. $\text{cov}(x_j, x_k) = 0$ ($j \neq k$).

Turning now to the interpretation of the computer output, the t-statistics for each independent variable are interpreted in exactly the same way as earlier. However, in the case of multiple regression, they have a t-distribution with $n - k - 1$ degrees of freedom.

If there are k independent variables, there will be $k + 1$ regression parameters (including the constant) thus the degrees of freedom will be $n - (k + 1)$ or $n - k - 1$.

As there are 51 sets of observations and four regression parameters, the degrees of freedom in the above example were 47. The 95% level of confidence (two-tailed) would require a t-value >2.02 and at the 99% level of confidence would require a t-value of >2.70. Thus in the above example the constant and the X_1 variable would not be signficantly different from zero, but the two other variables would be significant at the 95% level and X_2 would be significant at the 99% level of confidence.

Adjusted R^2: $\overline{R}^2$

In multivariate regression, adding additional explanatory variables will cause the coefficient of determination to increase. Consequently, the coefficient of determination should be adjusted to take account of the number of independent variables. The adjusted R^2, or $\overline{R}^2$, is calculated as

$$\overline{R}^2 = 1 - (1 - R^2)\frac{n-1}{n-k} \qquad (6.43)$$

where

n = the number of observations
k = the number of independent regressors.

To illustrate this recall the earlier R^2 of 0.52 with three regressors. We thus have

$$\bar{R}^2 = 1 - \left[(1 - 0.52) * \frac{50}{47} \right] = 0.49223$$

The adjusted R^2 will decrease in value if the additional variable is not significant. However, the reader is cautioned against adding or deleting variables simply according to their influence upon the adjusted R^2. The rational basis for inclusion or deletion is the theory behind the model that is being tested. Thus a variable that has a strong theoretical basis for inclusion should be added to the model even if the adjusted R^2 fails to improve.

Testing the significance of the adjusted R^2 is also a test of no significance between the dependent variable, Y, and any of the independent variables, the X_is. This is because if the regression model has a high explanatory power the variation in the dependent variable will be due to variations in the independent variables and the sums of squares due to regression (SSR) will be large relative to the sums of squares due to errors (SSE). If on the other hand the model has low explanatory power the variations in the dependent variable will be due to variation in the error term, and the SSE will be large relative to the SSR.

The test statistic is calculated as

$$\frac{R^2}{(1 - R^2)} \frac{(n - k)}{k - 1} \sim F_{k-1, n-k} \tag{6.44}$$

Thus, this test statistic has an F-distribution with $k - 1$ degrees of freedom in the numerator and $n - k$ degrees of freedom in the denominator.

To illustrate this

$$F = \frac{0.52}{1 - 0.52} * \left[\frac{(51 - 3)}{(3 - 1)} \right] = 1.0833 * 24 = 26$$

The 1% critical value of the F statistic for two degrees of freedom in the numerator and 48 in the denominator is 5.08. As the decision rule for testing the null hypothesis that $R^2 = 0$ is to reject H_0 if F is greater than the critical value, we reject the null hypothesis.

Chow test for equality of sub-period coefficients

There may be times when it is desirable to test a given hypothesis over different time periods. In such circumstances we need to know whether the resulting coefficients from the separate time periods are indeed statistically significantly different, or whether the differences are just due to chance. For this purpose we can apply the Chow test.

The Chow test is executed in three stages:

1. Run the regression model over complete data series, and derive the sums of squares due to error (SSE). We will label this SSE_1.
2. Run the regression model over the separate time periods and derive period-specific SSEs. We will assume two sub-periods, one with n observations and the other with m observations. Thus we get SSE_2 and SSE_3.

3. Calculate the Chow statistic as follows:

$$\frac{(SSE_1 - SSE_2 - SSE_3)/k}{(SSE_2 + SSE_3)/(n + m - 2k)} \tag{6.45}$$

where n and m = the observation in each respective sub-group.

The Chow statistic has an F-distribution with k degrees of freedom in the numerator and $m + n - 2k$ degrees of freedom in the denominator.

Breakdown of the OLS assumptions

It is important to determine whether the assumptions of OLS regression have been satisfied. In particular it is important to test for:

- **heteroscedasticity** the residuals do not have a constant variance
- **autocorrelation** the residuals are independent
- **multicollinearity** the independent variables are uncorrelated.

Heteroscedasticity

If the residuals have a constant variance they are said to be homoscedastic, but if they are not constant they are said to be heteroscedastic. The effects of heteroscedasticity are that the regression coefficients are no longer the best or minimum variance estimates, thus they are no longer the most efficient coefficients.

The consequence of heteroscedasticity on the prediction interval estimation and hypothesis testing is that although the coefficients are unbiased, the variances, and therefore the standard errors of those coefficients, will be biased. If this bias is negative, the estimated standard errors will be smaller than they should be and the test statistic will be larger than it is in reality. Thus we may be led to think that the coefficient is significant when it is not. Conversely, if the bias is positive, the estimated standard errors will be larger than they should be and the test statistics smaller. We may therefore accept the null hypothesis when in fact it should be rejected.

A test for heteroscedasticity is the Goldfeld–Quant test. This test requires that the residuals are divided into two groups of n observations, one group with small values and the other with large values. Usually the middle one-sixth of the observations is removed after sorting in ascending order so as to improve the discrimination between the two groups. Thus the number of residuals in each group is $(n - c)/2$, where c represents one-sixth of the observations.

The Goldfeldt–Quant test is the ratio of the sum of squares due to errors (SSE) of the high residuals divided by the SSE of the low residuals, that is

$$GQ = \frac{SSE_H}{SSE_L} \tag{6.46}$$

This statistic has an F-distribution with $(n - c)/(2 - k)$ degrees of freedom.

A solution to heteroscedasticity is to observe the relationship between the error terms and transform the regression model in a way that reflects that relationship. This may be achieved by regressing the error terms on various functional forms of the variable that causes the heteroscedasticity, for example

$$e_i = \alpha + \beta X_i^H \tag{6.47}$$

where X_i is the independent variable (or some function of the independent variable) that is assumed to be the cause of the heteroscedasticity and H reflects the power of the relationship between the errors and that variable, for example X^2 or $X^{1/n}$, etc.

Consequently the variance of the coefficients becomes

$$E(\sigma_i^2) = \sigma^2 X_i^H \tag{6.48}$$

Thus if $H = 1$ we would transform the regression model to

$$\frac{Y_i}{\sqrt{X_i}} = \frac{\alpha}{\sqrt{X_i}} + \beta_i \frac{e_i}{\sqrt{X_i}} \tag{6.49}$$

If $H = 2$, i.e. the variance increases in proportion to the square of the X variable in question, the transformation would be

$$\frac{Y_i}{X_i} = \frac{\alpha}{X_i} + \beta_i + \frac{e_i}{X_i} \tag{6.50}$$

Autocorrelation

Autocorrelation, also known as serial correlation, occurs when the residuals are not independent of each other because current values of Y are influenced by past values. The dependence between the residuals is described by an autoregressive scheme. For example, assume that the residual e_t is influenced by the residual in the previous time period e_{t-1} plus some current value of a random variable z_t. The residual e_t would be described by an autoregressive function

$$e_t = \rho e_{t-1} + z_t \tag{6.51}$$

This form of autoregressive function is referred to as a first-order autoregressive function or AR(1) because only one preceding time period is incorporated in the function.

If the current residual was influenced by, say, two or four previous residuals, the autoregressive functions AR(2) and AR(4) would be

$$\text{AR(2): } e_t = \rho_{t-1} e_{t-1} + \rho_{t-2} e_{t-2} + z_t \tag{6.52}$$

$$\text{AR(4): } e_t = \rho_{t-1} e_{t-1} + \rho_{t-2} e_{t-2} + \rho_{t-3} e_{t-3} + \rho_{t-4} e_{t-4} + z_t \tag{6.53}$$

The OLS regression model is a minimum variance, unbiased estimator only when the residuals are independent of each other. If autocorrelation exists in the residuals, the regression coefficients are unbiased but the standard errors will be underestimated and the tests of regression coefficients will be unreliable.

To test for first-order autocorrelation, the Durbin–Watson statistic must be calculated. This is calculated as

$$DW = \frac{\Sigma(e_t - e_{t-1})^2}{\Sigma e_t^2}$$ (6.54)

As a rule of thumb, if Durbin–Watson is two there is no positive autocorrelation, if it is zero there is perfect positive autocorrelation, and if it is four there is perfect negative autocorrelation. However, the Durbin–Watson statistic has a sampling distribution based on the table provided in their original paper (Durbin and Watson, 1950). This sampling distribution has two critical values d_L and d_U.

To test for autocorrelation in the data we test the following null hypothesis

H_0: no autocorrelation if $d_U \leqslant d \leqslant 4 - d_U$

H_1: positive autocorrelation $d < d_L$
negative autocorrelation if $d > 4 - d_L$

Unfortunately there are some grey areas within this distribution where the results are inconclusive. These are

$$d_L < d < d_U \quad \text{or} \quad 4 - d_U < d < 4 - d_U$$

Autocorrelation may be caused by omitted variables or the wrong functional form of the estimating equation, for example a linear model when it should be non-linear. Introducing lagged variables can also cause autocorrelation. As we will discover in Chapter 7, time-series data is also particularly susceptible to autocorrelation.

To solve the autocorrelation problem first consider the potential for omitted variables or wrong functional form. Then, if that is unsuccessful, use the Orcult–Cockrone procedure.

To apply this procedure, first calculate the autocorrelation coefficient, ρ, as follows:

$$\rho = \frac{\Sigma(e_t * e_{t-1})}{\Sigma e_t^2}$$ (6.55)

Then re-specify the equation, $Y_t = \alpha + \beta X_t$ as

$$Y_t - \rho Y_{t-1} = \alpha + \beta(X_t - \rho X_{t-1})$$ (6.56)

This has the effect of removing first-order autocorrelation from the data.

Multicollinearity

When some or all of the independent variables in a multiple regression are highly correlated, the regression model has difficulty untangling their separate explanatory effects on Y. In effect, the highly correlated independent variables act in unison, and they have insufficient independent variation to enable the model to isolate their separate influences. There is no exact boundary value of the degree of correlation between variables that cause the problem of multicollinearity, thus judgement is required.

Multicollinearity is particularly prevalent in macroeconomic data such as income and production where inflation, for example, can affect both series.

With multicollinearity the regression coefficients are unstable in the degree of statistical significance, magnitude and sign. Consequently they are unreliable. The R^2 may be high but the standard errors are also high and, consequently, the t-statistics are small, indicating lack of significance.

Multicollinearity can be remedied in several ways:

1. Add further sample data on the basis that more data mean lower variances of the OLS estimators. The problem with this solution is the difficulty of finding extra data.
2. Drop those variables that are highly correlated with the others. The problem here is that presumably the variables were included on theoretical grounds and it is inappropriate to exclude them just to make the statistical results "better".
3. Pooling of cross-section and time-series data. This technique takes a coefficient from, say, a cross-section regression and substitutes it for the coefficient of the time-series equivalent data. An example will illustrate this.

Assume that we want to fit the following time-series regression equation

$$Q_t = \alpha + b_1 P_t + b_2 Y_t + e_t \tag{6.57}$$

where Q is the quantity or value of pension fund money being invested in UK equities, P is the the level of the FTSE 100 index and Y is the pension fund income in the previous month. Both level and income are likely to be highly correlated over time because they are both influenced by inflation.

A cross-section study, from a survey, of the level of pension fund income and investment may be available. To that cross-section data we apply the following equation

$$Q = \lambda_1 + \lambda_2 Y \tag{6.58}$$

We then adjust the value of Q in the first equation as follows:

$$Q - \lambda_1 - \lambda_2 Y = Q* \tag{6.59}$$

Next we regress $Q*$ on P as follows:

$$Q* = \alpha_1 + \alpha_2 P \tag{6.60}$$

By this method we are able to get estimates of the parameters of P and Y from sources that will not be highly correlated.

Dummy variables

It may be necessary to incorporate one or more qualitative variables into a regression model, for example male or female investors. Alternatively it may be necessary to make a qualitative distinction between observations of the same data. For example, if a relationship between company size and monthly returns to shares is being tested, it may be desirable to incorporate a qualitative variable representing the month of January, because of the well-known "January effect" in security returns time-series. This "January effect" is the phenomenon that the average returns to particularly small companies are on average higher in January than in other months. Thus if we consider January observations to be qualitatively different to other observations a dummy variable enables such qualitative distinctions to be made.

Dummy variables are also used to account for the qualitative effects of changes in government policy on the data being analysed. For example, a study of the effects of currency of the level on the stock exchange index may have to take into account when the currency joins and/or leaves a currency block. For example, studies of the FTSE 100 index that incorporated August/September 1992 would have to account for sterling leaving the exchange rate mechanism of the European Monetary System.

A dummy variable is one of two types, a **shift dummy** or a **slope dummy**. A shift dummy is a variable which changes the intercept of the regression line, i.e. shifts the line up or down when the qualitative variable is applicable. A slope dummy is one that changes the slope of the regression line when the qualitative variable is applicable. Both types of dummy will have a value of plus one or minus one when the data observations coincide with the relevant qualitative variable, but have a value of zero when coinciding with observations where that variable does not apply. Figures 6.5 and 6.6 illustrate these points.

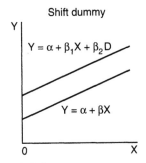 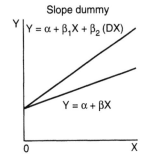

Figure 6.5 Shift dummy. **Figure 6.6** Slope dummy.

To illustrate a shift dummy consider the January effect again. To test the relationship between company size and investor returns, data regarding size of a company each month, returns to the company's shares that month and the qualitative variable (January) would be required. The January variable would be represented by a dummy with a value of one for the monthly observation of January and by a zero for all other months. The cross-section regression of monthly returns upon company size and the month of January would be

$$r_i = \alpha + \beta_1 size_i + \beta_2 D + e_i \tag{6.61}$$

where the value of D will be one when the return and size observations relate to January and zero when those observations refer to other months.

With regard to the FTSE 100 example, a dummy variable would be added to the model with a value of one when data observations coincided with sterling being in the ERM and zero when sterling was not in the ERM.

To illustrate a slope dummy consider the performance of investment managers. If they have market timing skills, the returns to their portfolios should rise faster than those of the market or fall less than those of the market.

This approach was used by Henriksson and Merton (1981) who used dummy variables in a linear regression of the following form

$$r_p - r_f = \alpha + \beta(r_m - r_f) + c[D(r_m - r_f)] + e_p \tag{6.62}$$

where

$r_p - r_f$ = the excess return to the portfolio
$r_m - r_f$ = the excess return to the market.

The dummy variable, D, is assigned a value of zero when the returns on the market are greater than the risk-free rate, and assigned a value of minus one when the return on the market is less that the risk-free return.

This equation has the following interpretation. When the returns to the market are higher than the risk-free rate $D = 0$ and therefore equation (6.62) reduces to

$$r_p - r_f = \alpha + \beta(r_m - r_f) + e_p \qquad (6.63)$$

and when the return on the portfolio is less than the risk-free rate, $D = -1$ so equation (6.62) reduces to

$$r_p - r_f = \alpha + (\beta - c)(r_m - r_f) + e_p \qquad (6.64)$$

Therefore, as β corresponds to the regression coefficient when the market gives returns higher than the risk-free rate, and $(\beta - c)$ corresponds to the portfolio β when the market gives returns that are lower than the risk-free rate, c is the difference between the two coefficients and indicates that the investment manager has market timing skills. With respect to the shape of the regression line, if c is positive, the slope of the line is less steep than if c is zero.

It should also be noted that in both equations (6.63) and (6.64) a positive α indicates that the investment manager also has positive stock selection skills.

Dummy variables can be applied to more than one qualitative difference in the variables. An example is seasonal variations in data. If we assume that in the United Kingdom, because of the tax year ending in April, security returns have a qualitative difference in January and in April, a study of the relationship between monthly returns and size would need two dummy variables. Indeed when there are n qualitative states of the data, it is necessary to have $n - 1$ dummy variables. In our example there are three qualitative states to the returns and size data, data relating to January, data relating to April and data relating to all other months. Thus two dummy variables will be required.

Non-linear regression

So far the discussion has centred on linear regression. However, it maybe that the relationship between the dependent variable and one or more of the independent variables is non-linear. Two ways of handling this problem are:

1. transform the data and apply linear regression
2. apply non-linear regression techniques.

Non-linear regression is beyond the scope of this book, thus we will concentrate on data transformations.

Data transformations

The following diagrams show a variety of relationships between Y and X that are non-linear. However, with suitable transformation of Y, α and X the relationship between Y and X can be transformed so that it is linear in terms of a and b. Thus OLS can be applied.

Consider the three non-linear forms given in Fig. 6.7. In the top diagrams the functional form is $Y = \alpha X^{\beta}$, where $0 < \beta < 1$ or $\beta > 1$. The transformation in these examples is to take the natural log of Y, α and X. The resulting regression equation will be

$$\ln Y = \ln \alpha + \beta \ln X \qquad (6.65)$$

The transformation of the bottom right-hand diagram is straightforward in that $1/X$ can be calculated as the independent variable.

With all of these transformations it is necessary to convert the output to its non-linear form in order to be able to interpret the output.

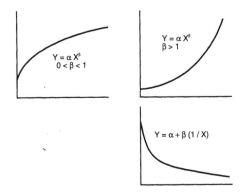

Figure 6.7

Application of regression analysis to hedging

The objective of hedging is to remove the risk from a portfolio of assets. Hedging a long position in a risky asset is achieved by going short a proportion of the portfolio value in a different, but highly correlated, risky asset. To illustrate this consider a portfolio that is long on asset A and it is desired to hedge this portfolio by going short a futures contract on asset A.

Two questions have to be asked before a hedge can be established:

1. What instrument should the short position be in?
2. What proportion of the value of the long position should the short position represent so as to minimize the variance of the overall portfolio?

The first question is simply answered by looking at the correlation coefficient of the changes in the price of the long position and the potential candidates for the short position. The candidate with the highest correlation with the long position should be chosen. In this example we assume that it is the futures contract.

With regard to the second question, the proportion to go short is known as the hedge ratio. OLS regression is often used to derive these hedge ratios.

To understand this, consider again our long bond position. The return to the long holding is given as R_B. It has been determined that the instrument that is most highly correlated is the bond future. The return to the future is given as R_F.

The returns to a hedged portfolio, i.e. one that is long one unit of the bond and short an appropriate amount h of the futures contract, is given as

$$R_p = (R_B - hR_F) \qquad (6.66)$$

where h is the headline ratio.

The variance of the portfolio returns is

$$\sigma^2 R_p = \sigma^2 R_b - 2h \text{ cov } (R_B, R_F) + h^2 \sigma^2 R_F \tag{6.67}$$

To derive the hedge ratio, h, which minimizes the variance of the hedged portfolio we must differentiate equation (6.68) with respect to h and set the derivative to zero, thus

$$\frac{d(\sigma^2 R_b - 2h \text{cov}(R_B, R_F) + h^2 \sigma^2 R_F)}{dh} = -2 \text{ cov } (R_B, R_F) + 2h \sigma R_F \tag{6.68}$$

To set this to zero, we need

$$2h\sigma^2 R_F = 2 \text{ cov } (R_B, R_F) \tag{6.69}$$

In such circumstances h is

$$h = \frac{\text{cov } (R_B, R_F)}{\sigma^2 R_F} \tag{6.70}$$

h will be recognized as analogous to β, the slope of a regression equation

$$R_B = \alpha + \beta R_F + e \tag{6.71}$$

Thus the slope coefficient β gives us the hedge ratio. In practice, the regression is set up by using $\Delta P_B/P_B$ and $\Delta P_F/P_F$, where P represents the prices of the bond and the future. This example can be generalized to using regression analysis to hedge any risk asset. However, alternative techniques are demonstrated in Chapter 7 regarding time-series analysis.

Exercises

Regression

1. What do you understand by the terms:

cross-section regression
time series regression
spurious regression.

The following exercises refer to the interpretation of the following output of a regression analysis. The figures in parentheses are t statistics.

$$\mathbf{Y = -0.01 + 0.04X_1 + 0.9X_2 - 0.1X_3}$$
$$(-0.2) \ (+1.75) \ (+3.2) \ (-2.6)$$

$R^2 = 0.85$ $DF = 51$
Adjusted $R^2 = 0.80$ $DW = 2.4$
$F = 27$.

2. Give a general explanation of the multiple regression equation

$$Y = -0.01 + 0.04X_1 + 0.9X_2 - 0.1X_3$$

including a discussion of the significance of each variable.

3. Explain the meaning of R^2 and adjusted R^2, and interpret the statistics given above.

4. What is autocorrelation? What are some of its causes? Explain how it affects the interpretation of the regression equation, and explain the role of the Durbin–Watson statistic in indicating the presence of autocorrelation.

5. What is the interpretation of the F statistic in the above data.

6. What is heteroscedasticity and what are some of its causes? What is its effect on the interpretation of the regression equation?

7. What test would you initiate to determine whether or not heteroscedasticity was a problem in the data being analysed? How would you solve the problem?

8. What is multicollinearity and what are some of its causes? What is the effect of multicollinearity on the interpretation of the regression equation, and how would you solve the problem if it existed?

9. What are dummy variables, and how are they applied in regression analysis?

10. Explain how to use regression analysis to derive the minimum risk hedge ratio when using financial futures to hedge risky assets.

Matrix algebra

11.

$$A = \begin{bmatrix} 2 & 3 & 1 \\ 6 & 0 & 5 \\ 9 & 2 & 1 \\ 1 & 1 & 3 \end{bmatrix}, \quad B = \begin{bmatrix} 1 & 0 \\ 5 & 2 \\ 7 & 1 \end{bmatrix}, \quad C = \begin{bmatrix} 0 & 8 \\ 5 & 0 \\ 1 & 1 \end{bmatrix}, \quad D = \begin{bmatrix} 2 & 3 \\ 0 & 1 \end{bmatrix}$$

(a) Find $B + C$ Find $A[B + C]$
   Find AB Find BC
   Find $AB + BC$

(b) The transpose of a matrix (denoted by X^T) is obtained by swapping its rows and columns.
   Find $B^T C$ and BC^T.
   Find $B^T(DC)$ and $[B^T C]D$.

(c) Find D^{-1} and $[B^T C]^{-1}$
   Show that $[BC^T]$ does not have an inverse.

(d) Show that $[B^T C]^T = C^T B$
   Show that $[[B^T C]D]^{-1} = D^{-1}[B^T C]^{-1}$

12. The returns on a security Y are thought to depend on the returns on underlying securities X_1 and X_2. Annualized monthly returns on the three securities are shown below for a six-month period:

Security	Returns					
X_1	0.054	0.053	0.049	0.049	0.054	0.060
X_2	0.063	0.062	0.061	0.058	0.057	0.057
Y	0.092	0.092	0.091	0.090	0.087	0.086

A regression model is to be built of the form $Y_i = \alpha_0 + \alpha_1 X_{i1} + \alpha_2 X_{i2} + e_i$, where the e_is are independent and normally distributed variables, all sharing the same variance.

(a) Construct a matrix X so that the model is specified by

$$\underline{Y} = X \begin{bmatrix} \alpha_0 \\ \alpha_1 \\ \alpha_2 \end{bmatrix} + \underline{e}$$

Note: $\underline{Y}$ and $\underline{e}$ are vectors of Y and e, respectively.

(b) Find $X^T X$.

(c) Describe how to compute $[X^T X]^{-1}$ by using a partitioned matrix. Confirm that the inverse is approximately

$$\begin{bmatrix} 189.11155 & -1084.56021 & -2200.26425 \\ -1084.56021 & 13338.36668 & 6291.68240 \\ -2200.26425 & 6291.68240 & 31269.66151 \end{bmatrix}$$

(d) Compute $X^T \underline{Y}$.

(e) Use $[X^T X]^{-1}$ and $X^T \underline{Y}$ to compute least squares estimates for α_i, $i = 0, 1, 2$.

(f) Compute the variance/covariance matrix for the securities X_1 and X_2.

Answers

11.

$$B + C = \begin{bmatrix} 1 & 8 \\ 10 & 2 \\ 8 & 2 \end{bmatrix} \qquad A[B+C] = \begin{bmatrix} 40 & 24 \\ 46 & 58 \\ 37 & 78 \\ 35 & 16 \end{bmatrix}$$

$$AB = \begin{bmatrix} 24 & 7 \\ 41 & 5 \\ 26 & 5 \\ 27 & 5 \end{bmatrix} \qquad AC = \begin{bmatrix} 16 & 17 \\ 5 & 53 \\ 11 & 73 \\ 8 & 11 \end{bmatrix}$$

$$AB + BC = \begin{bmatrix} 40 & 24 \\ 46 & 58 \\ 37 & 78 \\ 35 & 16 \end{bmatrix}$$

(b)

$$B^\mathrm{T}C = \begin{bmatrix} 32 & 15 \\ 11 & 1 \end{bmatrix} \text{ and } BC^\mathrm{T} = \begin{bmatrix} 0 & 5 & 1 \\ 16 & 25 & 7 \\ 8 & 35 & 8 \end{bmatrix}$$

$$B^\mathrm{T}[CD] = \begin{bmatrix} 64 & 111 \\ 22 & 34 \end{bmatrix} \text{ and } [B^\mathrm{T}C]D = \begin{bmatrix} 64 & 111 \\ 22 & 34 \end{bmatrix}$$

(c)

$$D^{-1} = \begin{bmatrix} 0.5 & -1.5 \\ 0.0 & 1.0 \end{bmatrix}, \ [B^\mathrm{T}C]^{-1} = \begin{bmatrix} -0.0075 & 0.1128 \\ 0.0827 & -0.2406 \end{bmatrix}$$

(d)

$$[B^\mathrm{T}C]^\mathrm{T} = \begin{bmatrix} 32 & 11 \\ 15 & 1 \end{bmatrix}$$

$$C^\mathrm{T}B = \begin{bmatrix} 0 & 5 & 1 \\ 8 & 0 & 1 \end{bmatrix} \begin{bmatrix} 1 & 0 \\ 5 & 2 \\ 7 & 1 \end{bmatrix} = \begin{bmatrix} 32 & 11 \\ 15 & 1 \end{bmatrix}$$

$$[[B^\mathrm{T}C]D]^{-1} = \begin{bmatrix} -0.1278 & 0.8346 \\ 0.0827 & -0.2406 \end{bmatrix}$$

$$D^{-1}[B^\mathrm{T}C]^{-1} = \begin{bmatrix} 0.5 & -1.5 \\ 0.0 & 1.0 \end{bmatrix} \begin{bmatrix} -0.0075 & 0.1128 \\ 0.0827 & -0.2406 \end{bmatrix} = \begin{bmatrix} -0.1278 & 0.8346 \\ 0.0827 & -0.2406 \end{bmatrix}$$

12.

(a)

$$X = \begin{bmatrix} 1 & 0.054 & 0.063 \\ 1 & 0.053 & 0.062 \\ 1 & 0.049 & 0.061 \\ 1 & 0.049 & 0.058 \\ 1 & 0.054 & 0.057 \\ 1 & 0.060 & 0.057 \end{bmatrix}$$

(b)

$$X^\mathrm{T}X = \begin{bmatrix} 6.000000 & 0.319000 & 0.358000 \\ 0.319000 & 0.017043 & 0.019017 \\ 0.358000 & 0.019017 & 0.021396 \end{bmatrix}$$

(d)

$$X^T\underline{Y} = \begin{bmatrix} 0.538000 \\ 0.028571 \\ 0.032132 \end{bmatrix}$$

(e)

$$\alpha = [X^TX]^{-1}[X^T\underline{Y}] = \begin{bmatrix} 0.056 \\ -0.239 \\ 0.774 \end{bmatrix}$$

(f) $\bar{X}_1 = 0.053167$; $\bar{X}_2 = 0.059667$

$$[\text{cov}(X_1, X_2)] = \frac{1}{6}\begin{bmatrix} 0.017043 & 0.019017 \\ 0.019017 & 0.021396 \end{bmatrix}\begin{bmatrix} \bar{X}_1 \\ \bar{X}_2 \end{bmatrix}[\bar{X}_1 \ \bar{X}_2]$$

$$= \begin{bmatrix} 0.002841 & 0.003170 \\ 0.003170 & 0.003566 \end{bmatrix} - \begin{bmatrix} 0.002827 & 0.003172 \\ 0.003172 & 0.003560 \end{bmatrix}$$

$$= \begin{bmatrix} 0.000014 & -0.00002 \\ -0.000002 & 0.000006 \end{bmatrix}$$

References and further reading

Henriksson, R. D. and Merton, R. C. (1981) On market timing and investment performance II. Statistical procedures for evaluating forecasting skills. *Journal of Business*, October, 513–33.

Mansfield, E. (1991) *Statistics for Business and Economics*, 4th edn. Norton, New York.

Gujarati, D. (1995) *Basic Econometrics*, 3rd edn. McGraw-Hill.

Pindyck, R. (1990) *Econometric Models and Economic Forecasting*, 3rd edn. McGraw-Hill.

Appendix 6.1: matrix algebra

Matrices are arrays of data set in a rectangular form. The data are arranged in rows and columns. The size of the matrix is given by the number of rows and the number of columns. This is said to be the order of the matrix, e.g. a matrix with five rows and four columns will be a matrix of order 5 × 4. For example, the following matrix has the order 3 × 3:

$$Y = \begin{bmatrix} 1 & 4 & 6 \\ 3 & 7 & 5 \\ 8 & 6 & 3 \end{bmatrix}$$

The individual cells in a matrix are identified first by their position in the row and then by their position in the column. Thus, the three at the left-hand end of the middle row in the above matrix is at cell number 2,1, whereas the three at the right-hand end of the bottom row is at cell number 3,3.

Vectors are rows or columns of data – in fact they are matrices with only one row or one column. Given the rule for identifying cells in matrices, a row vector with n elements will be a matrix of order $1 \times n$, whereas a column vector would be a matrix of order $n \times 1$.

Double-digit subscripts are often used to identify an element within a matrix. For example, the element at the juncture of the ith row and jth column of matrix Y would be depicted as y_{ij}.

Matrices can be multiplied if their shapes satisfy particular conditions.

They can be added and subtracted only if each matrix is of the same order.

Adding and subtracting matrices

Matrices are added together simply by adding each element in one matrix to the corresponding element in the other matrix. Subtraction of matrices is achieved simply by subtracting each element in the second matrix from the corresponding element in the first. This can be generalized to

$$X + Y = [x_{ij} + y_{ij}]$$

$$X - Y = [x_{ij} - y_{ij}]$$

where X and Y are matrices and x_{ij} and y_{ij} are individual elements in the respective matrices. To illustrate this consider the following two matrices:

$$X = \begin{bmatrix} 8 & 6 \\ 3 & 2 \end{bmatrix}, \quad Y = \begin{bmatrix} 9 & 4 \\ 6 & 1 \end{bmatrix}$$

$X + Y$ becomes

$$X + Y = \begin{bmatrix} (8+9) & (6+4) \\ (3+6) & (2+1) \end{bmatrix} = \begin{bmatrix} 17 & 10 \\ 9 & 3 \end{bmatrix}$$

$X - Y$ becomes

$$X - Y = \begin{bmatrix} (8-9) & (6-4) \\ (3-6) & (2-1) \end{bmatrix} = \begin{bmatrix} -1 & 2 \\ -3 & 1 \end{bmatrix}$$

Multiplication of matrices

Multiplying a matrix by a single number is known as scalar multiplication. Each cell in the matrix is simply multiplied by a single number or scalar. To illustrate this multiply the matrix X above by the scalar 5

$$5X = \begin{bmatrix} (5*8) & (5*6) \\ (5*3) & (5*2) \end{bmatrix} = \begin{bmatrix} 40 & 30 \\ 15 & 10 \end{bmatrix}$$

To multiply one matrix by another it is first essential to ensure that the matrices are conformable for multiplication. That means that each row in the first (or left-hand) matrix has the same number of cells as the columns of the second (or right-hand) matrix. In other words the number of columns in the first (or left-hand) matrix is equal to the number of rows in the second (or right-hand matrix). To illustrate this consider the following matrices

$$X = \begin{bmatrix} 6 & 2 & 1 \\ 8 & 9 & 4 \end{bmatrix}, \quad Y = \begin{bmatrix} 2 & 8 \\ 3 & 4 \\ 1 & 6 \end{bmatrix}$$

To multiply $X * Y$ it is necessary to multiply each element in the first row of X by the corresponding element in the first column of Y, sum the products and place the result in the first cell of matrix Z. This process is repeated taking the elements in the first row of X and multiplying by the elements in the second column of matrix Y and placing the result in the second cell of the first row of matrix Z. Next the elements in the first row of X are multiplied by those in the third column of Y and summed to find the value of the third cell in the first row of Z. This procedure is repeated using the second row of X to derive the second row in Z.

For example, the value of the first cell in matrix Z is derived as $(6 * 2) + (2 * 3)+(1 * 1) = 19$. The second cell in the first row of matrix Z is derived as $(6 * 8) + (2 * 4) + (1 * 6) = 62$.

The first cell in the second row of matrix Z is derived as $(8 * 2) + (9 * 3) + (4 * 1) = 47$ and finally the second cell in the second row of matrix Z is derived as $(8 * 8) + (9 * 4) + (4 * 6) = 124$.

$X * Y$ thus becomes

$$Z = \begin{bmatrix} 19 & 62 \\ 47 & 124 \end{bmatrix}$$

Note the order of matrix Z is only 2×2. This is because the order of the product matrix will be equal to the number of rows in the first matrix and the number of columns in the second matrix.

It is important to note that in matrix algebra XY is not the same as YX, that is matrix multiplication is not commutative. The reader can prove as an exercise, using the above X and Y matrices, that YX is

$$\begin{bmatrix} 76 & 76 & 34 \\ 50 & 42 & 19 \\ 54 & 56 & 25 \end{bmatrix}$$

Moreover it can be seen that the above YX matrix is not even the same shape as the XY matrix.

Inverting a matrix

Only square matrices, and then only some square matrices, have inverses. The inverse of a matrix is a matrix such that the product of the two is the identity matrix. An identity matrix is one in which every entry is zero except for the main diagonal which contains ones. For example the 3×3 identity matrix is

$$\begin{bmatrix} 1 & 0 & 0 \\ 0 & 1 & 0 \\ 0 & 0 & 1 \end{bmatrix}$$

The identity matrix has the property that when used in matrix multiplication it leaves the multiplied matrix unchanged.

Matrices cannot be divided. However, if a square matrix has an inverse then we can instead multiply by that inverse to achieve the desired results (see below for an application). The inverse of matrix X is denoted X^{-1}. To derive the inverse of a matrix, a partitioned matrix may be formed. This is achieved by positioning an identity matrix next to the matrix to be inverted, for example

$$\left[\begin{array}{ccc|ccc} 6 & 2 & 1 & 1 & 0 & 0 \\ 8 & 9 & 4 & 0 & 1 & 0 \\ 3 & 1 & 0 & 0 & 0 & 1 \end{array}\right]$$

The objective is then to convert the original matrix into an identity matrix by adding subtracting, multiplying or dividing each row. When the matrix on the left is an identity matrix, the resultant matrix on the right will be the inverse of the original matrix. This is illustrated below.

Take the following partitioned matrix X

$$\left[\begin{array}{ccc|ccc} 6 & 2 & 1 & 1 & 0 & 0 \\ 8 & 9 & 4 & 0 & 1 & 0 \\ 3 & 1 & 0 & 0 & 0 & 1 \end{array}\right]$$

To form a new matrix, Z, such that the identity matrix is on the left-hand side. We start by transforming cell x_{11} to one by dividing row one by six.

$$\left[\begin{array}{ccc|ccc} 1 & 0.33 & 0.17 & 0.17 & 0 & 0 \\ 8 & 9 & 4 & 0 & 1 & 0 \\ 3 & 1 & 0 & 0 & 0 & 1 \end{array}\right]$$

To convert the 8 in cell x_{21} to zero subtract 8 * row 1 from row 2.

$$\left[\begin{array}{ccc|ccc} 1 & 0.33 & 0.17 & 0.17 & 0 & 0 \\ 0 & 6.33 & 2.67 & -1.33 & 1 & 0 \\ 3 & 1 & 0 & 0 & 0 & 1 \end{array}\right]$$

To convert the 3 in cell x_{31} to zero subtract 3 * row 1 from row 3.

$$\begin{bmatrix} 1 & 0.33 & 0.17 & 0.17 & 0 & 0 \\ 0 & 6.33 & 2.67 & -1.33 & 1 & 0 \\ 0 & 0 & -0.5 & -0.5 & 0 & 1 \end{bmatrix}$$

To convert the 6.33 in cell x_{22} to 1 divide row 2 by 6.33.

$$\begin{bmatrix} 1 & 0.33 & 0.17 & 0.17 & 0 & 0 \\ 0 & 1 & 0.42 & -0.21 & 0.16 & 0 \\ 0 & 0 & -0.5 & -0.5 & 0 & 1 \end{bmatrix}$$

To convert the 0.33 in cell x_{12} to 0 subtract 0.33 * row 2 from row 1.

$$\begin{bmatrix} 1 & 0 & 0.03 & 0.24 & -0.05 & 0 \\ 0 & 1 & 0.42 & -0.21 & 0.16 & 0 \\ 0 & 0 & -0.5 & -0.5 & 0 & 1 \end{bmatrix}$$

To convert the –0.5 in cell x_{33} to 1 multiply row 3 by –2.

$$\begin{bmatrix} 1 & 0 & 0.03 & 0.24 & -0.05 & 0 \\ 0 & 1 & 0.42 & -0.21 & 0.16 & 0 \\ 0 & 0 & 1 & 1 & 0 & -2 \end{bmatrix}$$

To convert the 0.03 in cell x_{13} to 0 subtract 0.03 * row 3 from row 1.

$$\begin{bmatrix} 1 & 0 & 0 & 0.21 & -0.05 & 0.05 \\ 0 & 1 & 0.42 & -0.21 & 0.16 & 0 \\ 0 & 0 & 1 & 1 & 0 & -2 \end{bmatrix}$$

To convert the 0.42 in cell x_{23} to 0 subtract 0.42 * row 3 from row 2.

$$\begin{bmatrix} 1 & 0 & 0 & 0.21 & -0.05 & 0.05 \\ 0 & 1 & 0 & -0.63 & 0.16 & 0.84 \\ 0 & 0 & 1 & 1 & 0 & -2 \end{bmatrix}$$

Thus the inverse of

$$\begin{bmatrix} 6 & 2 & 1 \\ 8 & 9 & 4 \\ 3 & 1 & 0 \end{bmatrix}$$

is

$$\begin{bmatrix} 0.21 & -0.05 & 0.05 \\ -0.63 & 0.16 & 0.84 \\ 1 & 0 & -2 \end{bmatrix}$$

(to two decimal place accuracy).

Solving simultaneous equations

Matrix algebra can be used to solve simultaneous equations, and it is this function that makes it applicable to the calculation of the ordinary least-squares regression coefficients.

To show how matrix algebra can be used to solve simultaneous equations, consider the following

$$6a + 8b = 25$$

$$8a - 3b = 20$$

Our objective is to find the values of a and b that solve both equations.

This can be written in matrix form as

$$\begin{bmatrix} 6 & 8 \\ 8 & -3 \end{bmatrix} * \begin{bmatrix} a \\ b \end{bmatrix} = \begin{bmatrix} 25 \\ 20 \end{bmatrix}$$

We can solve for the simultaneous values of a and b as follows.

Pre-multiply both sides by the inverse of the matrix of coefficients

$$\begin{bmatrix} 6 & 8 \\ 8 & -3 \end{bmatrix}^{-1} \begin{bmatrix} 6 & 8 \\ 8 & -3 \end{bmatrix} \begin{bmatrix} a \\ b \end{bmatrix} = \begin{bmatrix} 6 & 8 \\ 8 & -3 \end{bmatrix}^{-1} \begin{bmatrix} 25 \\ 20 \end{bmatrix}$$

This gives

$$\begin{bmatrix} 1 & 0 \\ 0 & 1 \end{bmatrix} \begin{bmatrix} a \\ b \end{bmatrix} = \begin{bmatrix} 6 & 8 \\ 8 & -3 \end{bmatrix}^{-1} \begin{bmatrix} 25 \\ 20 \end{bmatrix}$$

i.e.

$$\begin{bmatrix} a \\ b \end{bmatrix} = \begin{bmatrix} 6 & 8 \\ 8 & -3 \end{bmatrix}^{-1} \begin{bmatrix} 25 \\ 20 \end{bmatrix}$$

Now

$$\begin{bmatrix} 6 & 8 \\ 8 & -3 \end{bmatrix}^{-1} = \begin{bmatrix} 0.036585 & 0.097561 \\ 0.097561 & -0.07317 \end{bmatrix}$$

so

$$\begin{bmatrix} a \\ b \end{bmatrix} = \begin{bmatrix} 0.036585 & 0.097561 \\ 0.097561 & -0.07317 \end{bmatrix} \begin{bmatrix} 25 \\ 20 \end{bmatrix} = \begin{bmatrix} 2.866 \\ 0.976 \end{bmatrix}$$

The resulting values of a and b are: $a = 2.866$ and $b = 0.976$.

Application to OLS regression

To see how this methodology applies to regression analysis, take the following estimated regression equation:

$$\hat{Y} = \hat{B}_0 + \hat{B}_1 X_1 + \hat{B}_2 X_2 + \ldots + \hat{B}_k X_k + e$$

For each of the observations of Y, and each set of observations of the X variables, a set of simultaneous equations can established as below

$$Y_1 = \hat{B}_0 + \hat{B}_1 X_{11} + \hat{B}_2 X_{21} + \ldots + \hat{B}_K X_{k1} + e_1$$

$$Y_2 = \hat{B}_0 + \hat{B}_1 X_{12} + \hat{B}_2 X_{22} + \ldots + \hat{B}_K X_{k2} + e_2$$

$$\vdots$$

$$Y_n = \hat{B}_0 + \hat{B}_1 X_{1n} + \hat{B}_2 X_{2n} + \ldots + \hat{B}_K X_{kn} + e_n$$

We can then draw up an n element vector of Y values, an $n * k$ matrix of the X values and an n element vector of e values, for example

$$Y = \begin{bmatrix} Y_1 \\ Y_2 \\ Y_3 \\ \cdot \\ \cdot \\ \cdot \\ Y_n \end{bmatrix} \quad X = \begin{bmatrix} 1 & X_{11} & X_{21} & \cdot & \cdot & X_{k1} \\ 1 & X_{12} & X_{22} & \cdot & \cdot & X_{k2} \\ \cdot & \cdot & \cdot & \cdot & \cdot & \cdot \\ \cdot & \cdot & \cdot & \cdot & \cdot & \cdot \\ \cdot & \cdot & \cdot & \cdot & \cdot & \cdot \\ 1 & X_{1n} & X_{2n} & \cdot & \cdot & X_{kn} \end{bmatrix} \quad e = \begin{bmatrix} e_1 \\ e_2 \\ \cdot \\ \cdot \\ \cdot \\ e_n \end{bmatrix}$$

The column of ones in the X matrix represents the intercept in the equation.

The objective is to solve for a vector of B_is

$$\begin{bmatrix} 1 & X_{11} & X_{21} & \cdot & \cdot & X_{k1} \\ 1 & X_{12} & X_{22} & \cdot & \cdot & X_{k2} \\ \cdot & \cdot & \cdot & \cdot & \cdot & \cdot \\ \cdot & \cdot & \cdot & \cdot & \cdot & \cdot \\ \cdot & \cdot & \cdot & \cdot & \cdot & \cdot \\ 1 & X_{1n} & X_{2n} & \cdot & \cdot & X_{kn} \end{bmatrix} * \begin{bmatrix} B_0 \\ B_1 \\ \cdot \\ \cdot \\ \cdot \\ B_k \end{bmatrix} = \begin{bmatrix} Y_1 \\ Y_2 \\ \cdot \\ \cdot \\ \cdot \\ Y_n \end{bmatrix}$$

If the above matrix is square, i.e. has the same number of observations as there are unknowns (i.e. $k = n$), then the matrix may have an inverse, in which case we get to

$$\begin{bmatrix} B_0 \\ B_1 \\ \cdot \\ \cdot \\ \cdot \\ B_K \end{bmatrix} = \begin{bmatrix} 1 & X_{11} & X_{21} & \cdot & \cdot & X_{k1} \\ 1 & X_{12} & X_{22} & \cdot & \cdot & X_{k2} \\ \cdot & \cdot & \cdot & \cdot & \cdot & \cdot \\ \cdot & \cdot & \cdot & \cdot & \cdot & \cdot \\ \cdot & \cdot & \cdot & \cdot & \cdot & \cdot \\ 1 & X_{1n} & X_{2n} & \cdot & \cdot & X_{kn} \end{bmatrix}^{-1} * \begin{bmatrix} Y_1 \\ Y_2 \\ \cdot \\ \cdot \\ \cdot \\ Y_n \end{bmatrix}$$

The problem in empirical work is that because there are usually more observations, n, than there are independent variables, k, the X matrix is not square, and it will be recalled that we can only invert square matrices. The OLS technique leads to the pre-multiplying of the X matrix and of the Y vector by the transpose of the X matrix, X^T. The transpose of the above X matrix is

$$\begin{bmatrix} 1 & 1 & \cdot & \cdot & \cdot & 1 \\ X_{11} & X_{12} & \cdot & \cdot & \cdot & X_{1n} \\ X_{21} & X_{22} & \cdot & \cdot & \cdot & X_{2n} \\ X_{k1} & X_{k2} & \cdot & \cdot & \cdot & X_{kn} \end{bmatrix}$$

The formula for the vector of regression coefficients is

$$\hat{B} = [X^T X]^{-1} X^T Y$$

$[X^T X]^{-1}$ is the inverse of the matrix resulting from multiplying the transpose of the X matrix by the X matrix. This inverse matrix is multiplied by the transpose of X again, and the product multiplied by the Y vector.

Worked example

The data for one dependent variable Y and three independent variables, V_1, V_2, V_3 are as follows:

Y	const	V_1	V_2	V_3
7	1	13	6	13
9	1	15	8	11
11	1	10	7	12
8	1	12	7	9
8	1	11	8	12
7	1	13	7	16
8	1	11	7	11

Expressed in matrix form we have

$$X = \begin{bmatrix} 1 & 13 & 6 & 13 \\ 1 & 15 & 8 & 11 \\ 1 & 10 & 7 & 12 \\ 1 & 12 & 7 & 9 \\ 1 & 11 & 8 & 12 \\ 1 & 13 & 7 & 16 \\ 1 & 11 & 7 & 11 \end{bmatrix}, \quad Y = \begin{bmatrix} 7 \\ 9 \\ 11 \\ 8 \\ 8 \\ 7 \\ 8 \end{bmatrix}$$

The transpose of X, X^T, is

$$X^T = \begin{bmatrix} 1 & 1 & 1 & 1 & 1 & 1 & 1 \\ 13 & 15 & 10 & 12 & 11 & 13 & 11 \\ 6 & 8 & 7 & 7 & 8 & 7 & 7 \\ 13 & 11 & 12 & 9 & 12 & 16 & 11 \end{bmatrix}$$

$X^T * X$ is

$$\begin{bmatrix} 7 & 85 & 50 & 84 \\ 85 & 1049 & 608 & 1023 \\ 50 & 608 & 360 & 598 \\ 84 & 1023 & 598 & 1036 \end{bmatrix}$$

$[X^T X]^{-1}$ is

$$\begin{bmatrix} 32.54 & -0.48 & -2.76 & -0.57 \\ -0.48 & 0.06 & -0.02 & -0.01 \\ -2.76 & -0.02 & 0.38 & 0.03 \\ -0.57 & -0.01 & 0.03 & 0.04 \end{bmatrix}$$

$[X^T X]^{-1} X^T$ is

$$\begin{bmatrix} 2.31 & -3.01 & 1.56 & 2.33 & -1.68 & -2.17 & 1.66 \\ 0.07 & 0.16 & -0.13 & 0.02 & -0.09 & 0.02 & 0.06 \\ -0.43 & 0.22 & -0.002 & -0.14 & 0.35 & 0.04 & -0.06 \\ -0.002 & -0.04 & 0.01 & -0.12 & 0.03 & 0.14 & -0.03 \end{bmatrix}$$

$[X^T X]^{-1} X^T Y$ is

$$\begin{bmatrix} 9.53 \\ -0.32 \\ 0.60 \\ -0.14 \end{bmatrix}$$

Thus $B_0 = 9.53$, $B_1 = -0.32$, $B_2 = 0.60$ and $B_3 = -0.14$, these being the constant and the regression coefficients for the variables V_1, V_2 and V_3.

Appendix 6.2

FTSE 100 returns	Gilts returns	S&P 500 returns	£/$ returns
−5.03865	0.97062	−0.81181	3.875111
−5.80479	−0.99772	−2.7947	−2.25009
6.756908	2.726713	2.726272	−1.01653
4.715902	−3.24295	0.786902	2.926117
−3.22561	−2.14766	−7.21813	4.249501
−4.68652	−3.13807	1.194216	−1.01554
−0.75337	−2.70719	1.775336	−2.09321
−4.78026	6.621534	−1.92271	0.641713
11.30555	3.500561	8.895581	2.585833
0.025298	0.276949	−1.00181	4.583169
−1.401	−0.35164	−1.12439	5.058423
−7.65644	0.330367	−9.72923	0.965673
−6.47293	4.124323	−2.3907	0.877082
−0.13797	3.771859	−2.54693	3.228606
6.430736	0.257748	5.413922	−1.67941
−0.89174	4.775062	1.870699	0.571579
1.030363	1.203434	3.811711	2.38076
9.725182	0.994964	7.689626	−4.05269
2.874212	0.189595	0.223789	−8.74905
2.090753	−0.1096	2.392376	−0.83731
0.294574	0.307949	2.022585	−1.57806
−2.91185	2.740759	−2.64774	−4.98216
5.884167	2.163633	2.405219	3.588102
3.33535	2.862377	2.123906	0.654959
−1.27696	0.204208	−1.58804	3.670671
−3.70006	−0.34128	0.543229	0.085776
−5.42394	1.512134	−2.56769	1.53151
3.186904	2.761412	8.945333	5.182791
2.655846	1.525032	−1.72668	−3.86864
−0.23072	−2.98952	0.607979	−2.50329
−5.87325	4.863675	−2.0131	−1.81065
9.920527	2.543311	2.032491	3.477124
1.411156	−0.58666	1.149646	2.106508
−7.85147	−0.24178	−1.06484	4.714317
−2.99976	−1.25275	2.914392	0.677969
−5.1637	3.832018	−2.14474	3.72397
11.25872	5.746209	0.052862	−13.9774
4.392257	−1.33108	1.539886	−12.5624
3.803517	2.737089	1.881653	0.325946
1.933198	1.197246	1.140232	−1.47519
0.179007	2.755257	1.548583	−4.07807
1.081242	1.040079	−0.11532	−0.93313
−0.14581	−1.80704	1.858152	5.767911
−2.29475	0.560659	−1.7564	2.811625
1.275118	4.08267	2.537261	−1.15386
1.380294	2.698462	−1.06552	−2.77841
1.814645	4.402739	0.251343	−1.26753
4.756401	0.078037	2.847012	1.201216
−1.49244	1.258159	−0.40457	−0.13276
4.033624	2.560811	1.681073	−1.64075
2.150889	4.099294	−1.54892	−0.03377

7

Time-series analysis

Introduction

Time-series analysis is a very wide subject. This chapter must, of necessity, be selective and will limit itself to four objectives. The first is to explain in a not too technical way some of the important terms used to define time-series processes. The second is to analyse the process generating the underlying time series as a **univariate stochastic process**, i.e. a stochastic process where the components of the process are functions of the underlying variable itself. The third and fourth objectives are to explain two econometric techniques of time-series analysis. Two such techniques which have become increasingly frequently applied to financial data are **cointegration** and **autoregressive conditional heteroscedasticity (ARCH)**, and its generalized form **GARCH**. However, before we analyse these concepts we must define some jargon and explain some basic forms of time-series analysis.

First some basics

There are a number of terms which describe the statistical characteristics of time series and the reader would be wise to become familiar with these terms, because they will be met frequently in this chapter and in wider reading of the financial literature. Indeed some jargon was used in the introductory section above.

In particular the reader will be interested in remembering the definitions of **random walks, martingales, stationarity** and **white noise**.

Random walks and martingales

A commonly used model of a financial time series is the **random walk**. A random walk defines the path of a random variable where each change or "**innovation**" is independent of all previous changes and each is drawn from the **identical** probability distribution.

By **independent** we mean that the inovation at one instant in time can have no effect on the subsequent innovation. This would be indicated by zero correlation between successive pairs of observations.

By **identical** we mean that each of the innovations has the same type of probability distribution with the same distributional parameters, e.g. the same mean and the same standard deviation.

A random walk is a stochastic process where the changes of level are given by the addition of a random variable, ϵ, which exhibits a zero mean and a constant variance, and where there is zero correlation between observations. This is given formally as

$$Y_t = Y_{t-1} + \epsilon \tag{7.1}$$

Note that it is the ϵ values that have zero mean and a constant variance of σ^2. Y_t has an expectation $E(Y_t) = \mu$ and a variance of $t\sigma^2$.

Sometimes a random walk can incorporate a drift element. This drift element actually refers to a time trend. Thus a random walk with drift is a random walk with a time trend. An example is

$$Y_t = Y_{t-1} + \alpha + \epsilon \tag{7.2}$$

where α is a constant.

Random walks exhibit two properties which have become particularly important in the analysis of financial time series. These properties are the **Markov property** and the **martingale property**.

The **Markov property** is simply that the only relevant information for determining conditional probability of a future (e.g. next) value of the random variable is the information contained in the current state of that variable, and not the historical probability distribution of that variable. For a random walk, this follows from the independence assumption because each change in the value of the random variable is independent of previous changes. However, the future level of the process does depend upon the current level.

The **martingale property** is that the **conditional expectation** of a future value of that random variable is the current value. This applies to a random walk since all innovations of a variable following a random walk without drift have zero mean.

In fact a **martingale** is a more general stochastic process than a random walk because in the case of a martingale the changes of level are given by a random variable which, although it must have zero mean, need not have constant variance. Nor need the innovations be independent.

A random walk with a positive drift is an example of a **sub-martingale**. With a negative drift it is an example of a **super-martingale**.

For example, consider the process defined by

$$Y_t = Y_{t-1} + 0.8 + \epsilon$$

This is an example of a sub-martingale.

This can be generalized to

$$Y_t = Y_{t-1} + \alpha + \epsilon \tag{7.3}$$

If $\alpha > 0$ we have a **sub-martingale**

$$Y_t = Y_{t-1} + \alpha + \epsilon \tag{7.4}$$

If $\alpha < 0$ we have a **super-martingale**

$$Y_t = Y_{t-1} + \alpha + \epsilon \tag{7.5}$$

If $\alpha = 0$, we have a random walk.

Diagrammatic examples of random walks, sub-martingales and super-martingales are given in Figs 7.1–7.3.

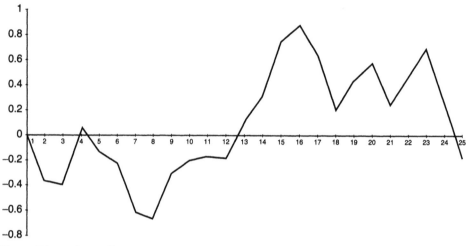

Figure 7.1 Random walk.

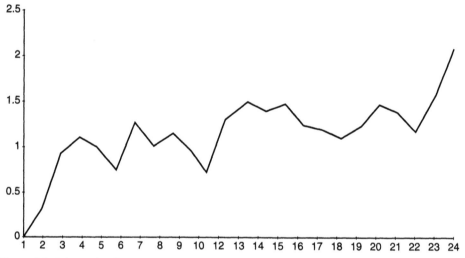

Figure 7.2 Sub-martingale.

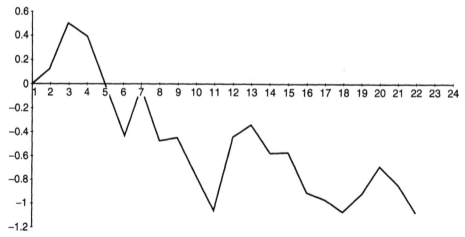

Figure 7.3 Super-martingale.

White noise

A time series is said to be "white noise" if the underlying variable has zero mean, a constant variance and zero correlation between successive observations, i.e. no autocorrelation. You will also recall from Chapter 6 that the assumptions surrounding the residual term in the OLS regression model are similar. Thus OLS residuals are assumed to exhibit "white noise".

When $E(\epsilon_t) \sim N(0,\sigma^2)$ we have Gaussian white noise. However, a white noise variable need not always be normally distributed.

An illustration of white noise is given in Fig. 7.4.

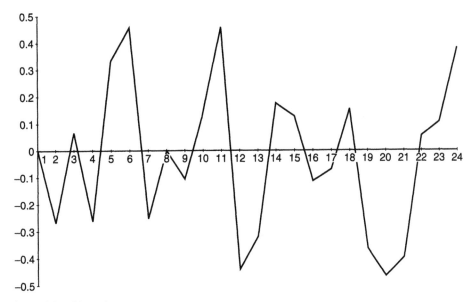

Figure 7.4 White noise.

Stationarity

A time series is said to be stationary if it has a constant mean, a constant variance and a covariance which depends only on the time between lagged observations.

For example, take the daily data of the FTSE 100 index. The index has grown from an opening level of 1000 to levels, at the time of writing, of around 3600, over a period of 12 years. Imagine that you were analysing the mean and standard deviation of the daily level in each calendar year. Because the level is generally rising year on year, the mean and standard deviation relating to year one will be lower than those in year two and so on, and clearly the mean level in a year when the index has risen to 3500 and fallen back to about 3100 will be higher than in the first year that the index started.

We noted in Chapter 2 that the magnitude of the variance and the standard deviation can be a function of the levels of the data used in the calculations. Thus the variance of an index oscillating around 1000 may very well be lower than one oscillating around 3000.

The covariance is similarly influenced by the levels of the data being analysed. In this case the covariance refers to that between successive observations of the data.

Intuitively, we would not expect many, if any, time series of prices, exchange rates or index levels to be stationary because rising or falling values over time are a feature of financial variables. However, the returns required by investors should be dependent upon the uncertainty surrounding the investment and independent of the level of the index. Thus returns data may have a constant mean and standard deviation, and a covariance between observations that depends only on the number of lags between those observations.

Figure 7.5 shows the time series of the daily levels of the FTSE 100 index from 1984 to 1992, while Fig. 7.6 shows the time series of the daily continuously compounded returns. Clearly the levels data are not stationary. However, the returns data may be – they seem to have a constant mean and a constant variance. Later in this chapter we will test to see if the returns data are indeed stationary.

Considering again the definition of **white noise** given earlier, a white noise series is also a stationary series. However, a stationary series will not be white noise if it has a mean that is different from zero or if there is covariance present.

Univariate stochastic models of time-series processes

Within the framework of **univariate stochastic models** the process generating the time series is considered to be made up of past components of the time-series itself. In other words, if the future values of the underlying variable are in any way predictable, those future values are some function of past values. In this section we will limit our analysis to analysing the components of the univariate stochastic process within the context of **autoregressive processes, moving average processes** and the degree

Figure 7.5 FTSE 100 index levels: 1984–1992.

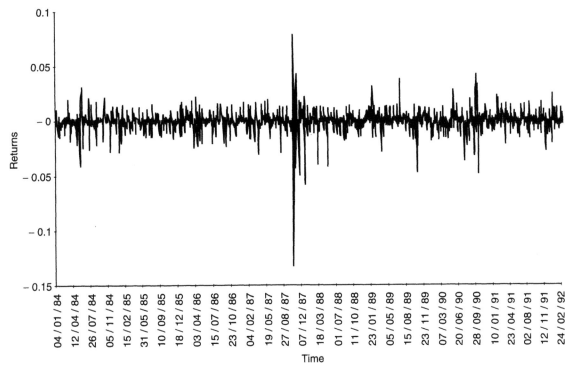

Figure 7.6 Daily returns FTSE 100 index: 1984–1992.

of **integration**. These three sub-processes can be conveniently grouped together under the heading of **autoregressive integrated moving average processes (ARIMA)**.

Autoregressive processes

We will begin our analysis of the ARIMA process by considering the **autoregressive process.** An autoregressive process is one in which the level of the process is a linear function of past levels. For example, if the current observation is a function of only the immediate past observation, i.e. the process relies upon just one lagged value of the underlying variable, the process is said to be autoregressive of order one, indicated as AR(1). This may be generalized to say that if the time-series process being analysed utilizes values lagged 1 to n periods, the process is autoregressive of order n, i.e. AR(n). To illustrate this, an AR(3) process would be given as

$$Y_t = \alpha_0 + \alpha_1 Y_{t-1} + \alpha_2 Y_{t-2} + \alpha_3 Y_{t-3} \tag{7.6}$$

Here the current value of Y is a linear function of the three most recent previous observations. Hence an autoregressive **model** is one in which the model values are given by a linear function of past observations. Readers will see the similarity with autocorrelation or serial correlation in regression analysis, where the residuals from the regression equation were correlated with previous residuals. Indeed the estimating equation given in equation (7.7) below will be recognized as a multivariate regression equation, where the past values of Y are the independent variables.

$$\hat{Y}_t = \alpha_0 + \alpha_1 Y_{t-1} + \alpha_2 Y_{t-2} + \alpha_3 Y_{t-3} + \epsilon_t \tag{7.7}$$

where ϵ_t is the residual or error term.

Later in this chapter we will explain how to determine the degree of autocorrelation in a time series by using the **autocorrelation coefficient** and the **partial auto-correlation coefficient.**

Integration

The autoregressive process, and indeed the moving average process which will be analysed in the next section, assumes that the data being analysed is stationary. The term integration refers to the degree of differencing that a data series requires for it to be transformed into a stationary series. Differencing is simply the process of finding the change in the value of a variable in successive time periods, i.e. $\Delta Y_t = Y_t - Y_{t-1}$. The ΔY series is the differenced series.

If a time series has to be differenced once in order to be transformed into a stationary series the original series is said to be "integrated of order one" or $I(1)$. If the original series has to be differentiated twice to become stationary, the original series is said to be integrated of order two or $I(2)$. If a series does not have to the differenced at all because it is already stationary, it is said to be integrated of order zero or $I(0)$. Recall the earlier definition of stationarity – we gave a chart of the levels and returns data, and suggested that the levels data were not stationary but the returns were more likely to be.

If a series is $I(0)$, i.e. stationary, the variance will be finite. A change in the underlying variable, often referred to as an innovation, will only have a transitory influence on the underlying time series. The autocorrelation coefficients decline steadily so that their sum is finite. If, on the other hand, the series is $I(1)$, any innovation in the time series has a permanent effect on that series, and the variance will increase without limit (tend to infinity) as the time horizon moves forward.

Recall the earlier discussion of stationarity. We suggested that the levels of the FTSE 100 index are unlikely to be stationary although returns may be. The index returns are analogous to the first difference of the levels data. If the returns data are stationary, the levels data will be $I(1)$.

Modelling with data that is not stationary can be problematical, for example it can give rise to **spurious correlation**. To understand this consider Fig. 7.7 which relates to the levels of the FTSE 100 and the S&P 500 indices. The correlation between these two series is 0.81. Yet it would be inappropriate to automatically assume that there is causality between the two indices. Certainly the two indices both rise over the long run, but does one cause the other to rise? Actually our economic theory would suggest that, given the integration of the national economies and positive inflation rates in each country, over the long run rising economic activity (measured in nominal values) in both economies would cause their stock markets to rise. Thus a third factor, nominal economic activity, is the causal effect of both markets rising in the long run.

The spurious nature of the correlation between the two levels series can be confirmed when analysing the returns series. If the two indices were highly correlated, we would expect negative (positive) returns in one to be accompanied by negative (positive) returns in the other. However, looking at Fig. 7.8, this is not always the case. Indeed the correlation between the returns series is only 0.3.

Thus modelling with $I(1)$ variables may give indications of correlation which may be interpreted as causality when none really exists.

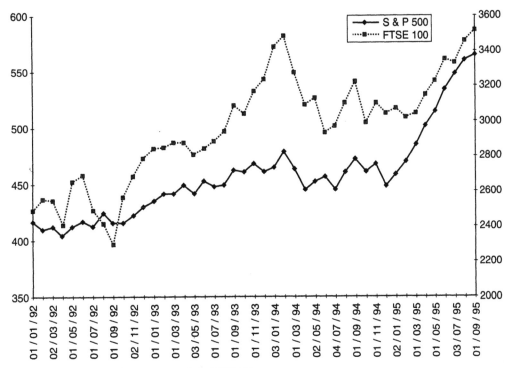

Figure 7.7 Monthly levels of the S&P 500 and the FTSE 100.

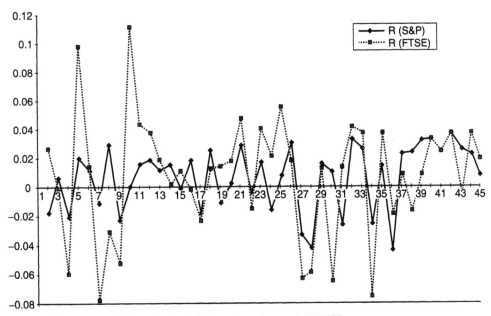

Figure 7.8 Monthly returns on the S&P 500 and the FTSE 100: 1/1/1992–1/9/1995.

Moving average models

A moving average model is one in which the model values are given by a linear function of past errors, i.e. the differences between past modelled values and past observations.

$$\hat{Y}_t = \beta_0 + \beta_1 \epsilon_{t-1} + \beta_2 \epsilon_{t-2} + \beta_3 \epsilon_{t-3} \tag{7.8}$$

where

$$Y_t = \hat{Y}_t + \epsilon_t \text{ or } \epsilon_t = Y_t - \hat{Y}_t \tag{7.9}$$

Note that the term moving average used in this chapter should not be confused with the process with the same name that relates to a technique of smoothing data.

Autoregressive moving average models

Time-series models have been developed that combine the autoregressive process with a moving average model. Not surprisingly, these models are known as **autoregressive moving average models** or **ARMA**. An ARMA (*pq*) model will have *p* lags in the autoregressive process and *q* lags in the moving model. To illustrate this an ARMA(3,2) model would look as follows:

$$\hat{Y}_t = \alpha_0 + \alpha_1 Y_{t-1} + \alpha_2 Y_{t-2} + \alpha_3 Y_{t-3} + \beta_1 \epsilon_{t-1} + \beta_2 \epsilon_{t-2} \tag{7.10}$$

Autoregressive integrated moving average models (ARIMA)

If differencing is needed to produce a stationary series before applying the ARMA process, then the degree of differencing will need to be known. Thus there are three parameters in the full ARIMA process. They are (Box Jenkins) *p*, the order of the autoregressive part of the model, *d*, the degree of preliminary differencing that is required, and *q*, the order of the moving average part of the model.

As ARIMA models encompass autoregressive processes, moving average models and integration, many time-series processes can be seen from an ARIMA standpoint. We have noted so far that data may have an autoregressive (AR) component. It may have a degree of integration, $I(1)$, $I(0)$ or even $I(2)$. If $I(1)$ or $I(2)$ the data will have to be differenced once or twice before it becomes stationary. Finally there may be a moving average (MA) component.

It is important to decompose the time series into these three components in order to be able to determine the structure of the process which is being modelled. The first stage in the analysis is the differencing in order to produce a stationary series. We then attempt to fit an ARMA model to that stationary series.

For example, take a completely random process where the Y_t depends only on the mean of the series and an error term, thus

$$Y_t = \mu + \epsilon_t \tag{7.11}$$

In this process there is no reliance on past values of Y_t, there is no differencing of Y_t and there is no reliance on past error terms. We can classify this process as ARIMA(0,0,0).

A white noise process is an ARIMA(0,0,0) process with a zero mean.

An ARIMA (1,0,0) process is of the form

$$Y_t = \alpha Y_{t-1} + \epsilon_t \tag{7.12}$$

where $-1 < \alpha < 1$ and ϵ_t is a white-noise process. The process depends on the immediate past value of Y and no differencing is required to make Y stationary. This is the same as an AR(1) process.

If $\alpha = 1$, the process would not be stationary and differencing would be required. This would be an example of an ARIMA(0,1,0) process, i.e.

$$Y_t = Y_{t-1} + \epsilon_t \tag{7.13}$$

because the first difference of Y_t has been taken in order to derive a stationary series. This will be recognized as a random walk.

An ARIMA(0,0,1) process is of the form

$$Y_t = \epsilon_t + \theta \epsilon_{t-1} \tag{7.14}$$

because Y_t is influenced only by error terms.

An ARIMA(1,0,1) process would be

$$Y_t = \alpha Y_{t-1} + \theta \epsilon_{t-1} + \epsilon_t \tag{7.15}$$

Vector autoregressive and vector moving average processes

So far we have considered only time series that have one underlying variable. In finance, as in many other applications, we may be concerned with the relationship between two or more variables. For example:

$$\hat{X}_t = \alpha X_{t-1} + \beta Y_{t-1}$$
$$\hat{Y}_t = \gamma X_{t-1} + \delta Y_{t-1} \tag{7.16}$$

This process can be restated in matrix format, with the X_t and Y_t forming a vector as follows:

$$\begin{bmatrix} \hat{X}_t \\ \hat{Y}_t \end{bmatrix} = \begin{bmatrix} \alpha & \beta \\ \gamma & \delta \end{bmatrix} \begin{bmatrix} X_{t-1} \\ Y_{t-1} \end{bmatrix} \tag{7.17}$$

Thus the coefficients α, β, γ and δ form a 2×2 matrix.

The process shown above is a vector AR(1) or VAR(1) process because only one lag of the underlying vector variable is included and there are no MA processes.

A VMA(1) process would have no lags of the underlying variable but only lags of the error terms, as follows:

$$\hat{X}_t = p\epsilon_{X_{t-1}} + q\epsilon_{Y_{t-1}}$$

$$\hat{Y}_t = r\epsilon_{X_{t-1}} + s\epsilon_{Y_{t-1}}$$

(7.18)

This process can also be stated in matrix form as

$$\begin{bmatrix} \hat{X}_t \\ \hat{Y}_t \end{bmatrix} = \begin{bmatrix} p & q \\ r & s \end{bmatrix} \begin{bmatrix} \epsilon_{X_{t-1}} \\ \epsilon_{Y_{t-1}} \end{bmatrix}$$

(7.19)

Clearly vector processes could incorporate elements of both the autoregressive and the moving average processes. An example of a vector ARMA incorporating one lag of levels and one lag of residuals, VARMA(1,1) would be

$$\hat{X}_t = \alpha X_{t-1} + \beta Y_{t-1} + p\epsilon_{X_{t-1}} + q\epsilon_{Y_{t-1}}$$

$$\hat{Y}_t = \gamma X_{t-1} + \delta Y_{t-1} + r\epsilon_{X_{t-1}} + s\epsilon_{Y_{t-1}}$$

(7.20)

This can be expressed in matrix form as

$$\begin{bmatrix} \hat{X}_t \\ \hat{Y}_t \end{bmatrix} = \begin{bmatrix} \alpha & \beta \\ \gamma & \delta \end{bmatrix} \begin{bmatrix} X_{t-1} \\ Y_{t-1} \end{bmatrix} + \begin{bmatrix} p & q \\ r & s \end{bmatrix} \begin{bmatrix} \epsilon_{X_{t-1}} \\ \epsilon_{Y_{t-1}} \end{bmatrix}$$

(7.21)

We have seen that time-series data may be generated by a number of processes and that some of the types of processes frequently met in analysis of financial time series come under the umbrella heading of autoregressive integrated moving average processes (ARIMA). These processes have as sub-processes: autoregressive processes (AR), integrated processes (I) and moving average processes (MA). Thus the analysis of such a time-series process can be achieved by breaking it down into three sub-processes: the autoregressive process, the moving average process and the degree of integration of the series. It is important to note that the ARMA process assumes that the time-series data are stationary in their mean and in variance. Thus when analysing data under the ARIMA process the first function is to determine the degree of integration, and if necessary difference the data so that they are stationary in their mean.

Later in this chapter we will examine cointegration and ARCH and GARCH, techniques developed because economic/financial time series are not stationary in mean or variance.

Cointegration has been developed in response to the growing need to analyse the relationship between groups of economic variables over time in a manner that gives a more conceptually and empirically valid measure of that relationship in the light of non-stationarity of the individual time series. Cointegration particularly addresses the issue of non-stationary means in data.

ARCH and GARCH have been developed to take account of the non-stationarity of the variance, in particular because of the need for improved forecasts of the volatility of financial time series, the advent of financial options and the generally greater volatility of financial markets in the last 20 or so years.

Tools for analysing time series

Clearly the process generating time-series data can take many forms. We have limited our discussion to three elements and have shown that when analysing time-series it is necessary to look for the **degree of autocorrelation**, the **degree of integration** and the extent of the **moving average component**. In this section we will describe the use of the **autocorrelation coefficient (ACC)** and the **partial autocorrelation coefficient (PACC)** to identify the AR and MA elements in the process generating the time series. Then we shall utilize **augmented Dickey–Fuller** tests to identify the degree of integration.

Test for autocorrelation: the autocorrelation coefficient or function

In order to identify the degree of autocorrelation in time series we must determine the strength of association between the current and lagged values of the underlying variable. One such measure is the autocorrelation coefficient (ACC) which gives rise to the autocorrelation function (ACF); it measures the correlation between the current and lagged observations of the time series and is calculated as follows:

$$\rho_k = \frac{\sum_{t=1}^{n-k}\left(Y_t - \overline{Y}\right)\left(Y_{t+k} - \overline{Y}\right)}{\sum_{t=1}^{n}\left(Y_t - \overline{Y}\right)^2} \tag{7.22}$$

where k represents the number of lags in question. Thus the first-order autocorrelation coefficient will have just a one-period lag, the second-order autocorrelation coefficient will have a two-period lag and so on. The autocorrelation coefficient is calculated for all lags and significance tests are carried out to see which lags are statistically significant. Only the lags that are statistically significant are retained in any modelling.

The significance tests for the autocorrelation coefficient are the standard error test and the Box–Pierce Q-test. There are two tests because there are two approaches to testing for autocorrelation. One approach, which utilizes the standard error test, tests the autocorrelation coefficient for individual lags to see which ones are significant. The second approach utilizes the Box–Pierce Q-test to test whether a whole set of autocorrelation coefficients, as a group, are significant.

The standard error of the correlation coefficient is calculated as

$$SE_{r_k} = \frac{1}{\sqrt{n}} \tag{7.23}$$

The autocorrelation coefficients of random data have a sampling distribution which is approximately normally distributed with a mean of zero and a standard deviation of $1/\sqrt{n}$.

To illustrate this application consider daily data on UK long-term government bond levels and bond returns. The first-order autocorrelation coefficients are calculated from a sample of 900 observations. The standard error is $1/\sqrt{900} = 1/30 = 0.0333$.

If r_1 is within the limits

$$-1.96 * 0.0333 \leqslant r_1 \leqslant + 1.96 * 0.0333 = -0.065 \leqslant r_1 \leqslant + 0.065$$

The data are considered not to exhibit first-order autocorrelation.

The calculated autocorrelation coefficients for bond levels were all very much greater than 0.065. This is not surprising given that the time series is in levels. However, the data relating to bond returns displayed a weak autocorrelation structure in which only the first, third, seventh and eighth lags show significant autocorrelation. This is evidenced in Table 7.1 below.

Table 7.1

Lag	ACC	Lag	ACC	Lag	ACC
1	0.095*	2	0.012	3	0.074*
4	−0.009	5	0.022	6	0.031
7	0.080*	8	0.068*	9	0.011

*Significant at the 5% level.

The Q statistic is calculated as

$$Q = n\sum_{i=1}^{m} r_i^2 \sim \chi_m^2$$

where m is the maximum lag being considered.

To illustrate this in the nine-lag case

$$Q = 900 * 0.027 = 24.5800$$

$$\chi_9^2(0.005) = 23.59$$

Thus, as a group, the nine lags are significant.

The partial autocorrelation coefficient and function

A partial autocorrelation coefficient (PAC) giving rise to the partial autocorrelation function (PAF) measures the relationship between the current observation of a variable X_t and successive lagged values of that variable $X_{t-1}, ..., X_{t-k}$ when the effects of intervening lags have been removed. Thus the first partial autocorrelation coefficient is the same as the autocorrelation coefficient as there are no intervening lags. However, the second and subsequent partial autocorrelation coefficients will be different.

The partial autocorrelation coefficient is used to identify the degree of autocorrelation in a time series. For example, a series is AR(m) when the last statistically significant partial autocorrelation coefficient is associated with lag m. Thus in a series that is AR(2), only lags 1 and 2 will be significantly correlated with the current variable. If the series is AR(4) lags 1 to 4 inclusive will be significantly different from zero, but all higher lags will not be significantly different from zero.

In an AR(m) process, the partial autocorrelation coefficients, are significantly different from zero for lags 1 to m, and then fall abruptly to zero for lag $m + 1$ and above.

Test for a moving average process

Knowledge of the behaviour of the autocorrelation coefficients and the partial autocorrelation coefficients can also be used to determine whether a series contains a moving average component in the process. If the series is MA rather than AR, the autocorrelations will not indicate the order of the MA process. However, if the partial autocorrelation coefficients fall to zero exponentially rather than being significant for some lags and then abruptly zero, the series can be assumed to be an MA process rather than an AR one.

Test for ARMA processes

To test for autocorrelation in time series that have both autoregressive and moving average processes, the Ljung–Box (1978) test is used. The Ljung–Box (LB) statistic is calculated as

$$\text{LB} = n(n+2)\sum_{k=1}^{m}\left(\frac{1}{n-k}\right)r_k^2 \approx \chi_{m-p-q}^2 \tag{7.24}$$

where m is the maximum lag being considered, p is the order of the autoregressive process and q is the order of the moving average process.

Tests for degree of integration and for stationarity

Recall that integration refers to the degree of differencing required to make a time series stationary. Also recall that stationarity is very important because many methods of time-series analysis assume that the data being analysed are in fact stationary. Tests of stationarity are often referred to as unit root tests. If the data exhibit a unit root it will be integrated $I(1)$.

The early approach to testing for the degree of integration and of stationarity was called the Dickey–Fuller test. This approach tests for the value of α in equation (7.25) having a value of one or a value less than one

$$Y_t = \alpha_1 Y_{t-1} + e_t \tag{7.25}$$

If α has a value of one, the data is said to have a "unit root" and is integrated of degree one, i.e. $I(1)$. If α is less than one, the series is integrated of order zero, $I(0)$. In finance, α is unlikely to be greater than one as that implies an explosive series. Such series are unlikely because economic pressures would stop the values becoming infinite.

There are some theoretical problems with equation (7.25) because the potential of non-stationarity breaks the assumptions of OLS regression which assumes a constant variance in the residuals. As an example consider $Y_t = Y_{t-1} + u_t = (Y_{t-1} + u_{t-1}) + u_t = \ldots = u_t + Y_0 + u_1 + u_2 + \ldots + u_t$. Since the u_ts are independent with a constant variance, this shows that the variance of Y_t grows without limit (tends to infinity) as t tends to infinity.

Thus the equation has to be re-specified in terms of changes in Y_t as follows:

$$\Delta Y_t = \beta Y_{t-1} + e_t \tag{7.26}$$

where $\beta = (\alpha_1 - 1)$.

If β is zero, α is equal to one, the Y series is said to exhibit a unit root and is $I(1)$, and the ΔY series will be stationary. If β is negative, α is less than one and the Y series is itself stationary in the mean, $I(0)$.

Equation (7.25) assumes a zero mean and no time trend. In financial time series it is often appropriate to include a positive mean, because over time risky assets are expected to offer a positive rate of return. The resulting equation with a positive mean is

$$Y_t = \alpha_0 + \alpha_1 Y_{t-1} + e_t \tag{7.27}$$

which may be similarly transformed to

$$\Delta Y_t = \alpha_0 + \beta Y_{t-1} + e_t \tag{7.28}$$

A third form of equation that is appropriate in finance incorporates a time trend as follows:

$$Y_t = \alpha_0 + \alpha_1 Y_{t-1} + \gamma T + e_t \tag{7.29}$$

which may be transformed into

$$\Delta Y_t = \alpha_0 + \beta Y_{t-1} + \gamma T + e_t \tag{7.30}$$

It is not appropriate to use the traditional t-test of the significance of β, because in using regression to estimate β we are assuming that β is less than zero (i.e. $\alpha < 1$). It can be shown that when $\beta = 0$, too large a percentage of those estimates will be rejected by the t-test. Thus the **null** hypothesis of the existence of the unit root will be rejected too often.

In addition Phillips (1987) has shown that such "unit-root tests", as they are also known, are robust against reasonable degrees of heteroscedasticity, but autocorrelation causes problems. The problem of testing for **stationarity** when there is autocorrelation in the residuals is solved by using the augmented Dickey–Fuller test. This approach incorporates lagged values of the dependent variable in the regression equation, with the number of lags being chosen simply to be sufficient to remove the autocorrelation in the residuals. For example, the equation could be

$$\Delta Y_t = \alpha_0 + \beta_1 Y_{t-1} + \gamma_1 \Delta Y_{t-1} + \gamma_2 \Delta Y_{t-2} + \ldots + \gamma_n \Delta Y_{t-n} + e_t \tag{7.31}$$

The exact form of the significance tests depends upon the form of the model being tested, i.e. without a positive mean (equation (7.26)), with a mean (equation (7.28)) and with a mean and time trend (equation (7.30)).

Null hypothesis without a mean. Testing equation (7.26) that assumes no mean (i.e. no constant) but is adapted to account for the autocorrelation as in equation (7.31), the null hypothesis is

$$H_0: \beta_1 = 0$$

If $\beta_1 = Y$ is $I(1)$, and if β_1 is significantly less than zero, Y is stationary, i.e. $I(0)$.

The **null** hypothesis will be rejected if the test statistic, $\beta_1/\text{SE}(\beta_1)$, has a larger negative value than the critical number in the tables contained in Dickey and Fuller (1979). The 1% and 5% critical values from Dickey and Fuller are −2.58 and −1.95, respectively.

If the null hypothesis is accepted, the Y series is a random walk without drift.

This test has been generalized to take account of the sample size T. This is achieved by computing modified critical values given by the formula

$$\phi_\infty + \frac{\phi_1}{T} + \frac{\phi_2}{T^2}$$

where ϕ_∞ is −2.57 (1%) or −1.94 (5%), ϕ_1 is −1.96 (1%) or −0.398 (5%) and ϕ_2 is −10.04 (1%) or 0 (5%). (These ϕ values are tabulated in MacKinnon (1991).)

Null hypothesis with a mean. Testing equation (7.28), which incorporates a mean (i.e. a constant), involves using the same test statistic, $\beta_1/\text{SE}(\beta_1)$, and the same formula for critical values. However, the ϕ values are now given as

$$\phi_\infty = -3.43 \text{ (1\%) or } -2.86 \text{ (5\%)}$$
$$\phi_1 = -6.00 \text{ (1\%) or } -2.74 \text{ (5\%)}$$
$$\phi_2 = -29.25 \text{ (1\%) or } -8.36 \text{ (5\%)}$$

Null hypothesis with a mean and a time trend. Testing equation (7.30), which incorporates a mean and a time trend, again involves using the same process but the ϕ values are given as

$$\phi_\infty = -3.96 \text{ (1\%) or } -3.41 \text{ (5\%)}$$
$$\phi_1 = -8.35 \text{ (1\%) or } -4.04 \text{ (5\%)}$$
$$\phi_2 = -47.44 \text{ (1\%) or } -17.83 \text{ (5\%)}$$

Illustration of stationary foreign exchange returns. To illustrate a test for stationarity, we apply the augmented Dickey–Fuller test to both daily levels and daily returns of the US\$–sterling exchange rate for the period from 1992 to 1995.

Below we give the test statistics for the regression of the exchange rate levels with a mean (equation (7.28)) and with a mean and time trend (equation (7.30)), each for one, two and three lags.

Lags	Test statistic (mean)	Critical value	Test statistic (mean & trend)	Critical value
1	−2.1307	−2.86	−1.8430	−3.41
2	−2.0508	−2.86	−1.8443	−3.41
3	−2.7975	−2.86	−2.0365	−3.41

The critical values relate to the 95% level of confidence.

Recall that the null hypothesis of the exchange rate levels being $I(1)$ is rejected in favour of the alternative hypothesis that they are $I(0)$ if the test statistic is a larger negative number than the critical value. In each of the examples above, this is not the case, thus the null hypothesis is accepted, i.e. that the levels are $I(1)$.

Next we look at the tests on the returns data

Lags	Test statistic (mean)	Critical value	Test statistic (mean & trend)	Critical value
1	−20.2910	−2.86	−20.3090	−3.41
2	−16.5143	−2.86	−16.5463	−3.41
3	−13.8929	−2.86	−13.9255	−3.41

Here we can clearly see that in relation to the foreign exchange returns data, the null hypothesis of being $I(1)$ is rejected in favour of being $I(0)$. Thus the returns data are stationary.

Cointegration

Intuitive introduction

Earlier, in Chapter 2, we discussed correlation as a measure of the linear association between pairs of variables. Now that we have discussed the concept of stationarity, it is clear that for the correlation coefficient to be a statistically meaningful concept the relationship between the two time-series must be stationary. Indeed we say that the two time series must be **jointly covariance stationary**. A single variable is covariance stationary if both $E(X_t)$ and $\sigma^2(X_t)$ are finite constants for all t, the correlation coefficient between X_t and X_{t-n} is the same for all t, and thus the covariance between two observations of X depends only on the time between the observations. For a pair of variables to be jointly covariance stationary, the individual series must be covariance stationary, and the correlation between X_t and Y_t must be the same for all t so that $cov(X_t, Y_t)$ is independent of t.

The problem with using the correlation coefficient in finance is that, in many cases, there is little reason to expect financial variables to be covariance stationary. For example, currencies or equity market indices between countries with weak economic links are unlikely to have a stationary relationship between them. However, there may be a long-run relationship which needs identifying. Consequently, we need another measure of the degree of association between variables that accommodates the practical realities of a series not being jointly covariance stationary in the short run, but which exhibits a long-run equilibrium. Cointegration is that concept.

Cointegration describes the long-run linear relationship between a group of variables which exhibit an equilibrium relationship with each other. Consider a two-variable example, the level of the FTSE 100 index and the level of the FTSE 100 future, which we will designate X and Y, respectively. There are economic reasons to expect that in the long run these two variables have an equilibrium relationship with each other. To understand this, consider the cash and carry arbitrage model introduced in Chapter 1 to explain the pricing of financial futures. What would happen if the levels on the future moved much above or below the theoretical level? If the future were above its fair value arbitragers would sell the future and buy the index, forcing the prices back to their equilibrium level. Conversely, if the future were below its fair value, arbitragers would buy the future and sell the index.

Let us assume that a fair futures price of 1.1 times the equity index level is the equilibrium relationship. This can be expressed as

$$Y = 1.1X \tag{7.32}$$

this is the same as saying

$$Y - 1.1X = 0 \tag{7.33}$$

This relationship only holds in equilibrium; in the short run each of these variables will vary over time in its own way, and will probably be $I(1)$. Even though each series is $I(1)$, if there is a long-run equilibrium relationship between them, a third variable, z, given by

$$Y - 1.1X = z \tag{7.34}$$

will be stationary and will measure the extent to which the variables are out of equilibrium. z is termed the equilibrium error because the forces that set up the equilibrium process will force z towards its mean value.

Thus, if an equilibrium relationship exists, it will be possible to find some combination of the two variables, i.e. $aY + bX = z$, such that z is stationary. Remember that if z is indeed stationary, it moves around a constant mean, and that when it deviates from the mean value it tends to revert to that value. However, in the short run the two variables may not move together, and thus will not be in short-term equilibrium, but any short-term deviation will be reversed by market forces. This is depicted in Fig. 7.9.

If there exists a and b such that $aY + bX$ is $I(0)$, then $\gamma + (b/\alpha)X$ will be stationary. If we have a function, $Y + \beta X = z$, where z is stationary and both X and Y are $I(1)$, we have cointegration between X and Y. β is then referred to as the coefficient of cointegration.

Thus cointegration describes the long-run relationship between pairs or groups of variables and results from those pairs or groups of variables exhibiting a common stochastic trend over time.

One function of cointegration analysis is to provide a complementary analysis of the benefits of portfolio diversification in addition to the correlation analysis incorporated in mean-variance portfolio structures. In particular, the cointegration analysis will indicate whether there are long-term relationships between variables and the speed at which short-term deviations from the long-term equilibrium will return to that equilibrium (Clare et al., 1995).

A second function of cointegration analysis has been to identify weak form inefficiencies in markets. See, for example, Chelley-Steeley and Pentacost (1994) and Choudhry (1994).

A third function has been to provide a forecasting capability through the modelling of what is termed "an error correction model" (Alexander, 1992).

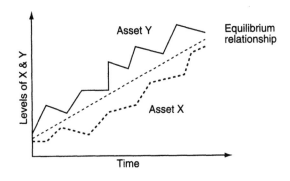

Figure 7.9

Cointegration between two variables

Cointegration between two variables occurs when two series are each integrated of order b, but some linear combination of the two series results in a third series which is integrated of order a, $a < b$. In such circumstances the two $I(b)$ series are said to be cointegrated. For practical purposes in finance $b = 1$ and $a = 0$. Thus the cointegrated series will be $I(1)$ and the linear combination of the two series will be $I(0)$. To illustrate this consider the £/DM and £/FFR exchange rates. Assume that each is integrated of order 1, $I(1)$, but if from the two series we can deduce a variable, z, in $z_t = Y_t - \lambda X_t$, which is $I(0)$, i.e. so that the value of z oscillates around a constant mean, Y and X are said to be cointegrated. The parameter λ is referred to as the constant of cointegration.

Cointegration characterizes an equilibrium relationship between the variables because, for z to be stationary, when X and Y move away from their equilibrium relationship with each other, they will return towards their equilibrium relationship so that z oscillates about its (constant) mean value. This tendency to revert to the equilibrium characterizes what is known as error correction. The model of this process of reverting to the equilibrium is the **error correction model** and corresponds to an interesting feature of stationarity, which is that the stationary series, z, is mean reverting.

To understand the equivalence of stationarity and mean reversion, consider equation (7.35) used to test for stationarity ignoring the effects of autocorrelation

$$Y_t = \alpha Y_{t-1} + \epsilon_t \tag{7.35}$$

Let us assume that the constant mean is 0, α is 0.8 and that there has been a sudden deviation from this mean such that the value of Y_{t-1} was 2. Ignoring the random effects, the ϵs, Y_t and Y_{t+1} would be

$$Y_t = 0.8(2) = 1.6$$

$$Y_{t+1} = 0.8(1.6) = 1.28$$

If on the other hand the value of Y_{t-1} was -3, Y_t and Y_{t+1} would be

$$Y_t = 0.8(-3) = -2.4$$

$$Y_{t+1} = 0.8(-2.4) = -1.92$$

Thus we see that after sudden positive deviations from the constant mean, the changes are negative, and the magnitude of those changes is a function of the size of the positive deviation. Similarly if the sudden deviation from the mean is negative the changes will be positive. Thus irrespective of the sign of the deviation from the mean, subsequent changes tend to pull the variable back to its mean.

Thus we can generalize this by saying that if $\alpha < 1$, the series has a tendency to revert back towards its mean value. Now in the case of a cointegrating regression, the equilibrium errors, the zs, are stationary, implying that the two cointegrated variables tend to move back towards their equilibrium relationship.

Tests for cointegration between two variables

The first stage of any cointegration analysis is to estimate the **constant of cointegration** (assuming one exists). In the two-variable case OLS regression can be used. Alternatively a more sophisticated approach using **maximum likelihood** estimation techniques to determine the cointegrating vector may be used. This latter approach is more complex but has the advantage of being generalizable to the multivariate case. In this section we will test using OLS regression. In subsequent sections we will develop the error correction model. This will be followed by the approach developed by Johansen (1988) and Johansen and Jesulius (1990) which use the maximum likelihood technique.

Recall that for a pair of variables to be cointegrated:

1. their individual time series must be $I(1)$, and
2. the linear combination must be $I(0)$.

It should be stressed that not all $I(1)$ variables are cointegrated, only those where a linear relationship between them can be shown to be $I(0)$. Thus testing for cointegration is in two stages.

Stage one

Determine that the variables in question are $I(1)$ using the augmented Dickey–Fuller test. Thus the researcher must arbitrarily introduce sufficient lagged values of Y and X to convert the residuals in the following regressions into white noise

$$\Delta X_t = \beta X_{t-1} + \Sigma \theta_i \Delta X_{t-i} + \epsilon_t$$
$$\Delta Y_t = \gamma Y_{t-1} + \Sigma \phi_i \Delta Y_{t-i} + \epsilon_t \tag{7.36}$$

where both β and γ are analogous to $(\alpha - 1)$ in equation (7.3).

As this is the **augmented Dickey–Fuller** test, the tests for the significance of both β and γ use the t-statistic on Y_{t-1} and X_{t-1}, respectively. If any of these parameters are not significantly different to zero, the appropriate series (i.e. the X or the Y series) will be $I(1)$.

The researcher may also wish to introduce a mean and/or a time trend into equation (7.36) as discussed in the section on integration.

Stage two

Having satisfied ourselves that the original variables are $I(1)$, we apply OLS regression in the following form, known as the **cointegrating regression**:

$$Y_t = \lambda_0 + \lambda X_t + u_t \tag{7.37}$$

This isolates the residuals u_t in order that we can test whether or not they are stationary. If we add λ_0 and u_t to get z_t, then we have $Y_t - \lambda X_t = z_t$ and we shall need to test for the stationarity of z. If the zs are indeed stationary, λ will be the cointegrating vector as discussed earlier

Note: ordinarily regressing Y on X and X on Y will be different, but in the case of cointegrated time series the "long-run" correlation between the variables is 1, and the regressions are essentially the same.

Stage three

Determine that the residuals are $I(0)$. This is achieved by running the following regression

$$\Delta u_t = \beta u_{t-1} + e_t \tag{7.38}$$

This is analogous to equation (7.26), but with u rather than Y as the independent variable. The test for the null hypothesis of no cointegration is that $\beta = 0$. The reasoning is that if β is not significantly different from zero, u_t is $I(1)$ and therefore Y and X are not cointegrated.

We cannot use the same significance test on the coefficient of u_t as we did for Y_{t-1} in equation (7.25) because the residuals are themselves the result of an estimation. We must therefore use the tables given in Mackinnon (1991).

This test has a number of drawbacks. Firstly, the test suffers from small-sample bias in the same way as regression analysis. Secondly, it is based on minimizing the variance of the residuals rather than identifying maximum stationarity. However, small-sample bias is unlikely to be a problem with financial time series as samples are usually large.

Error correction model

Granger (1986) and Engle and Granger (1987) have shown that if variables are cointegrated an **error correction model** is embodied within them. The error correction model describes the process by which the $I(1)$ variables return to equilibrium once they have departed from it. For example, consider the relationship between the two exchange rates introduced earlier. Over short periods of time investors' enthusiasm for one may cause it to rise relative to the other. At other times the alternative currency may be more attractive. In both these situations the currencies will move away from their long-run equilibrium relationship with each other. However, economic forces will cause them to revert to that long-run equilibrium. In the example cited, if a currency gets too strong economic and political pressures may lead to a lower level of interest rates in the relevant country. If a currency gets too weak, interest rates may rise and or the country's exports may grow and imports fall until the long-run equilibrium was again reached. The model of the process of returning to the long-run equilibrium is the **error correction model**.

It should be clear now why and how cointegration implies an error correction model. The z variable has to be stationary if the $I(1)$ variables are cointegrated. To be stationary, z has to oscillate about a constant mean with a constant variance. This implies that when X and Y deviate from their equilibrium relationship there must be forces (an error correction process) bringing them back into line. Thus we can see that error correction models are modelling cointegrated processes, and this fact is embodied in a formal result known as the **Granger representation theorem**.

Engle and Granger two-stage process

Regression is useful when there are only two series to be analysed, because there can be at most one cointegrating coefficient. Engle and Granger (1987) developed the two-stage process of estimating the error correction model.

The first stage was the estimation of the cointegrating regression as described earlier. The second step is the construction of the following error correction model:

$$\Delta X_t = \alpha_1 Z_{t-1} + \sum_{i=1}^{n} B_{1i} \Delta Y_{t-i} + \sum_{i=1}^{n} C_{1i} \Delta X_{t-i} + \epsilon_t$$

$$\Delta Y_t = \alpha_2 Z_{t-1} + \sum_{i=1}^{n} B_{2i} \Delta Y_{t-i} + \sum_{i=1}^{n} C_{2i} \Delta X_{t-i} + \epsilon_t$$

(7.39)

with α_1 and $\alpha_2 \neq 0$.

Thus we can use OLS regression to estimate the current ΔYt and ΔXt from past observations of Y and X and the value of the cointegrated variable.

However, in the multivariate case there may be more than one cointegrating vector. We therefore need a methodology that identifies the structure of all the cointegrating vectors. Such a process has been developed by Johansen (1988) and Johansen and Jesulius (1990) which specifies the set of time series as a vector autoregressive (VAR) process. The error correction model is developed as follows.

Vector autoregressive specification of the error correction model

To illustrate this derivation of the **error correction model**, consider a two-variable vector (AR3) process (i.e. one in which the levels of the variables are linear combinations of three past observations). This would be expressed thus

$$X_t = a_1 X_{t-1} + a_2 X_{t-2} + a_3 X_{t-3} + b_1 Y_{t-1} + b_2 Y_{t-2} + b_3 Y_{t-3}$$

$$Y_t = d_1 X_{t-1} + d_2 X_{t-2} + d_3 X_{t-3} + e_1 Y_{t-1} + e_2 Y_{t-2} + e_3 Y_{t-3}$$

(7.40)

This somewhat cumbersome equation can be expressed in matrix form as

$$\begin{bmatrix} X_t \\ Y_t \end{bmatrix} = A_1 \begin{bmatrix} X_{t-1} \\ Y_{t-1} \end{bmatrix} + A_2 \begin{bmatrix} X_{t-2} \\ Y_{t-2} \end{bmatrix} + A_3 \begin{bmatrix} X_{t-3} \\ Y_{t-3} \end{bmatrix}$$

(7.41)

where the As and Bs are all 2×2 matrices. For example, A_1 is a 2×2 matrix as follows:

$$A_1 = \begin{bmatrix} a_1 & b_1 \\ d_1 & e_1 \end{bmatrix}$$

(7.42)

Recall that the analysis of stationarity required us to convert

$$Y_t = \alpha Y_{t-1}$$

(7.43)

into first differences, i.e.

$$\Delta Y_t = (1 - \alpha)Y_{t-1} \tag{7.44}$$

So here we also have to reorganize the model into differences. We can achieve this as follows.

Firstly, we subtract

$$\begin{bmatrix} X_{t-1} \\ Y_{t-1} \end{bmatrix} \tag{7.45}$$

from both sides to turn the left-hand side into differences, noting that

$$A_1 \begin{bmatrix} X_{t-1} \\ Y_{t-1} \end{bmatrix} - \begin{bmatrix} X_{t-1} \\ Y_{t-1} \end{bmatrix} = [A_1 - I] \begin{bmatrix} X_{t-1} \\ Y_{t-1} \end{bmatrix} \tag{7.46}$$

This gives

$$\begin{bmatrix} \Delta X_t \\ \Delta Y_t \end{bmatrix} = [A_1 - I] \begin{bmatrix} X_{t-1} \\ Y_{t-1} \end{bmatrix} + [A_2] \begin{bmatrix} X_{t-2} \\ Y_{t-2} \end{bmatrix} + A_3 \begin{bmatrix} X_{t-3} \\ Y_{t-3} \end{bmatrix} \tag{7.47}$$

remembering that the A_is on the right-hand side are actually 2×2 matrices.

Now we employ a little algebraic sleight of hand by rewriting the expression as follows:

$$\begin{bmatrix} \Delta X_t \\ \Delta Y_t \end{bmatrix} = [A_1 - I] \begin{bmatrix} X_{t-1} \\ Y_{t-1} \end{bmatrix} + [(A_1 + A_2 - I) - (A_1 - I)] \begin{bmatrix} X_{t-2} \\ Y_{t-2} \end{bmatrix} + A_3 \begin{bmatrix} X_{t-3} \\ Y_{t-3} \end{bmatrix} \tag{7.48}$$

We open our brackets to give

$$\begin{bmatrix} \Delta X_t \\ \Delta Y_t \end{bmatrix} = [A_1 - I] \begin{bmatrix} X_{t-1} \\ Y_{t-1} \end{bmatrix} + [A_1 + A_2 - I] \begin{bmatrix} X_{t-2} \\ Y_{t-2} \end{bmatrix} - [A_1 - I] \begin{bmatrix} X_{t-2} \\ Y_{t-2} \end{bmatrix} + A_3 \begin{bmatrix} X_{t-3} \\ Y_{t-3} \end{bmatrix} \tag{7.49}$$

and pair off terms with matching coefficients, for example

$$[A_1 - I] \begin{bmatrix} X_{t-1} \\ Y_{t-1} \end{bmatrix} - [A_1 - I] \begin{bmatrix} X_{t-2} \\ Y_{t-2} \end{bmatrix} \tag{7.50}$$

becomes

$$[A_1 - I] \left[\begin{bmatrix} X_{t-1} \\ Y_{t-1} \end{bmatrix} - \begin{bmatrix} X_{t-2} \\ Y_{t-2} \end{bmatrix} \right] = [A_1 - I] \begin{bmatrix} \Delta X_{t-1} \\ \Delta Y_{t-1} \end{bmatrix} \tag{7.51}$$

A similar procedure is applied to the A_3 matrices. The result is

$$\begin{bmatrix} \Delta X_t \\ \Delta Y_t \end{bmatrix} = [A_1 - I] \begin{bmatrix} \Delta X_{t-1} \\ \Delta Y_{t-1} \end{bmatrix} + [A_1 + A_2 - I] \begin{bmatrix} \Delta X_{t-2} \\ \Delta Y_{t-2} \end{bmatrix} + [A_1 + A_2 + A_3 - I] \begin{bmatrix} X_{t-3} \\ Y_{t-3} \end{bmatrix} \tag{7.52}$$

Remember that the resulting $[A_1 + A_2 + A_3 - I]$ matrix is actually a 2×2 matrix, and we relabel this Π. The $[A_1 - I]$ matrix is relabelled A_1^* and the $[A_1 + A_2 - I]$ matrix, A_2^*. The result is

$$\Delta\begin{bmatrix} X_t \\ Y_t \end{bmatrix} = A_1^* \Delta\begin{bmatrix} X_{t-1} \\ Y_{t-1} \end{bmatrix} + A_2^* \Delta\begin{bmatrix} X_{t-2} \\ Y_{t-2} \end{bmatrix} + \Pi\begin{bmatrix} X_{t-3} \\ Y_{t-3} \end{bmatrix} \tag{7.53}$$

so we see that the VAR process in levels can be written as a VAR process in differences, save for one term

$$\Pi\begin{bmatrix} X_{t-3} \\ Y_{t-3} \end{bmatrix} \tag{7.54}$$

The rank of the Π matrix gives the number of cointegrating vectors in the system (recall that the rank of a matrix is the number of linearly independent rows). Thus the rank of the Π matrix tells us about what we should be doing. If the rank of Π is zero, then Π is the null matrix (i.e. zeros in every entry) and we do in fact have a VAR process in differences. This indicates that there is no cointegration, and that differencing is necessary to achieve stationarity.

If Π is of full rank, the levels data are already stationary. (Π has an inverse and so the expression can be solved for levels in terms of differences. This can only be true if the levels are $I(0)$.)

If Π is of rank $0 < m < n$ (in this case $n = 2$ and $m = 1$) then there are m cointegrating vectors. These vectors describe the long-run equilibrium relationships between the variables. The error correction model **embodies** the short-run dynamics which maintain that long-run equilibrium relationship.

To understand this, the Π matrix can be decomposed into two matrices α and γ such that

$$\Pi = \begin{bmatrix} \alpha_1 \\ \alpha_2 \end{bmatrix} [\gamma_1 \; \gamma_2] \tag{7.55}$$

so that

$$\Pi\begin{bmatrix} X_{t-3} \\ Y_{t-3} \end{bmatrix} = \begin{bmatrix} \alpha_1 \\ \alpha_2 \end{bmatrix} [\gamma_1 \; \gamma_2] \begin{bmatrix} X_{t-3} \\ Y_{t-3} \end{bmatrix} \tag{7.56}$$

If

$$[\gamma_1 \; \gamma_2] \begin{bmatrix} X_{t-3} \\ Y_{t-3} \end{bmatrix} = Z \tag{7.57}$$

is stationary, we have cointegration, and α_1 and α_2 are interpreted as the **rate of adjustment of the process towards equilibrium**.

Thus the error correction model is

$$\Delta\begin{bmatrix} X_t \\ Y_t \end{bmatrix} = A_1^* \Delta\begin{bmatrix} X_{t-1} \\ Y_{t-1} \end{bmatrix} + A_2^* \Delta\begin{bmatrix} X_{t-2} \\ Y_{t-2} \end{bmatrix} + \begin{bmatrix} \alpha_1 Z \\ \alpha_2 Z \end{bmatrix} \tag{7.58}$$

where Z is $I(0)$.

The method considered to give the most accurate derivation of the Π matrix is the maximum likelihood procedure developed by Johansen (1988) and Johansen and Jesulius (1990) which is applied to multivariate cointegration in the next section.

Cointegration amongst several variables

We are now in a position to extend the analysis of cointegration to several variables, for example X, Y and W. There are four possible linear combinations of variables, for example $X\&Y$, $X\&W$, $Y\&W$, $X,Y\&W$. However, we are only concerned with the independent combinations, as it is only the independent combinations that may be cointegrated. Any linear combination of cointegrating vectors will itself be a cointegrating vector. Thus we can only have a maximum of $n - 1$ cointegrating vectors. As we have three variables we can have only two independent combinations.

To illustrate this, consider the four combinations above.

We can show that if X and Y are cointegrated, and Y and W are cointegrated, then it follows that X, Y and W must be cointegrated, and that X and W must be cointegrated.

If X and Y are cointegrated there exist a, b and c such that $aX + bY + c$ is $I(0)$.

If Y and W are cointegrated there exist p, q and r such that $pY + qW + r$ is $I(0)$.

Adding gives $aX + (p + b)Y + qW + (c + r)$ is $I(0)$, so X, Y and W are cointegrated.

Multiplying by p and b respectively and subtracting gives $apX - bqW + (cp - br)$ is $I(0)$, so X and W are cointegrated.

Thus there are, at a maximum, only two independent cointegrating vectors.

To identify the cointegrating vectors in the multivariate case and to construct the error correction model we apply the maximum likelihood approach of Johansen. The error correction model in the multivariate case is simply a generalization of the two-variable model discussed above. Again we begin by constructing a VAR model and manipulating it into differences. However, now the vectors will be $n \times 1$ vectors instead of 2×1, and the matrices will be $n \times n$ instead of 2×2.

We can write this in matrix shorthand where the underlined variables are vectors, as

$$\Delta \underline{X}_t = A_1^* \Delta \underline{X}_{t-1} + A_2^* \Delta \underline{X}_{t-2} + \Pi \underline{X}_{t-3} \tag{7.59}$$

In this example we have assumed three AR elements so that the final equation has lags over two periods, just as in the earlier example. However, the Π matrix is $n \times n$.

It is the rank of the Π matrix which will determine the number of distinct cointegrating vectors amongst the variables. If the rank is m ($m < n$), then there are m cointegrating vectors.

When cointegrating vectors are present, Π can be decomposed into two matrices, one $n \times m$ the other $m \times n$. We will call these α and γ, and Π is the product of α and γ, i.e. $\Pi = \alpha\gamma$. Rows of γ are such that for each row of γ, $\gamma_i * X_{t-3}$ is $I(0)$. It is the rows of the γ matrix which form the cointegrating vectors. Thus in matrix shorthand we have

$$\gamma \underline{X}_{t-3} = \underline{Z}_{t-3}$$

Z_{t-3} will be vector $I(0)$ of order m if there are m cointegrating vectors. Again the α matrix represents the speed of adjustment back towards equilibrium.

Test for cointegration between several variables

The use of ordinary least-squares regression is not suitable for identifying the various cointegrating vectors which may exist within the multivariate framework. A more powerful test of the components of the cointegrating vector is the Johansen likelihood ratio or "trace" test (Johansen, 1988; Johansen and Jeseulius, 1990), which draws on the error correction model to specify the independent cointegrating vectors and to test for their stationarity.

The Johansen procedure has two functions. One is to identify the number of co-integrating vectors in a group of time series. The other is to provide maximum likelihood estimations of the cointegrating vectors and the speed of adjustment vectors. The process utilizes canonical correlation and maximum likelihood estimation – both techniques are briefly explained in Appendices 7.1 and 7.2, respectively. However, many econometric packages contain cointegration procedures. We have used Microfit 3.0 to produce the results for the illustration given below.

To illustrate the multivariate test for cointegration and the estimation of an error correction model, we have chosen daily exchange rates between sterling and the Hong Kong dollar (HK$), Malaysian ringgit (MR), Thai bhat (TB) and Filipino peso (FP) for the years 1991–5 inclusive.

The **first stage** is to test that the exchange rate data is $I(1)$. Here we report the augmented Dickey–Fuller test where a time trend is allowed for as in equation (7.30) above.

$X1$	$X2$	$X3$	$X4$
−1.7976	−1.7447	−1.8849	−1.9562

We would reject the null hypothesis of no stationarity if the test statistic was more negative than the critical value. As the critical value of the test statistic is −3.4168, we conclude that the data are $I(1)$.

The **second stage** is to test for the rank of the Π matrix. As we have four currencies there can be a maximum of three cointegrating vectors. The procedure is to test first the null hypothesis that there is one cointegrating vector, then the null that there are two and so on. We reject the null hypothesis that m, the number of cointegrating vectors, is less than n if the test statistic is greater than the critical values specified. The details of the likelihood ratio test or **trace test** on our four currencies are given in Table 7.2 below.

Table 7.2

Null	Alternative	Statistic	95% Critical
$m = 0$	$m = 1$	62.1827	47.2100
$m \leqslant 1$	$m = 2$	19.5523	29.6800
$m \leqslant 2$	$m = 3$	8.6202	15.4100
$m \leqslant 3$	$m = 4$	2.4095	3.7620

To identify the number of cointegrating vectors we first test the null hypothesis that there are no cointegrating vectors, $m = 0$, against the alternative hypothesis that there is one cointegrating vector. We reject the null hypothesis if the test statistic is greater than the critical value. With a test statistic of 62.1827 against a critical value of 47.2100, there is clearly at least one cointegrating vector. Next we test the null hypothesis that there is one cointegrating vector against the alternative hypothesis that there are two cointegrating vectors. As the test statistic is smaller than the critical value we accept the null hypothesis. In similar tests for three and four cointegrating vectors, the null hypothesis is rejected. Thus we conclude that this data exhibits one cointegrating vector.

Next we show the long-run Π matrix in Table 7.3.

Table 7.3

	X1	*X2*	*X3*	*X4*
X1	−0.046042	−0.076484	0.0003787	0.019572
X2	−0.023645	−0.039278	0.0001945	0.058003
X3	−0.13644	−0.22666	0.0011223	0.058003
X4	−0.085460	−0.14196	0.0007030	0.036329

This Π matrix can be decomposed into the matrix of estimated cointegrated vectors which is given by the 1×4 vector in Table 7.4 and the vector of adjustment parameters which is given by the 4×1 vector in Table 7.4.

Table 7.4

$$
\begin{bmatrix}
-0.26189 \\
(0.04606) \\
-0.13449 \\
(0.023645) \\
-0.77610 \\
(0.13644) \\
-0.4861 \\
(0.085460)
\end{bmatrix}
\begin{bmatrix}
0.17581 & 0.29205 & -0.0014461 & -0.074736 \\
(-1.0000) & (-1.6612) & (0.0082255) & (-0.42510)
\end{bmatrix}
$$

The multiplication of the standardized variables in the 4×1 vector by the standardized variables in the 1×4 vector results in the 4×4 Π matrix in Table 7.3 above.

Multiplying the elements of a 1×4 vector of appropriately lagged changes to the exchange rates by the unstandardized elements of the 4×1 vector gives Z, as follows:

$$
Z = [0.17581 \quad 0.29205 \quad -0.0014461 \quad -0.074736]
\begin{bmatrix}
\text{HK\$}_{t-2} \\
\text{MR}_{t-2} \\
\text{TB}_{t-2} \\
\text{FP}_{t-2}
\end{bmatrix}
\tag{7.60}
$$

$$
= 0.17581 \text{ HK\$} + 0.29205 \text{ MR} - 0.0014461 \text{ TB} - 0.074736 \text{ FP}
$$

Estimating the multivariate error correction model

In order to model

$$
\Delta \underline{X}_t \tag{7.61}
$$

We must estimate the A_1^* and A_2^* matrices. This is achieved by regressing the

$$
\Delta \underline{X}_t - \Pi \underline{X}_{t-3}
$$

on

$$\Delta \underline{X}_{t-1} \text{ and } \Delta \underline{X}_{t-2}$$

Recall that the estimating equations for the two-variable error correction model were given in equation (7.53). In that example ΔX_t was regressed on one lagged value of Z and the lagged changes of X and Y. Here we are generalizing this process to four variables and a vector process. Thus to estimate error correction models for each of the four currencies we run four separate estimating regressions. We will illustrate the process with an example of just one of the four, the HK dollar (again assuming an AR3 process).
The estimating equation is

$$\Delta HK\$_t - \alpha Z_{t-2} = B_0 + B_1 \Delta HK\$_{t-1} + B_2 \Delta MR_{t-1} + B_3 \Delta TB_{t-1} +$$
$$B_4 \Delta FP_{t-1} C_1 \Delta HK\$_{t-2} + C_2 \Delta MR_{t-2} + C_3 \Delta TB_{t-2} + C_4 \Delta FP_{t-2}$$

The regression results are

$$\Delta HK\$_t - \alpha Z_{t-2} =$$

$$0.0806 - 0.3353\ \Delta HK\$_{t-1} + 0.3079\ \Delta MR_{t-1} + 0.0274\ \Delta TB_{t-1} + 0.0877\ \Delta FP_{t-1}$$
$$(28.88) \qquad (-3.45) \qquad\quad (1.20) \qquad\qquad (3.19) \qquad\qquad (4.03)$$

$$-\ 0.2524\ \Delta HK\$_{t-2} + 0.3489\ \Delta MR_{t-2} + 0.0161\ \Delta TB_{t-2} + 0.0429\ \Delta FP_{t-2}$$
$$(-2.65) \qquad\qquad (1.36) \qquad\qquad (1.87) \qquad\qquad (2.01)$$

The parameters relating to the first and second lags of the Malaysian ringgit and the second lag of the Thai bhat are insignificant at the 5% level. Consequently these could be removed and the error correction model estimated with fewer variables.

Generalized autoregressive conditional heteroscedasticity (GARCH)

As financial markets have become more volatile in the last two decades and because of the growing importance of options in financial risk management there has developed a growing interest in the volatility of financial markets. Mandlebrot (1963) noted that large changes in asset prices were followed by large changes of either sign, while small changes were followed by small changes. In particular, financial variables often exhibit quiet periods followed by volatile periods, that is volatility is not constant but **time varying**. Techniques developed in the 1980s (Engle, 1982; Bollerslev, 1986; Nelson, 1991) give us the econometric tools for making predictions of future time-varying volatility. Engle coined the term autoregressive conditional heteroscedasticity (ARCH). Bollerslev generalized the process to generalized ARCH or GARCH.

Given the time-varying volatility it seems reasonable for risk-averse economic agents to require time-varying risk premiums as reward for bearing financial risk. The ARCH in mean model developed by Engle *et al.* (1987) and developed further by Nelson (1991) provides a framework within which to analyse the influence of time-varying risk on the risk premiums required by those agents.

This section will begin by introducing the conditional moments of a time series. We will then proceed to an analysis of ARCH and GARCH and illustrate the application of univariate GARCH to forecasting volatility of the US dollar exchange rate. Next we will apply the E-GARCH method to the same exchange rate data. Following that we will explain the GARCH-M model and its application to time-varying risk premiums in financial assets. We will then proceed to explain how bivariate GARCH techniques can provide time-varying variances and correlations as inputs to the risk management process.

The conditional moments of a time series

The starting point for us is to define some basic concepts. The traditional representation of the expected value of a random variable, y_t, is

$$E(y_t) = \mu \tag{7.62}$$

μ in this case is referred to as the **unconditional mean**. It is simply the probabilistically weighted mean of the expected outcomes of the random variable. Similarly the **unconditional variance** of a random variable is given as

$$\sigma^2 = E[(y_t - \mu)^2] \tag{7.63}$$

However, we are interested in the **conditional mean**, m_t, and the **conditional variance**, which we shall label h_t^2. The conditional mean is the expected value of a random variable when the expectation is influenced (conditioned) by knowledge of other random variables. The mean is usually a function of these other variables. Similarly, the conditional variance is the variance of a random variable conditioned by knowledge of other random variables.

The conditional mean can be formally stated as

$$m_t = E[y_t \mid F_{t-1}] \tag{7.64}$$

The vertical line $\mid$ should be interpreted as saying "given".

This equation says that m_t is the expected value of y_t, conditioned on the set of information, F, available in the previous time period. The information set F could include any sources of information. A natural choice is past values of y_t. Thus we might, for example, find the conditional mean, m_t, by first regressing y_t on past values of y_t, lagged one period.

$$y_t = \alpha_0 + \alpha_1 y_{t-1} + \epsilon \tag{7.65}$$

then m_t is derived as the sum of $\alpha_0 + \alpha_1 y_{t-1}$ and is thus a time-varying estimate of the mean. This particular equation will be recognized as an AR(1) process.

The mean is a conditional mean because it is conditional on previous values of y. We could extend the number of lags either because of *a priori* knowledge of the lag structure, or until there is no autocorrelation in the residuals.

The conditional variance is defined as

$$h_t^2 = E[(y_t - m_t)^2 \mid F_{t-1}] \tag{7.66}$$

Recall from the conditional mean equation that the difference between y_t and the mean is ϵ_t. Therefore, we can derive the conditional variance, h_t^2, as a function of past **squared residuals** of the conditional mean equation. Thus, for example, we can derive the value of h_t^2 from the following equation

$$h_t^2 = \alpha_0 + \alpha_1 \epsilon_{t-1}^2 \qquad (7.67)$$

provided that one lag only is appropriate.

ARCH and GARCH models

We now have the building blocks to develop ARCH and GARCH models. We will begin with univariate ARCH to be followed by univariate GARCH before proceeding to the multivariate equivalents.

Univariate ARCH

It will be recalled that when, in Chapter 2, we calculated the standard deviation from historical data we were in effect calculating the unconditional standard deviation. We were actually assuming that variability was constant over the history of the data. Thus the data was assumed to be stationary in its variance. In fact it is well known that many financial time series exhibit time-varying variance. Recall from Chapter 6 that such a condition was termed **heteroscedasticity**. The ARCH process was first developed by Engle (1982) in order to accommodate the time-varying variance. Among its many uses has been the modelling of the volatility of asset returns which is well known to be time varying.

In order to identify the ARCH characteristics in a time series, we start by modelling the **conditional mean**. To do this we will specify an autoregressive model of returns. An AR(p) model would hypothesize that p lags of the independent variable are required. We will illustrate this with an AR(1) model where we hypothesize that the current level of return is a function of just one lagged observation of return, for example

$$r_t = \alpha_0 + \alpha_1 r_{t-1} + \epsilon_t \qquad (7.68)$$

We could of course include any number of lagged values of r_t until there is no autocorrelation in the ϵ_ts. If r_t is the value of a random variable $\alpha_0 + \alpha_1 r_{t-1}$ is a conditional mean because it is conditioned on the previous value (or values of r_t). The formula can be generalized to take account of p lags as follows:

$$r_t = \alpha_0 + \sum_{i=1}^{p} \alpha_i r_{t-i} + \epsilon_t \qquad (7.69)$$

The objective of modelling the conditional mean is to construct a series of squared residuals (ϵ_t^2) from which to derive the conditional variance. Recall from Chapter 6 that the residuals from the OLS regression were assumed to have a constant (zero) mean and a standard deviation of ϵ (homoscedastic). Thus the variance of the residuals is simply ϵ_t^2. In the ARCH methodology, the ϵs are assumed to have a non-constant variance, denoted by h^2. Thus $\epsilon_i = \sqrt{h_i^2} * z$, where z is unit normal.

Thus from a time series of the squared residuals of the conditional mean equation we develop the equation for the conditional variance as follows:

$$h_t^2 = \beta_0 + \beta_1 \epsilon_{t-1}^2 \qquad (7.70)$$

where h_t^2 is the conditional variance again in this case assuming one lag. Thus h_t^2, the **conditional variance**, is an autoregressive process of the squared residuals, hence the term **autoregressive conditional heteroscedasticity**. This approach can be generalized to any number (q) of the lagged residuals being included as follows:

$$h_t^2 = \beta_0 + \sum_{i=1}^{p} \beta_i \epsilon_{t-i}^2 \tag{7.71}$$

This representation of ARCH is adapted from Engle (1982) and referred to as the linear ARCH(p) model because p lags of the squared residuals are included.

Tests for univariate ARCH

To test for the existence of ARCH it is necessary to square the error term from the original conditional mean equation. This squared series is regressed on a constant and p lagged squared series of itself. The test statistic is $T * R^2$, where T is the sample size and the R^2 is the multiple regression coefficient from the regression of squared errors. This test statistic has a χ^2 distribution. The degrees of freedom are equal to the number of lags in the regressors. If the test statistic is greater than the critical value in the χ^2 tables the null hypothesis of no ARCH is rejected.

Univariate GARCH

One problem that arises from the original ARCH formulation is that the α_is have to be constrained to be non-negative in order to ensure that the conditional variance is always positive. However, when many lags are required to model the process correctly, the non-negativity constraints may be violated. The early practice was to ensure that the number of lags was often constrained arbitrarily by imposing an *ad hoc* linearly declining structure of coefficients.

Bollerslev (1986) generalized the ARCH model by the inclusion of lagged values of the conditional variance in order to avoid the long-lag structure of the ARCH(q) developed by Engle (1982). Thus the generalized ARCH or GARCH (p, q) specifies the conditional variance to be a linear combination of p lags of the squared residuals from the conditional mean equation and q lags of the conditional variance as follows:

$$h_t^2 = \beta_0 + \sum_{i=1}^{p} \beta_i \epsilon_{t-i}^2 + \sum_{i=1}^{q} \gamma_i h_{t-i}^2 \tag{7.72}$$

where β and γ are constrained to be non-negative in order to avoid the possibility of negative conditional variances.

This is the GARCH equation. What it is saying is that the current value of the conditional variance is a function of a constant, some value of the squared residual from the conditional mean equation, plus some value of the previous conditional variance. For example, if the conditional variance is best described by a GARCH (1,1) equation it will be because the series is AR(1), i.e. the epsilons lagged 1, and has one lag of the conditional variance.

To illustrate the application of GARCH we use the technique to forecast the volatility of the US dollar returns to a sterling holder.

The conditional mean model was an AR(2) model and the regression parameters with t-statistics in parentheses were as follows:

$$r_{t_{USS}} = \alpha_0 + \alpha_1 r_{t-1} + \alpha_2 r_{t-2} + \epsilon$$

$$r_{t_{USS}} = 0.00005 + 0.01927 r_{t-1} - 0.0571 r_{t-2}$$

$$(0.285) \quad (0.502) \quad (-1.526)$$

The conditional variance equation with t-statistics is

$$h_t^2 = 0.0 + 0.04643 \epsilon_{t-1}^2 + 0.9429 h_{t-1}^2$$

$$(2.062) \quad (3.572) \quad (57.178)$$

This result shows that the conditional variance at time t is explained very significantly in terms of one lag of the squared residual of the conditional mean equation and one lag of the conditional variance itself.

Exponential GARCH: E-GARCH

In the GARCH (p, q) model, the conditional variance depends upon the size of the residuals, not on their sign. Yet there is evidence, e.g. Black (1976), that volatility and asset returns are negatively correlated. Thus when security prices rise, giving positive returns, volatility falls, and when asset prices fall giving negative returns volatility rises. Indeed casual empiricism shows that periods of high volatility are associated with falls in security markets and periods of low volatility are associated with rises in markets.

Nelson (1991) developed E-GARCH to accommodate such a situation. It is as follows:

$$\log h_t^2 = \alpha_0 + \sum_{i=1}^{p} \alpha_i \frac{|\epsilon_{t-i}|}{h_{t-i}} + \sum_{i=1}^{p} \gamma_i \frac{\epsilon_{t-i}}{h_{t-i}} + \sum_{i=1}^{q} \beta_i \log h_{t-i}^2 \tag{7.73}$$

Notice that the ϵs are included in this equation both as raw observations, ϵ, and in modulus form, $|\epsilon|$, where the modulus of ϵ is simply its size, i.e. ignoring the signs. Thus E-GARCH models the conditional variance as an asymmetric function of the ϵs. It allows positive and negative lagged values to have different impacts on volatility. The logarithmic formulation accommodates negative residuals without the conditional variance itself being negative.

Note also that the conditional standard deviations (h_{t-i}) are included as denominators on the right-hand side.

We have applied the E-GARCH model to the same US dollar data as in the GARCH illustration. The regression parameters and the appropriate t-statistics are given below

$$\log h_t^2 = -0.01097 + 0.118 \frac{|\epsilon_{t-1}|}{h_{t-1}} + 0.2496 \frac{\epsilon_{t-1}}{h_{t-1}} + 0.9885 \log h_{t-1}^2$$

$$(-1.807) \quad (5.306) \quad (2.070) \quad (164.077)$$

This result shows the significance of the asymmetric treatment of the residuals from the conditional mean equation. It also highlights again the significance of the GARCH variable.

The GARCH-M model

If the riskiness of financial assets varies over time, it would be appropriate to assume that the return that investors require also varies over time. As all assets, including risk-free assets, earn at least the risk-free rate of return, typically proxied by the return to a short-term government zero coupon bond such as a Treasury bill, the appropriate variable to model is the risk premium. This risk premium is proxied by the difference between the risky asset return and the return to the risk-free asset.

The GARCH-M model developed by Engle *et al.* (1987) formulates the conditional mean as a function of the conditional variance as well as an autoregressive function of past values of the underlying variable. A GARCH adaptation of the original ARCH model is as follows:

$$y_t = \beta + \delta h_t + \epsilon_t$$

$$h_t^2 = \gamma + \alpha \sum_{i=1}^{i} \epsilon_{t-i}^2 + \sum_{i=1}^{i} h_{t-i}^2$$

$$(7.74)$$

Note that in the conditional mean equation the variance is transformed into a conditional standard deviation so that it is in the same unit of measurement as the risk premium being modelled.

Engle *et al.* (1987) applied the ARCH formulation of the above model to the risk premium of one- and six-month Treasury bills in relation to the assumed risk-free asset of a three-month bill, and 20-year corporate bonds versus the three-month Treasury bill. In the latter case the mean regression equation incorporated a third variable to accommodate the yield spread between the three-month and 20-year bonds.

The above model was applied by French *et al.* (1987) on the US equity risk premium for the period 1928–84. They used a GARCH (1,2) model on the conditional variance.

Testing the GARCH model

To test the adequacy of the GARCH model it is necessary to examine the standardized residuals, which are ϵ/h, where h is the conditional standard deviation as calculated by the GARCH model and ϵ is the residuals of the conditional mean equation. If the GARCH model is well specified, the standardized residuals will be independent and identically distributed. This test is in two stages.

The first stage entails calculating the Ljung–Box (LB) statistic on the **squared observations** of the raw data. This entails calculating k autocorrelation coefficients using T observations. Next the autocorrelation coefficients, γ, are squared giving γ^2. The LB statistic is calculated as

$$LB = T(T+2)\sum_{k=1}^{m}\left(\frac{1}{T-k}\right)\gamma_k^2 \approx \chi_m^2 \tag{7.75}$$

where m is the maximum lag of the autocorrelation coefficients.

The second stage entails calculating the LB statistic on the squared standardized residuals. Thus each residual is divided by the appropriate observation of the conditional standard deviation. Next the autocorrelation coefficients of the squared standardized residuals (γ') are calculated and squared. The LB statistic is then calculated as

$$LB = T(T+2)\sum_{k=1}^{m}\left(\frac{1}{T-k}\right)\gamma_k'^2 \approx \chi_{m-p-q}^2 \tag{7.76}$$

where m is the maximum lag in the autocorrelation as before, p is the number of lags of the squared residuals from the conditional mean equation and q is the number of lags of the conditional variance.

If the GARCH model is well specified the LB statistic of the standardized residuals will be less than the critical value of χ_{m-p-q}^2.

A question arises as to what is the best GARCH model and also the most appropriate GARCH parameters. The answer to that is found by trial and error, i.e. by comparing the LB statistics on models of alternative types or parameter structures.

GARCH volatility

We noted earlier that volatility is not constant but time varying. Consequently GARCH volatility, which is time varying by definition, is the correct statistical measure to use. Of course, this is only a satisfactory statement if the correct GARCH specification is applied. Financial theory says little if anything about the correct specification and therefore it must be a subject of continuing empirical investigation.

However, assuming the correct specification is applied, to find the annualized volatility one simply has to take the square root of the conditional variance and multiply that by the square root of the number of data observations per year. This volatility measure is time varying in the sense that the current volatility is a function of the past volatility.

To forecast volatility using GARCH one may use a recursive model of the following form

$$\begin{aligned}
h_{t+1}^2 &= \beta_0 + \beta_1\epsilon_t^2 + \gamma_1 h_t^2 \\
h_{t+j}^2 &= \beta_0 + (\beta_1 + \gamma_1)h_{t+j-1}^2
\end{aligned} \tag{7.77}$$

Note that in the second equation, ϵ^2, which is not known when forecasting, is replaced by its conditional estimate h^2. Thus the second equation allows us to forecast h^2 at time $t + 2$ (by putting $j = 2$), thence to forecast h^2 at time $t + 3$ ($j = 3$), etc. The result of each calculation is the single-period forecast of the conditional variance, j periods ahead. To derive a volatility estimate for the overall period the single-period estimates have to be summed and square rooted.

In addition, it is possible to obtain the standard errors of the coefficients and develop time varying confidence intervals around our forecasts.

Bivariate GARCH

Bivariate GARCH can be used to determine the conditional variances and the conditional covariance and correlation between the variables. In addition it is possible to include a cointegration parameter in the conditional mean equation, which then makes the GARCH parameters suitable for deriving more effective hedge ratios, because the hedging process will be most effective when the variables are cointegrated.

We will first illustrate bivariate GARCH for the returns on a share, s_t, and the returns to a future, f_t. Later we will incorporate the cointegration parameter and derive hedge ratios.

To illustrate this application assume that the two models for the conditional means are

$$s_t = a_0 + a_1 s_{t-1} + \epsilon_{s_t}$$
$$f_t = b_0 + b_1 f_{t-1} + \epsilon_{f_t}$$

(7.78)

and that the conditional variance and covariance equations are

$$h_{s_t}^2 = \alpha_0 + \alpha_1 \epsilon_{s_{t-1}}^2 + \alpha_2 h_{s_{t-1}}^2$$
$$h_{f_t}^2 = \beta_0 + \beta_1 \epsilon_{f_{t-1}}^2 + \beta_2 h_{f_{t-1}}^2$$
$$\text{cov}_{sf_t} = h_{sf_t}^2 = \gamma_0 + \gamma_1 \epsilon_{s_{t-1}} \epsilon_{f_{t-1}} + \gamma_2 h_{sf_{t-1}}^2$$

(7.79)

The conditional covariance in this case is a symmetrical 2×2 matrix as follows:

$$\begin{bmatrix} h_{s,t}^2 & h_{sf_t}^2 \\ h_{sft}^2 & h_{f,t}^2 \end{bmatrix} = \begin{bmatrix} \sqrt{h_{s_t}^2} & 0 \\ 0 & \sqrt{h_{f_t}^2} \end{bmatrix} \begin{bmatrix} 1 & \rho \\ \rho & 1 \end{bmatrix} \begin{bmatrix} \sqrt{h_{s_t}^2} & 0 \\ 0 & \sqrt{h_{f_t}^2} \end{bmatrix}$$

(7.80)

where the diagonals are the conditional variances and the off diagonals are the conditional covariances.

When the GARCH volatilities are obtained by annualizing the square root of the conditional variances as discussed above, the GARCH correlations are derived as follows:

$$\rho_{sf_t} = \text{cor}_{sf_t} = \frac{h_{sf_t}^2}{\sqrt{h_{s_t}^2 h_{f_t}^2}}$$

(7.81)

Thus bivariate GARCH can be used to derive time varying correlations and covariances, with their attendant applications in portfolio construction and the derivation of minimum variance hedge ratios.

Alternatively one may wish to use an error correction model that results from co-integration. An example of such a model is

$$s_t = a_0 + a_1(s_{t-1} - \lambda f_{t-1}) + \epsilon_{s_t}$$
$$f_t = b_0 + b_1(s_{t-1} - \lambda f_{t-1}) + \epsilon_{f_t}$$

(7.82)

The residuals from the above equations are incorporated into the conditional variance equations as described earlier.

If the bivariate GARCH is to be used to derive hedge ratios, the hedge ratio is calculated as

$$b_t^* = \frac{\mathrm{cov}_{sf_t}}{h_{f_t}^2}$$ (7.83)

where cov_{sf} is the conditional covariance between s and f, and h_f^2 is the conditional variance of f. This is analogous to the slope coefficient β in the OLS regression. The advantage of the bivariate GARCH hedge ratio is that it is derived from time-varying variances and covariance, whereas the OLS slope coefficient is based on assumed stationarity of the variances and covariance.

Exercises

1. Distinguish between:

   (a) random walks and martingales
   (b) the Markov property and the martingale property
   (c) stationarity and white noise.

2. Within an ARIMA framework explain what is meant by an AR process, an MA process and the degree of integration.

3. Explain how the autocorrelation coefficient and the partial autocorrelation coefficient are used to analyse the structure of time-series data.

4. Explain how to test for the degree of integration in data.

5. Give an intuitive explanation of cointegration.

6. Explain why it is that cointegrated variables will embody an error correction model.

7. Explain how to run the Engle–Granger two-stage process to identify cointegration.

8. Explain how to derive the conditional variance under:

   (a) a univariate ARCH process
   (b) a univariate GARCH process.

9. Explain how E-GARCH and GARCH-M models differ from the original GARCH model.

10. Explain how to derive the volatility of the underlying variable using GARCH models.

11. What is bivariate GARCH, and how can it be used to derive hedge ratios?

12. What is maximum likelihood estimation? Explain how it operates.

References

Black, F. (1976) Studies in stock price volatility changes. *Proceedings of the 1976 Meetings of the American Statistical Association*, August.

Bollerslev, T. (1986) Generalized autoregressive conditional heteroskedasticity. *Journal of Econometrics*, **31**, 307–27.

Bollerslev, T., Chou, R.Y. and Kroner, K.F. (1992) ARCH modeling in finance: a review of the theory and empirical evidence. In Engle, R.F. and Rothchild, M. eds, *ARCH Models in Finance*. Supplement to the *Journal of Econometrics*, **52**.

Clare, A.D., Maras, M. and Thomas S.H. (1995) The integration and efficiency of international bond markets. *Journal of Business Finance and Accounting*, **22**(2), 313–22.

Cheeley-Steeley, P.L. and Pentacost, E.J. (1994) Stock market efficiency, the small firm effect and cointegration. *Applied Financial Economics*, **4**, 405–11.

Choudhry, T. (1994) Stochastic trends and stock prices: an international enquiry. *Applied Financial Economics*, **4**, 383–90.

Dickey, D.A. and Fuller, W.A. (1979) Distribution of estimators for autocorrelated time series with a unit root. *Journal of the American Statistical Association*, **74**, 427–31.

Engle, R.F. (1982) Autoregressive conditional heteroskedasticity with estimates of the variance of UK inflation. *Econometrica*, **50**, 987–1008.

Engle, R.F. and Granger, C.W.J. (1987) Cointegration and error correction: representation, estimation and testing, *Econometrica*, **55**, 251–76.

Engle, R.F., Lillian, D.M. and Robins, R.P. (1987) Estimating time varying risk premia in the term structure: the ARCH-M model. *Econometrica*, **55**, 391–408.

Engle, R.F. and Rothchild, M. (1992) ARCH models in finance. Supplement to the *Journal of Econometrics*, **52**.

French, K.R., Swert, G.W. and Stambaugh, R.F. (1989) Expected stock returns and volatility. *Journal of Financial Economics*, **19**, 3–29.

Granger, C. (1986) Developments in the study of cointegrated variables. *Oxford Bulletin of Economics and Statistics*, **43**, 213–28.

Johansen, S. (1988) Statistical analysis of cointegrating vectors. *Journal of Economic Dynamics and Control*, **12**, 231–54.

Johansen, S. (1991) Estimation and hypothesis testing of cointegrating vectors in Gaussian vector autoregressive models. *Econometrica*, **59**, 1551–81.

Johansen, S. and Jesulius, K. (1990) Maximum likelihood estimation and inference on cointegration – with applications to the demand for money. *Oxford Bulletin of Economics and Statistics*, **52**, 169–210.

Mackinnon, J.J. (1991) Critical values for cointegrating tests. In Engle, R.F. and Granger, C.W.J. eds, *Long Run Economic Relations*. Oxford University Press.

Mandelbrot, B.B. (1963) New methods in statistical economics. *Journal of Political Economy*, **71**, 421–40.

Nelson, D.B. (1991) Conditional heteroskedasticity in asset returns: a new approach. *Econometrica*, **2**, 347–70.

Phillips, P.C. (1987) Time series regressions with a unit root. *Econometrica*, **55**, 277–301.

Phillips, P.C. and Perron P. (1988) Testing for a unit root in time series regression. *Biometrika*, **75**.

Appendix 7.1: maximum likelihood estimation

There are two forms of maximum likelihood estimation, full information maximum likelihood and limited information maximum likelihood. The latter is a single equation technique and the former is a multivariate technique and is the technique which will be described here.

In Chapter 6 covering regression analysis, we used the method of ordinary least squares to estimate unknown parameters from a sample. The parameters specify a particular model which we are using to describe observed data. Thus, in the regression chapter we wished to explain the observation of a dependent variable Y in terms of independent variables X_i ($i = 1, ..., n$). The model that we constructed was of the form

$$\hat{Y} = \alpha_0 + \alpha_1 X_1 + ... + \alpha_k X_k \tag{A.7.1}$$

The parameters are the α_is, and our task was to choose them so that $\hat{Y}$ was as close as possible to Y over the range of our observations. We achieved this by finding the values of the α_is which minimized

$$\Sigma(Y - \hat{Y})^2 \tag{A.7.2}$$

A more general approach to estimating parameters is provided by the method of **maximum likelihood.** In this approach the data is regarded as evidence relating to the parameters of the distribution. That evidence is expressed as a function of the unknown parameters – the **likelihood function** as follows:

$$L(X_1, X_2, X_3, ..., X_n; \Phi_1, \Phi_2, ... \Phi_k) \tag{A.7.3}$$

where the X_is are the data and the Φ_is denote the parameters that we want to estimate. For example, if the population that we are sampling from is normally distributed, the parameters that we want to estimate will be μ and σ^2.

This likelihood function represents the joint probability of observing the sample

$$L(X_1, X_2, X_3, ..., \Phi_1, \Phi_2, ..., \Phi_k)$$
$$= P(X_1 \wedge X_2 \wedge X_3 ... \wedge X_n) \tag{A.7.4}$$

The objective of the maximum likelihood technique is to maximize the likelihood function. This is achieved by differentiating the likelihood function with respect to each of the parameters to be estimated and setting those partial derivatives to zero. The values of the parameters which maximize the value of the likelihood function are then taken to be our estimates.

It is usual to take the logarithm of the likelihood function first, thus simplifying the subsequent working.

The OLS regression parameter estimations are the same as the maximum likelihood estimations if the OLS residuals are normally distributed. Thus it is convenient to illustrate the maximum likelihood approach with regard to the OLS estimates.

Recall from Chapter 6 that the OLS model is

$$Y = \alpha + \beta X + \epsilon \tag{A.7.5}$$

where the ϵs are assumed to be distributed $N(0, \sigma^2)$, i.e.

$$Y - (\alpha + \beta X) \sim N(0, \sigma^2) \tag{A.7.6}$$

For each mutual observation of X and Y there will be, assuming normality, a probability density function of the following form

$$f(X_i, Y_i) = \frac{1}{\sqrt{2\Pi\sigma^2}} e^{-\frac{1}{2}\left(\frac{Y_i - (\alpha + \beta X_i)}{\sigma}\right)^2} \tag{A.7.7}$$

Given n joint observations of X and Y the total probability of observing all the values in the sample is the product of all the individual probability density function values. Thus the likelihood function is given by the product of these probability density function values as follows:

$$L(\alpha, \beta) = \prod_{i=1}^{n} \frac{1}{\sqrt{2\Pi\sigma^2}} e^{-\frac{1}{2}\left(\frac{Y_i - (\alpha + \beta X_i)}{\sigma}\right)^2} \qquad (A.7.8)$$

As it is easier to differentiate summations rather than products, the log of the likelihood function is often taken, thus

$$\ln L(\alpha, \beta) = \sum_{i=1}^{n} \left(\ln\left(\frac{1}{\sqrt{2\Pi\sigma^2}}\right) - \frac{1}{2\sigma^2}(Y_i - (\alpha + \beta X_i))^2 \right) \qquad (A.7.9)$$

This useful transformation does not affect the final result because $\ln L$ is an increasing function of L. Thus the values of the parameters α and β that maximize $\ln L$ will also maximize L.

Next we take the first derivatives of (A.7.9) with respect to α and β and set to zero, thus

$$\frac{\partial \ln L(\alpha, \beta)}{\partial \alpha} = \sum_{i=1}^{n} \frac{1}{\sigma^2}(Y_i - (\alpha + \beta X_i)) \qquad (A.7.10)$$

which becomes

$$\frac{\partial \ln L(\alpha, \beta)}{\partial \alpha} = \frac{1}{\sigma^2}\left(\sum_{i=1}^{n} Y_i - \left(n\alpha + \beta \sum_{i=1}^{n} X_i^2 \right) \right) \qquad (A.7.11)$$

and

$$\frac{\partial \ln L(\alpha, \beta)}{\partial \beta} = \frac{1}{\sigma^2}\left(\sum_{i=1}^{n} X_i Y_i - \left(\alpha \sum_{i=1}^{n} X_i + \beta \sum_{i=1}^{n} X_i^2 \right) \right) \qquad (A.7.12)$$

Setting these derivatives to zero we require

$$\sum_{i=1}^{n} Y_i = n\alpha + \beta \sum_{i=1}^{n} X_i \qquad (A.7.13)$$

and

$$\sum_{i=1}^{n} X_i Y_i = \alpha \sum_{i=1}^{n} X_i + \beta \sum_{i=1}^{n} X_i^2 \qquad (A.7.14)$$

These are the OLS regression equations that we developed in Chapter 6, thus showing that the OLS solution is a maximum likelihood solution when the OLS residuals are normally distributed.

Johansen used the maximum likelihood technique to derive the parameters of the cointegrating vector.

Maximum likelihood estimation in ARCH and GARCH modelling

Recall that in ARCH and GARCH modelling we start with an equation for the conditional mean, e.g.

$$r_t = \alpha_0 + \sum_{i=1}^{m} \alpha_i r_{t-i} + \epsilon_t \tag{A.7.15}$$

From this the residuals are given as

$$\epsilon_t = r_t - \left(\alpha_0 + \sum_{i=1}^{m} \alpha_i r_{t-i} \right)$$

Furthermore, $\epsilon_t = h_t^2 z$, where h^2 is the conditional variance and $z \sim N(0,1)$.
Thus $\epsilon_t \sim N(0, h_t^2)$ where

$$h_t^2 = \beta_0 + \sum_{i=1}^{p} \beta_i \epsilon_{t-i}^2 + \sum_{i=1}^{q} \gamma_i h_{t-i}^2 \tag{A.7.16}$$

(Note that this is GARCH – for ARCH put gammas = zero).

So we have $m + 1 + p + q + 1$ parameters to estimate $(m + 1)$ alphas from the conditional mean equation, $(p + 1)$ betas and q gammas from the conditional variance equation.

The probability density for our observation is

$$f(r_t; \alpha_0, \alpha_1, ..., \alpha_m, \beta_0, \beta_1, ..., \beta_p, \gamma_1, \gamma_2, ..., \gamma_q)$$

$$= \frac{1}{\sqrt{2\pi h_t}} e^{-\frac{1}{2} \left(\frac{r_t - \left(\alpha_0 + \sum_{i=1}^{m} \alpha_i r_{t-i} \right)}{h_t} \right)^2}$$

where

$$h_t^2 = \beta_0 + \sum_{i=1}^{p} \beta_i \epsilon_{t-i}^2 + \sum_{i=1}^{q} \gamma_i h_{t-i}^2$$

Our likelihood function is the product of these fs for our various observations of r (given enough past observations to be able to compute

$$\alpha_0 + \sum_{i=1}^{m} \alpha_i r_{t-i}$$

$$\beta_0 + \sum_{i=1}^{p} \beta_i \epsilon_{t-i}^2 + \sum_{i=1}^{q} \gamma_i h_{t-i}^2$$

for given values of the alphas and betas).

Hence

$$\ln L(\alpha_0, \alpha_1, ..., \alpha_m, \beta_0, \beta_1, ..., \beta_p, \gamma_1, \gamma_2, ..., \gamma_q)$$

$$= -\frac{n}{2}\ln(2\pi) - \frac{1}{2}\sum_t \ln(h_t) - \frac{1}{2}\sum_t \left(\frac{r_t - \left(\alpha_0 + \sum_{i=1}^{m}\alpha_i r_{t-i}\right)}{h_t}\right)^2$$

This expression is dependent not only on the alphas (which can be seen), but also on the betas and gammas which are hidden in the sense that they are subsumed within the *h*s.

The task is to find the values of the alphas, betas and gammas which maximize ln *L*. This is achieved by numerically based search routines.

Maximum likelihood estimation in cointegration

The outline model given in the chapter excluded the stochastic element in the interests of clarity. Thus equation (7.59), for instance, should read

$$\Delta X_t = A_1^* \Delta X_{t-1} + A_2^* \Delta X_{t-2} + \Pi X_{t-3} + \epsilon_t \qquad (A.7.17)$$

The ϵ_t in this expression is a vector. The assumption is that its components are independent and normally distributed with unknown variances. Thus the likelihood function will be a function of those unknown variances together with the elements of the matrices A_1^*, A_2^* and Π.

In theory we could expand (A.7.17) to obtain explicit expressions for the components of ϵ_t in terms of those parameters and hence construct and maximize the likelihood function. In practice that would be very cumbersome and efficient routines are required exploiting matrix algebra to achieve this. Johansen's routine was developed for this purpose.

Appendix 7.2: canonical correlation and regression

In regression analysis we seek to model a dependent variable as a linear combination of a set of independent variables. Our task is to choose the "best" linear combination – that is the linear combination which achieves the best match between modelled values and observed values. When our independent variable is itself an unknown linear combination of other variables then we need canonical analysis.

In our analysis of multivariate cointegration we showed that by simple algebraic manipulation we could express a vector-autoregressive process $\underline{X}_t = A_1\underline{X}_{t-1} + A_2\underline{X}_{t-2} + A_3\underline{X}_{t-3}$ in the form

$$\Delta\underline{X}_t = A_1^* \Delta\underline{X}_{t-1} + A_2^* \Delta\underline{X}_{t-2} + \Pi X_{t-3} \qquad (A.7.18)$$

We said that if Π is of full rank then we can solve for $\underline{X}_{t-3}$. This would imply that the components of $\underline{X}$ are $I(0)$, since they would be expressed in terms of differences. But this would contradict the initial assumption that they are $I(1)$. The conclusion is that the components of X are in fact stationary, that $\underline{X}$ is over-differenced and that the correct model is the VAR model in levels.

We said that if the rank of Π is zero, then we have a VAR in differences – the standard approach for modelling a non-stationary process.

Our interest is therefore focused on the possibility that Π is neither of full nor null rank. This indicates the existence of cointegration. Our problem is that, given the presence of background noise, we will not know exactly what Π is. Thus we must build a statistical procedure for estimating it and its components (α and γ, and an associated statistical test for its rank.

One such procedure was provided by Johansen (1988). In this approach equation (A.7.18) is re-expressed in the form

$$\Delta \underline{X}_t - \Pi \underline{X}_{t-3} = A_1^* \Delta \underline{X}_{t-1} + A_2^* \Delta \underline{X}_{t-2} \tag{A.7.19}$$

In this form, and bearing in mind that the noise terms have been omitted for convenience, we can see that we are looking to explain a linear combination of $\Delta \underline{X}t$ and $\underline{X}_{t-3}$ in terms of a linear combination of lagged differences.

If we knew Π then we could regress $\Delta \underline{X}_t - \Pi \underline{X}_{t-3}$ on $\Delta \underline{X}_{t-1}$ and $\Delta \underline{X}_{t-2}$ to find A_1^* to find A_2^* (note that this would involve running separate regressions for each component). We could thus try this for different hypothesized Π matrices, choosing that for which the resulting regressions give the best fit. The best fit occurs when the two linear combinations representing the left-hand side and the right-hand side of equation (A.7.19) have the highest correlation, so the analysis is called canonical correlation. The regressions are called canonical regressions.

The reader will recognize that the philosophy of this approach is similar to that of maximum likelihood. In this context the parameters (the elements of the matrix Π) are chosen so as to maximize, not the likelihood function, but the correlation function.

Of course, we do not have repeatedly to make guesses at Π. We can proceed through the algebra, using maximum likelihood estimators for the regressions. The details of that algebra are not appropriate to this text, but a general outline of the resulting procedure is as follows:

1. Regress $\Delta \underline{X}_t$ on the lagged differences and record the residuals R_{0t}.

2. Regress $\underline{X}_{t-k}$ on the $k-1$ lagged differences and record the residuals R_{kt} (note: $k = 3$ in our example).

3. Construct the four matrices

$$S_{00} = \frac{1}{n}\sum_{t=1}^{n} R_{0t}R_{0t}^{T}, \quad S_{0k} = \frac{1}{n}\sum_{t=1}^{n} R_{0t}R_{kt}^{T}, \quad S_{k0} = \frac{1}{n}\sum_{t=1}^{n} R_{kt}R_{0t}^{T}, \quad S_{kk} = \frac{1}{n}\sum_{t=1}^{n} R_{kt}R_{kt}^{T}$$

4. The estimate of Π is given by $S_{0k}S_{kk}^{-1}$.

5. The eigenvalues of Π are the solutions to the equation $|\lambda S_{kk} - S_{k0}S_{00}^{-1}S_{0k}| = 0$, where the vertical lines denote "determinant".

6. Eigenvalues (squared canonical correlation coefficients) which are insignificantly different from zero indicate a reduced rank of P. Thus tests for the rank of P are based on examining the smallest of them when arranged in order (the eigenvalue test), or on the sum of the smallest (the trace test).

7. The corresponding eigenvectors are the estimates of the cointegrating vectors.

8. Estimates of the A_i^* are provided by OLS once the estimate of Π is known.

8

Numerical methods

Introduction

The term "numerical methods" collectively describes ways of solving mathematical problems by repetitive application of a mathematical procedure, either to search for a solution or to aggregate many approximations into a final solution. An example of the former is the use of an iterative procedure to solve an equation for which no convenient formula exists. An example of the second is the aggregation of many small areas under the normal distribution curve to find the total area, since this cannot be found by analytic integration. A third form of "numerical method" is known as Monte Carlo simulation. As its name implies, this process solves problems by imitating the random process, i.e. running the mathematical model many, many times and taking the average outcome.

Equation solving

The process of mathematical modelling – expressing a real-world problem in terms of a mathematical problem – very often leads to a further problem, that of solving the equation or equations of the mathematical model. This particularly applies to non-linear equations. Non-linear equations are equations where one of the variables is raised to a power greater or less than one. For example

$$2x^2 - 2 = 4$$

is a non-linear equation, because x is raised to a power, in this case the power two. If the equation were of the form

$$2x - 2 = 4$$

the equation would be linear.

All linear and some non-linear equations can be solved quite simply because we have formulae for doing so. There are other equations, which are non-linear in form, that cannot be solved that way, not because the equations are particularly complex, but because there is no formulae that can be applied to give the solution.

Even where a formula exists it may be more convenient, or necessary, to obtain an approximate solution. As an example, take the value of x in the following non-linear equation

$$x^2 - 2 = 0$$

The answer is the square root of two. However, there is no rational number that squares to two. The answer is an irrational number. To 10 decimal places it is 1.4142135624, but in fact the answer has an infinite number of decimal places.

We have to search, by trial and error, for a solution that has an acceptable degree of accuracy. Searching by trial and error is known as iteration.

An example, very common in finance, of a non-linear equation that cannot be solved by a formula is the equation for the internal rate of return (IRR) of a series of five or more cash flows.

We learned in Chapter 1 that the IRR is that rate of discount which equates a stream of future cash flows to their present value. For example, the rate that discounts the future coupon and redemption payments to the present market price is the IRR. It is referred to as the yield to maturity or the gross redemption yield.

In order to calculate the yield to maturity we have to solve a polynomial equation derived from the bond price calculation. For example, consider a two-year bond paying annual coupons of 10 and currently priced at 100. The price of that bond would be given by the equation

$$P = \frac{10}{1+r} + \frac{110}{(1+r)^2}$$

In order to determine the yield to maturity (the IRR) we must know the price. Thus if the current price is 100, this equation becomes

$$100 = \frac{10}{1+r} + \frac{110}{(1+r)^2}$$

This equation is the basis of a polynomial equation of degree two because one of the arguments is raised to the power of two. The polynomial is

$$100x^2 - 10x - 110 = 0$$

where $x = 1 + r$. This type of equation is also known as a quadratic equation. We have a formula for solving quadratic equations. It is

$$x = \frac{-b \pm \sqrt{b^2 - 4ac}}{2a} \tag{8.1}$$

which gives solutions to a quadratic equation of the form

$$\boldsymbol{ax^2 + bx + c = 0} \tag{8.2}$$

Applying this to our polynomial where $a = 100$, $b = -10$ and $c = -110$, we get two solutions

$$\frac{+10 + \sqrt{10^2 + 4*100*110}}{2*100} = 1.1$$

$$\frac{+10 - \sqrt{10^2 + 4*100*110}}{2*100} = -1.0$$

As economic theory precludes negative interest rates only the positive solution is economically significant. Thus $(1 + r) = 1.1$.

It is possible, though not particularly helpful, to solve cubic and quartic equations in a similar way to quadratics, although the procedure needs an algorithm rather than a formula. (The cubic is actually more difficult than the quadratic, and needs several pages of closely typed text to describe it.) Mathematicians struggled for several centuries to develop similar algorithms for quintics (i.e. equations of degree five – those where the polynomial can be written in the form $ax^5 + bx^4 + cx^3 + dx^2 + ex + f = 0$) or for polynomial equations of higher degree. However, in 1846, it was shown that no such algorithms could exist.

However, such equations are very important in finance. Take for example a bond with a maturity of just 2.5 years paying half-yearly coupons. To find the IRR we will have to solve the following equation

$$P = \frac{C_1}{(1+r)^{0.5}} + \frac{C_2}{(1+r)} + \frac{C_3}{(1+r)^{1.5}} + \frac{C_4}{(1+r)^2} + \frac{C_5}{(1+r)^{2.5}} \tag{8.3}$$

To convert this into a polynomial equation of degree five, we start by converting the required annual IRR into its semi-annual equivalent, i.e. $(1 + r)^{0.5}$. If we call this x, the above equation is transformed as follows:

$$P = \frac{C_1}{x} + \frac{C_2}{x^2} + \frac{C_3}{x^3} + \frac{C_4}{x^4} + \frac{C_5}{x^5} \tag{8.4}$$

Multiplying through by x^5 we get

$$Px^5 = C_1x^4 + C_2x^3 + C_3x^2 + C_4x + C_5 \tag{8.5}$$

Note: $(C_1/x) * x^5 = C_1x^4$.

After collecting all the arguments on the left-hand side, the final outcome is the following polynomial

$$Px^5 - C_1x^4 - C_2x^3 - C_3x^2 - C_4x - C_5 = 0 \qquad (8.6)$$

A numerical example will clarify this. A 2.5-year bond pays semi-annual coupons of 5 and is currently priced at 98. The bond equation is

$$98 = \frac{5}{(1+r)^{0.5}} + \frac{5}{(1+r)} + \frac{5}{(1+r)^{1.5}} + \frac{5}{(1+r)^2} + \frac{105}{(1+r)^{2.5}}$$

To derive the polynomial equation, we need to convert $(1 + r)$ into the semi-annual compounded equivalent, as discussed above. The polynomial equation is then

$$98x^5 - 5x^4 - 5x^3 - 5x^2 - 5x - 105 = 0$$

To solve this polynomial equation and thus find $(1 + r)$ we have to use trial and error methods. Remember that x is $(1 + r)^{0.5}$, therefore, we need to square x in order to get $(1 + r)$.

The question is how do we go about our iterative procedures to search for the elusive x. There are several approaches. All start with a guess at the correct value, and then a search for that solution that has the required degree of accuracy. Below we will explain two methods of iteration. We will begin with **bisection** and then go on to the **Newton–Raphson** method.

Bisection

Recall that we are trying to solve the equation $f(x) = 0$. A simple, although not a financial, example to illustrate bisection is to consider the problem of finding an approximation to the square root of two (i.e. an approximation to a solution to the equation $x^2 - 2 = 0$). We know that one is too small since $1^2 < 2$ (note that $1^2 - 2 = 1 - 2 = -1 < 0$), and that two is too large since $2^2 > 2$ (note that $2^2 - 2 = 2 > 0$). So our solution must lie between one and two. Clearly 1.5 bears examination – and we find that it is too big because $1.5^2 = 2.25 > 2$ (i.e. $1.5^2 - 2 = 2.25 - 2 = 0.25 > 0$). Thus we now know that the solution lies between one and 1.5. We have halved the interval of uncertainty at a cost of one calculation $(1.5^2 - 2)$.

We may now continue this process by bisecting the range within which the solution lies.

In this example we evaluate at $x = (1.5 + 1)/2 = 1.25$, find that $1.25^2 - 2 < 0$, and thus decide that the interval within which the solution lies (the interval of uncertainty) is between 1.25 and 1.5. We repeat this until we have reduced the interval of uncertainty sufficiently to satisfy our accuracy requirements.

The process of **bisection** is portrayed diagrammatically in Fig. 8.1.

We can formally specify this procedure as follows. To begin we must guess two values of x, x_1 and x_2, such that the function $f(x_1) < 0$ and the function $f(x_2) > 0$. There must be a solution somewhere between these two values. Evaluate $f(x)$ at the midpoint, $(x_1 + x_2)/2$, i.e. find $f[(x_1 + x_2)/2]$. If $f[(x_1 + x_2)/2] < 0$ then replace x_1 by $(x_1 + x_2)/2$. If $f[(x_1 + x_2)/2] > 0$, then replace x_2 by $(x_1 + x_2)/2$ and repeat.

We can lay this out in tabular form. Initially x_1 has been chosen as one and x_2 as two.

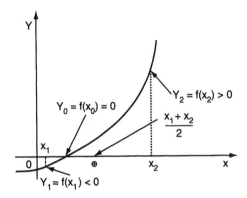

Figure 8.1

Table 8.1 shows 20 iterations to find the answer.

Table 8.1			
x_1	x_2	$(x_1 + x_2)/2$	$f[(x_1 + x_2)/2]$
1	2	1.5	0.25
1	1.5	1.25	−0.4375
1.25	1.5	1.375	−0.1094
1.375	1.5	1.4375	0.06641
1.375	1.4375	1.40625	−0.0225
1.4063	1.4375	1.42188	0.02173
1.4063	1.4219	1.41406	−0.0004
1.4141	1.4219	1.41797	0.01064
1.4141	1.418	1.41602	0.0051
1.4141	1.416	1.41504	0.00234
1.4141	1.415	1.41455	0.00095
1.4141	1.4146	1.41431	0.00026
1.4141	1.4143	1.41418	−9E-05
1.4142	1.415	1.41461	0.00112
1.4142	1.4146	1.4144	0.00052
1.4142	1.4144	1.41429	0.00021
1.4142	1.4143	1.41424	6.3E-05
1.4142	1.4142	1.41421	−1E-05
1.4142	1.4142	1.41422	6.4E-06
1.4142	1.4142	1.41422	2.5E-05

We knew intuitively that 1^2 was too small and that 2^2 was too large, so we take these two values as the initial values of x_1 and x_2, respectively

$$f\left(\frac{x_1 + x_2}{2}\right) = f\left(\frac{1+2}{2}\right) = f(1.5) \qquad (8.7)$$

Next we evaluate our function, $x^2 - 2$, at $x = 1.5$. The result is

$$f(1.5) = 2.25 - 2 = 0.25 > 0$$

As this is greater than zero, $(x_1 + x_2)/2$, 1.5 on this occasion, is substituted for x_2. If, as on the next line, the function is less than zero, $(x_1 + x_2)/2$ would be substituted for x_1.

Now we will apply the same procedure to find the IRR of the 2.5-year bond introduced earlier. To apply this method we have to find a solution to the equation below

$$f(x) = 98x^5 - 5x^4 - 5x^3 - 5x^2 - 5x - 105$$

In this example x is $(1 + IRR)$. We have guessed that appropriate values of x_1 and x_2 are 1.0 and 1.1, respectively. At $x = 1.0$, $f(x) = -27$, thus 1.0 is too small. At $x = 1.1$, $f(x) = +27.3$, thus 1.1 is too large. So the next step is to evaluate the function

$$f\left(\frac{x_1 + x_2}{2}\right) = \left(\frac{1.0 + 1.1}{2}\right) = F(1.05)$$

$$f((98 * 1.05^5 - 5 * 1.05^4 - 5 * 1.05^3 - 5 * 1.05^3 - 5 * 1.05^2 - 5 * 1.05 - 105)) = -2.5526$$

To explain the procedure, consider the first line of Table 8.2. In this example the function $f[(x_1 + x_2)/2]$ has a negative value of -2.5526, so $(x_1 + x_2)/2$, i.e. 1.05, is substituted for x_1. On the next line (the next iteration) the function, $f[(x_1 + x_2)/2]$, is positive (11.6497), so $(x_1 + x_2)/2$, i.e. 1.075, is substituted for x_2 in the next iteration. This procedure is repeated, with the appropriate substitutions made before each iteration until the required degree of accuracy is achieved. In this case where $f[(x_1 + x_2)/2]$ equals zero to four decimal places.

This example has calculated the semi-annual IRR. Squaring 1.054679 gives 1.112348 or an annualized IRR of 11.2348%.

This approach is simple and powerful. Halving the interval of uncertainty at each stage very quickly reduces it to infinitesimal size, and at each stage we are certain that we have the solution "trapped" so that the process is under control. Nevertheless, it does have one disadvantage – it requires that we begin by having a solution "trapped" in the first place. This is not always easy. It requires thought, and that means that it is difficult to automate fully.

Table 8.2

x_1	x_2	$(x_1 + x_2)/2$	$f[(x_1 + x_2)/2]$
1	1.1	1.05	-2.5526
1.05	1.1	1.075	11.6497
1.05	1.075	1.0625	4.37346
1.05	1.0625	1.05625	0.86746
1.05	1.0563	1.05313	-0.8532
1.0531	1.0563	1.05469	0.00445
1.0531	1.0547	1.05391	-0.425
1.0539	1.0547	1.0543	-0.2105
1.0543	1.0547	1.05449	-0.103
1.0545	1.0547	1.05459	-0.0493
1.0546	1.0547	1.05464	-0.0224
1.0546	1.0547	1.05466	-0.009
1.0547	1.0547	1.05468	-0.0023
1.0547	1.0547	1.05468	0.00109
1.0547	1.0547	1.05468	-0.0006
1.0547	1.0547	1.05468	0.00025
1.0547	1.0547	1.05468	-0.0002
1.0547	1.0547	1.05468	4.2E-05
1.0547	1.0547	1.05468	-6E-05

Newton–Raphson

The Newton–Raphson method is the most widely used of the iterative procedures and has the advantage of converging more rapidly than the bisection method. Like the bisection method, the Newton–Raphson method begins with a guess of the value of x. Let x_n be our current guess. The function $f(x_n)$ is evaluated at x_n. The value of $f(x_n)$, is represented by the height of the vertical line on Fig. 8.2 below. The point where the tangent to the curve at $(x_n, f(x_n))$ cuts the x-axis should provide a better guess. We will call this value of x, x_{n+1}.

Recall from Chapter 3 which dealt with calculus that the slope = height/distance. Height in this example is $f(x_n)$, distance is $x_n - x_{n+1}$. Therefore, we can see that knowing the lengths of the sides of the triangle, the slope is given by

$$\frac{f(x_n)}{x_n - x_{n+1}} = f'(x_n) \tag{8.8}$$

Thus

$$f(x_n) = f'(x_n)(x_n - x_{n+1}) \tag{8.9}$$

so

$$x_{n+1} = x_n - f(x_n)/f'(x_n) \tag{8.10}$$

Thus each iteration is found by taking the last iteration and deducting the ratio of the function, $f(x)$, and its first derivative, $f'(x)$.

This process only fails if, by chance, one of our x_ns happens to be a point for which $f'(x_n)$ is exactly or nearly zero (in this case re-starting the procedure from a new point when it fails will nearly always solve the problem).

It might be argued that this technique suffers from the disadvantage of required preliminary work, since $f'(x)$ must be found. However, even this may be avoided. It is possible to replace $f'(x)$ by the finite difference approximation

$$\frac{f(x+h) - f(x)}{h} \tag{8.11}$$

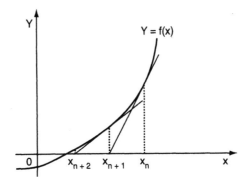

Figure 8.2

where h is small and may be made smaller as the iteration proceeds. Thus each iteration will involve our computer in calculating $f(x_n)$ and $f(x_n + h)$ (for the current values of x_n and h), and then evaluating

$$x_n - \frac{f(x_n)}{\left[\dfrac{f(x_n + h) - f(x_n)}{h}\right]} \tag{8.12}$$

Alternatively, given the increasing sophistication of computer algebra packages (which ought to be called algebra/calculus packages!), the exact differentiation can be performed automatically.

Now we will apply the Newton–Raphson procedure to valuing the 2.5-year bond analysed earlier.

We noted earlier that in the case of the 2.5-year bond, $f(x) = 0$ was a polynomial of degree five. The exact format was

$$f(x) = 98x^5 - 5x^4 - 5x^3 - 5x^2 - 5x - 105 = 0 \tag{8.13}$$

To start the iterative process we guess that $x = 1.055$. Replacing x by 1.055 in expression (8.13) to give the following result

$$f(x) = 98 * 1.055^5 - 5 * 1.055^4 - 5 * 1.055^3 - 5 * 1.055^2 - 51.055 - 105 = 0.176625$$

We must adjust 1.055 by the ratio of 0.176625, and the first derivative of expression (8.13), i.e. 551.2939. Thus x_1, the value of x in the second iteration, is

$$x_1 = 1.055 - \frac{0.176625}{551.2939} = 1.054680$$

Running equation (8.12) again but with 1.054680 as x gives the following

$$f(x) = 98 * 1.054680^5 - 5 * 1.054680^4 - 5 * 1.054680^3 - $$
$$5 * 1.054680^2 - 5 * 1.054680 - 105 = 0.0001125 \tag{8.14}$$

We adjust 1.054680 by the ratio of 0.0001125 and the first derivative of equation (8.14), i.e. 550.5916. So x_2, the value of x in the third iteration, is

$$x_2 = 1.054680 - \frac{0.001125}{550.5916} = 1.054679$$

Running equation (8.12) again, this time with $x = 1.054679$, gives

$$f(x) = 98 * 1.054679^5 - 5 * 1.054679^4 - 5 * 1.054679^3 - $$
$$5 * 1.054679^2 - 5 * 1.054679 - 105 = 0.000017$$

Thus $x = 1.054679$ with the required degree of accuracy of $f(x) = 0$ to four decimal places.

However, recall that we are trying to find the annual IRR of a 2.5-year bond but so far we have found only the IRR in half-year cash flow periods. We therefore need the value of x^2 to give us the required 1 + IRR. So 1 + IRR = 1.054679^2 = 1.112348. This shows that the gross redemption yield of this particular bond is 11.2348%. This is the same result as that achieved by bisection, but here the result was achieved with far fewer iterations.

Numerical methods for integration

It will be recalled from Chapter 3 that the process of integration finds the area under a curve, and that the first stage of this process is to find the primitive of the function to be integrated. We then evaluate that primitive at the end points of the integral in order to find the area. Unfortunately, there are many functions for which no primitive exists – though this does not mean that there is no integral.

You will recall from Chapter 4 that the function

$$\int_1^2 \frac{1}{\sqrt{2\pi}} e^{-\frac{x^2}{2}} dx \tag{8.15}$$

is the probability of a variable with a standard normal distribution taking a value between one and two. The value of this integral is represented by the shaded area in Fig. 8.3.

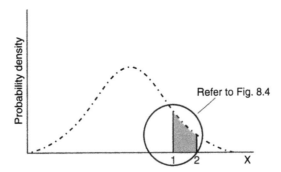

Figure 8.3

Since there is no function $f(x)$ which has the property that

$$f'(x) = \frac{1}{\sqrt{2\pi}} e^{-\frac{x^2}{2}} \tag{8.16}$$

the methods of integral calculus learned in Chapter 3 cannot be used. Consequently a numerical method is required. Two methods will be demonstrated here. Both entail dividing the area under the curve into small vertical strips, each with a height equal to the distance between a particular point on the horizontal axis and the curve. The area of each strip is calculated and the sum of the calculated areas is approximately the area under the curve.

The trapezium rule

The first method that we shall demonstrate uses the trapezium rule. In this approach the area under the curve which is to be evaluated is split into a number of vertical strips of equal width, as shown in Fig. 8.4 below. If the tops of adjacent lines are connected by another straight line as indicated by a, the enclosed strip forms a trapezium which approximates the true area under the curve between the vertical lines.

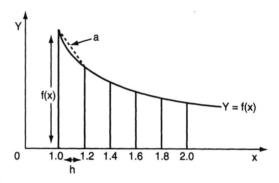

Figure 8.4

In this case, because the curve is convex along the relevant range, it can be seen that the area of each trapezium is slightly greater that the true area under the curve between the respective lines. Thus the sum of the areas of the trapezia will overestimate the required area. If the curve were concave, the area of the trapezia would understate the true area under the curve. This type of error can be reduced by introducing more intervals. We will demonstrate this later.

Each vertical line is at a point value of x where the function $f(x)$ is evaluated. The height of the line is given by the value of the function $f(x)$. The area of the trapezium is then found by evaluating each pair of $f(x)$, for example $f(x = 1.0)$ and $f(x = 1.2)$ in Fig. 8.4, taking the mean of the two function values (heights) and multiplying by the width of the trapezium, 0.2 in this example. The area under the curve is then approximated by the sums of the areas of trapezia, as illustrated.

To demonstrate this approach, we will find the area under the curve between points 1 and 2. We have chosen to divide the space between point 1 and point 2 into five intervals as in Fig. 8.4.

The area of each trapezium is given by its mean height times its width, so the total area is

$$\left[\frac{f(1.0)+f(1.2)}{2}*0.2\right]+\left[\frac{f(1.2)+f(1.4)}{2}*0.2\right]+\left[\frac{f(1.4)+f(1.6)}{2}*0.2\right]$$

$$+\left[\frac{f(1.6)+f(1.8)}{2}*0.2\right]+\left[\frac{f(1.8)+f(2.0)}{2}*0.2\right]$$

A more compact way of expressing this function is

$$\frac{0.2}{2}[f(1.0)+f(2.0)+2(f(1.2)+(1.4)+f(1.6)+f(1.8))]$$

This generalizes to the formula

$$h/2(\text{``ends''} + 2 * \text{``middles''}) \tag{8.17}$$

where h is the width of each trapezium, "ends" means the **sum** of the function values at the extreme of the integral, and "middles" means the **sum** of the function values at the intermediate values of x.

Thus to integrate

$$\int_1^2 \frac{1}{\sqrt{2\pi}} e^{-x^2/2}$$

between one and two, we first evaluate the equation for the six values of x that encompass the five spaces as set out below in order to derive the appropriate values of $F(x)$.

x		1	1.2	1.4	1.6	1.8	2
$f(x) = \frac{1}{\sqrt{2\pi}} e^{-x^2/2}$		0.2420	0.1942	0.1497	0.1109	0.0790	0.0540

$h = 0.2$
"ends" = 0.2420 + 0.0540 = 0.2960
"middles" = 0.1942 + 0.1497 + 0.1109 + 0.0790 = 0.5338

So

$$\int_1^2 \frac{1}{\sqrt{2\pi}} e^{-x^2/2} \approx \frac{0.2}{2}(0.2960 + 2*0.5338) = 0.1364$$

We noted earlier that the accuracy of this method can be improved by increasing the number of strips. So now let's run this example again, but this time we will use 10 intervals.

x		1	1.1	1.5	1.2	1.3	1.4	1.5
$f(x) = \frac{1}{\sqrt{2\pi}} e^{-x^2/2}$		0.2420	0.2179	0.1295	0.1942	0.1714	0.1497	0.1295

1.6	1.7	1.8	1.9	2
0.1109	0.0940	0.0790	0.0656	0.0540

$h = 0.1$
"ends" = 0.2960 (as before)
"middles" = 0.2179 + 0.1942 + 0.1714 + 0.1497 + 0.1295 + 0.1109 + 0.0940 + 0.0790 + 0.0656 = 1.2122

So

$$\int_1^2 \frac{1}{\sqrt{2\pi}} e^{-x^2/2} \approx \frac{0.1}{2}(0.2960 + 2*1.2122) = 0.1360$$

Simpson's rule

Simpson's rule is an improvement upon the trapezium rule. It divides the area under the curve into an **even** number of intervals of equal width. The function $f(x)$ is evaluated

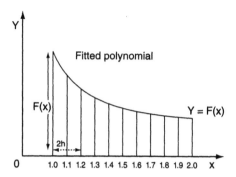

Figure 8.5

at the two ends and the middle x values within each pair of intervals, giving three points on the curve, as illustrated in Fig. 8.5.

A unique quadratic curve can be fitted to these three points and the area beneath that curve can be evaluated. Under most circumstances this will give a closer approximation to the true area within these two intervals than that obtained by evaluating and summing the areas of two trapezia.

We need not concern ourselves with the details of fitting the quadratic and finding the area underneath it – it is tedious, but the result is easy. The approximating area turns out to be the total width of the two intervals times a weighted mean of the heights (i.e. function values), with weights of 1, 4 and 1.

Thus, the first two intervals are defined by x values of 1, 1.1 and 1.2. The approximating area is

$$0.2 * \frac{f(1.0) + 4f(1.1) + f(1.2)}{6}$$

Thus the approximation to the whole area is

$$\left[0.2 * \frac{f(1.0) + 4f(1.1) + f(1.2)}{6} \right] + \left[0.2 * \frac{f(1.2) + 4f(1.3) + f(1.4)}{6} \right] +$$

$$\left[0.2 * \frac{f(1.4) + 4f(1.5) + f(1.6)}{6} \right] + \ldots + \left[0.2 * \frac{f(1.8) + 4f(1.9) + f(2.0)}{6} \right]$$

$$(8.18)$$

This equals

$$\frac{0.2}{6} * [f(1.0) + f(2.0) + 4(f(1.1) + f(1.3) + \ldots + f(1.9)) + 2(f(1.2) + f(1.4) + \ldots + f(1.8))]$$

As the 0.2 represents $2h$, this may be written as $h/3 * [\text{"ends"} + 4x\text{"evens"} + 2x\text{"odds"}]$, the "evens" and "odds" being so labelled since $x = 1.1, 1.3, 1.5$, and so on, are the second, fourth and sixth, etc., points at which the function is evaluated, whilst $x = 1.2, 1.4, 1.6$, and so on, are the third, fifth, seventh, etc. points at which it is evaluated.

The table of function values from the previous example can be used to give

$$h = 0.1$$
$$\text{"ends"} = 0.2960$$
$$\text{"evens"} = 0.6784$$
$$\text{"odds"} = 0.5338$$

So

$$\int_1^2 \frac{1}{\sqrt{2\pi}}e^{-x^2/2}dx \approx \frac{h}{3}(\text{"ends"} + 4\text{"evens"} + 2\text{"odds"})$$

(8.19)

$$= \frac{0.1}{3}[(0.2960) + (4*0.6784) + (2*0.5338)] = 0.1359$$

Accurate tables of the standard normal distribution give the probability of a result between one and two to be 0.1359. Thus we see that Simpson's rule gives a more accurate answer than the trapezium rule. The computation effort in each case is the same (it is the number of function evaluations which is potentially expensive in terms of computational time).

Polynomial approximations to the cumulative normal curve

Although it is not possible to find an analytic integral for the standard normal density function, it is possible to approximate, numerically, the area under the curve. We could use the trapezium rule or Simpson's rule, and both of these techniques come under the heading of numerical integration. An alternative approximation is to fit a polynomial to the cumulative normal curve.

A description of such an approach, is as follows:

1. Firstly, consider the area under the standardized normal curve as shown in Fig. 8.6(a). The associated cumulative function, or ogive, is shown in Fig. 8.6(b).
   Each cumulative extent of area in Fig. 8.6(a) is plotted as a height in Fig. 8.6(b). Figure 8.6(b) is in effect the "area so far" curve.
2. Next, fit a polynomial approximation to the top half of the ogive.
3. To fit the polynomial to the ogive, apply the following function

$$N(z) = 1 - \frac{(0.319382x - 0.356564x^2 + 1.78148x^3 - 1.82126x^4 + 1.33027x^5)e^{-\frac{1}{2}z^2}}{\sqrt{2\pi}}$$

(8.20)

where $N(z)$ is the cumulative standard normal density function evaluated at z, and x is derived from the following equation

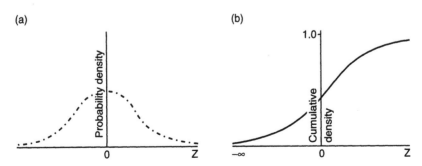

(a)   (b)

Figure 8.6

$$x = \frac{1}{(1 + 0.231642z)} \tag{8.21}$$

where z is the standardized normal variable being evaluated.

This method assumes that z is positive. It only evaluates the top half of the curve and, therefore, only relates to cumulative probabilities in the right-hand half of the normal density function.

If z is negative, we have to invoke the symmetric property of the normal density function. For example, if we want to know the cumulative probability of $z = -2$, we calculate for $z = 2$ and subtract the result from 1.0.

To illustrate the application, we will first calculate the area under the curve where $z = 2$, and then when $z = -0.8$

$z = 2.0$

$$x = \frac{1}{(1 + (0.231642 * 2))} = 0.683394$$

$$1 - \frac{\left[\begin{array}{l} (0.319382 * 0.683394) - (0.356564 * 0.683394^2) + (1.78148 \times 0.683394^3) \\ -(1.82126 * 0.683394^4) + (1.33027 * 0.683394^5) \end{array} \right] e^{-0.5 * 2^2}}{\sqrt{2\pi}} = 0.97725$$

$z = -0.8$

We first put $z = +0.8$ and calculate x, thus

$$x = \frac{1}{(1 + (0.231642 * 0.8))} = 0.8436585$$

$$1 - \frac{\left[\begin{array}{l} (0.319382 * 0.8436585) - (0.356564 * 0.8436585^2) + (1.78148 * 0.8436585^3) \\ -(1.82126 * 0.8436585^4) + (1.33027 * 0.8436585^5) \end{array} \right] e^{-0.5 * 0.8^2}}{\sqrt{2\pi}}$$

$$= 0.788328$$

as we are calculating the area under the curve for $z = -0.8$, we subtract the value 0.788328 from 1.0 to get 0.211672.

If we wished to find out the area between, say, $z = 2$ and $z = 1.5$, we solve equation (8.20) for $z = 2$, and then again for $z = 1.5$, and subtract the second value from the first.

Numerical methods to solve stochastic problems

Models of the behaviour of a variable are either **deterministic** or **stochastic**. When the models assume a fixed (known and constant) relationship between the variables, the models are said to be deterministic. For example, a model to determine the salary of a derivatives trader that was based on a predetermined proportion of net profit of the trading department would be deterministic, once profits are known.

When the models assume some uncertain component such as a random variable, the models are said to be stochastic. For example, a model that tries to forecast the future level of a stock market or the future price of a financial asset, and introduces chance or uncertainty, is known as a stochastic model. The stochastic model will not produce an unambiguous outcome; it produces a probability distribution of possible answers. The model of the trader's salary would also be stochastic if it were being used to forecast the salary using past levels of profits as an indicator of future profits.

We noted in Chapter 4 that those variables that are assumed to exhibit uncertainty are known as random variables. When we construct deterministic models, we assign a single value to a variable. In stochastic models we assign probability distributions to those variables.

The numerical techniques dealt with so far in this chapter have assumed deterministic processes, that is processes in which chance, or randomness, plays no part. In this section we will look at three widely used numerical methods for dealing with stochastic problems. They are:

- the binomial process
- the trinomial process
- Monte Carlo simulation.

One of the major applications of all three methods is in finding the value of options for which there is no formula, or at best one which is difficult to apply. We will therefore illustrate the use of these methods by way of application to option pricing. Accordingly we will digress at this point to introduce the reader to the basics of option pricing.

A fourth group of numerical methods, the finite difference methods, is applied to solving continuous-time partial-differential equations. As these equations are analysed in Chapter 10, the discussion of finite difference methods is deferred until that chapter.

Basics of option pricing

A call option gives the buyer "the right but not the obligation to purchase an asset at a previously agreed price, on or before a particular date" and for this right with no obligation the option price or premium is paid. A put option gives "the right but not the obligation to sell an asset at a previously agreed price, on or before a particular date".

From this we note that the option has a fixed maximum life, it is exercisable at a fixed price known as the **exercise price** or **strike price,** and that it can be abandoned by the purchaser without penalty. The current value of an option depends on the probability distribution of the underlying asset at the time the option expires. In particular the value of a call option depends on the probability of the asset price being above the exercise price of the option when it expires whilst the value of a put option depends upon the probability of the asset price being below the exercise price at expiry.

At expiry nobody would pay more for an option to buy a particular asset than the difference between what they would have to pay to buy the asset in the market and what they will have to pay to buy under the option, i.e. exercise price. Thus the upper limit to the value of a call is the asset price itself minus the exercise price.

In the case of a put at expiry nobody would pay more for an option to sell an asset than the difference between the price at which they can sell that asset in the open market and the price at which they can sell it under the option. Thus the upper limit to the value of a put is the exercise price minus the current value of the asset.

As the option can be abandoned by the holder without further liability, the price of the option cannot be negative. Thus the lower limit for both calls and puts is zero.

Both of these boundary conditions may be expressed formally as:

$$C = \max [0, S - X]$$
$$P = \max [0, X - S]$$

(8.22)

Equation (8.22) gives the value of the option at expiry.

A number of models have been developed for pricing options during their lifetime (i.e. before expiry). The most widely noted model is that developed by Black and Scholes (1973) which uses continuous-time stochastic calculus to find the values and is thus deferred to Chapter 10. The most popular discrete time model is the binomial model developed by Cox *et al.* (1976) and Rendelman and Barter (1979). More recently there has been a growing interest in trinomial models. This section will discuss the binomial and trinomial models.

Binomial models

These models assume that the underlying random variable (the price of the security underlying the option) has a binomial distribution, as discussed in Chapter 4. The application of binomial models to option pricing entails modelling the price of the underlying asset as a binomial process, be it a security price, an exchange rate or an interest rate, in order to determine the distribution of that variable when the option expires. Then, using the boundary conditions noted above, the future value of the option is determined and discounted back to the present to get the current option price.

The binomial model assumes the possibility of **creating a risk-free portfolio by hedging a long position in the underlying asset with a short position in a number of fairly priced call options on that asset**. Consequently only the risk-free rate of interest need be used for discounting. The rationale behind this thinking is that if the portfolio is perfectly hedged, it will be risk free and thus should only earn the risk-free rate of interest.

Two approaches are demonstrated in this section. The first, and by far the most popular, is the construction of a binomial lattice or recombining tree. The second approach is to use binomial algebra to derive the sum of the present values of each of the probabilistically weighted future values of the option. Both of these approaches were introduced in Chapter 4. We will begin with the binomial lattice.

The binomial lattice approach

The cost of establishing the risk-free portfolio is the cost of buying the underlying asset minus the premiums from the written (sold) options. The following example illustrates how the risk-free portfolio is created.

Assume that the price of the underlying asset $(S) = 35$, the exercise price of the option $(X) = 35$, the risk-free rate of interest $(r) = 10\%$ or 0.1 in decimals, and $R = 1 + r$ or 1.1. The time to expiry of the option is one year. In addition assume that at the end of the one-year period the asset price will have either risen by 25% from 35 to 43.75 or fallen 25% from 35 to 26.25. This can be illustrated graphically as shown in Fig. 8.7, where u

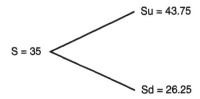

Figure 8.7

is the multiplicative upward movement in the asset price (1 + the percentage rise in the asset price), 1.25 in this example, and d is the multiplicative downward movement in the asset price (1 + the percentage fall in the asset price), 0.75 in this example. It is also a requirement that $d < (1 + r) < u$. The reason is that if d and u are less than the risk-free rate, the risk-free asset would always show higher returns than the risky asset, which of course, is contrary to financial theory. If d and u are greater than the risk-free rate, the risky asset would always show higher returns than the risk-free one.

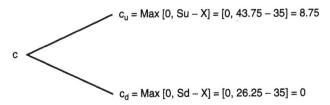

Figure 8.8

It is important that the percentage up and down movements are expressed in multiplicative form, i.e. 1.25 * 35 or 0.75 * 35. If they were expressed in additive form, i.e. plus or minus 25%, negative asset prices could result.

From the above **one-period binomial tree** of the asset price, and the boundary conditions for the price of an option, we develop a similar looking tree that relates to the value of an option, in this case a call.

Figure 8.8 clearly shows that if the asset rises to 43.75 at the end of the one-year period the option must be worth 8.75 – we will label this value as c_u. If the asset falls to 26.25 the option will be worth nothing – we will label this value c_d.

To create a fully-hedged portfolio, one unit of the underlying asset is purchased and H call options are sold against it. Being fully hedged, the portfolio value will be the same whether the price of the asset rises or falls. In terms of a binomial tree of this portfolio, $Su - Hc_u = 26.25 = Sd - Hc_d = 26.25$ (see Fig. 8.9).

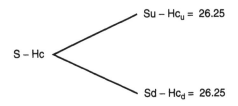

Figure 8.9

At this point we have two problems to solve. The **first problem is to find out how many options to sell that makes the portfolio riskless,** i.e. what is the magnitude of H? The **second problem is to determine the fair price at which these options should be sold.**

Problem 1: the number of options to sell
To determine the number of options to be sold so as to create the riskless portfolio, note that from the binomial trees presented above we can see that range of asset prices is given by $S(u - d)$ or $35(1.25 - 0.75) = 17.5$. The range of the option values is given $(c_u - c_d)$ or $8.75 - 0 = 8.75$. From this information we are able to calculate the number of options that have to be written over each unit of the underlying asset in order to achieve a perfectly hedged portfolio.

The hedge ratio is calculated as follows:

$$H = \frac{S(u-d)}{(c_u - c_d)} \tag{8.23}$$

or in numerical form

$$H = \frac{35(1.25 - 0.75)}{(8.75 - 0)} = \frac{17.5}{8.75} = 2$$

This hedge ratio can be explained intuitively by noting that $S(u - d)$ was the range of the potential prices of the asset, and $(c_u - c_d)$ was the range of the potential prices of the option. Thus as the asset prices have a range twice that of the option prices, a short position in two options is required to achieve the profit that fully offsets the loss on one unit of the long position in the asset.

Thus in the above example a riskless hedge consists of buying one unit of the underlying asset and writing two options, each with an exercise price of 35. The two outcomes of this strategy at the end of the time period will be as follows:

Asset rises to 43.75: $Su - Hc_u = 1.25(35) - 2(8.75) = 26.25$

Asset falls to 26.25: $Sd - Hc_d = 0.75(35) - 2(0) = 26.25$

Problem 2: the fair price
To solve our second problem of determining the fair price at which to sell the options today, note that as the strategy is riskless, it should only earn the risk-free rate of return. The current value of a portfolio, which is long one unit of S and short two options, earning the risk-free rate for one year must equal the present value of the end of year pay-off. That end of year pay-off is 26.25. Therefore, the present value must be

$$26.25/1.1 = 23.86$$

As the current value of S is 35 the value of the two options sold short must be $35 - 23.86 = 11.14$. Therefore a single option must be fairly priced at $11.14/2 = 5.57$.

This process may be expressed in a more general form. Firstly note that $R = 1 + r$, where r is the risk-free rate of interest. Then it follows that

$$R(S - Hc) = (Su - Hc_u) \tag{8.24}$$

This can be rearranged as follows:

$$RS - RHc = Su - Hc_u \tag{8.25}$$

Collecting RHc on the left-hand side gives

$$-RHc = -RS + Su - Hc_u \tag{8.26}$$

Multiplying both sides by -1 gives

$$RHc = RS - Su + Hc_u$$
$$\text{i.e. } RHc = S(R - u) + Hc_u \tag{8.27}$$

Dividing through by $RH = HR$ gives

$$c = \frac{S(R-u) + Hc_u}{HR} \tag{8.28}$$

Remembering that

$$H = \frac{S(u-d)}{(c_u - c_d)} \tag{8.29}$$

and substituting into equation (8.28) and rearranging, this becomes

$$c = \left[c_u \frac{(R-d)}{(u-d)} + c_d \frac{(u-R)}{(u-d)} \right] / R \tag{8.30}$$

which is the equation for a call option with one period to expiry.

Putting numbers to this example and remembering that $c_u = 8.75$, $c_d = 0$, $u = 1.25$, $d = 0.75$ and $R = 1.1$

$$c = \left[8.75 \frac{(1.1 - 0.75)}{(1.25 - 0.75)} + 0 \frac{(1.25 - 1.1)}{(1.25 - 0.75)} \right] / 1.1$$

$$c = \left[8.75 \left(\frac{0.35}{0.5} \right) + 0 \right] / 1.1 = 5.57$$

We can simplify this procedure by making $p = (R - d)/(u - d)$ and $(1 - p) = (u - R)/(u - d)$. Then equation (8.30) becomes

$$c = \frac{[pc_u + (1-p)c_d]}{R} \tag{8.31}$$

$$c = [(8.75 * 0.7) + (0.3 * 0)] / 1.1 = 5.57$$

Thus the fair value of the call will be 5.57. We can test whether indeed that is the fair value for this option because this portfolio should earn the risk-free rate of interest which is 10%.

The portfolio cost is $(35 - (2 * 5.57))$ or 23.86 to set up. That sum invested at the risk-free rate for one year will accrue to: $23.86 * 1.1 = 26.25$, which is exactly the value of the hedged portfolio one year hence.

The pricing of the options on the basis of creating a risk-free hedge of the underlying asset obviates the need for the price of an option to be dependent on investors' expectations of the future price of the underlying asset. All that is required is that the effectiveness of the hedge can be maintained so that the portfolio remains risk free.

The multi-period binomial model

The example illustrated above assumed that the time between now and the expiry of the option was divided into just one period, in our example one year. However, the binomial approach can be generalized so that the life of the option can be divided into any number of time periods, i.e. binomial trials. The more trials over a given period, i.e. the smaller the time interval represented by each trial, the more accurate the option valuation will be.

Indeed as the time interval between trials becomes infinitesimally small, so that trading is in effect continuous, the binomial model converges to the Black–Scholes model which we look at in Chapter 10.

Irrespective of the number of binomial trials, the same principles are used to solve the value of the option at each node of the tree, working back from the expiry of the option to the present time period and, therefore, the current price of the option. This will be illustrated in the following example, which assumes an asset price (S) of 35, an exercise price (X) of 35, the annual risk-free rate (r) is 10%, an annual volatility (σ) of 20% and the one-year time period is divided into four quarterly sub-periods or binomial trials.

Before we increase the number of binomial trials, the annual risk-free interest rate must be adjusted to reflect the shorter time period between trials. For example, in this four quarterly model, the quarterly compounded equivalent of the annual rate would be used. The quarterly compounded equivalent of the annual rate, r, is $(1 + r)^{0.25} - 1$. In our example $(1.1)^{0.25} - 1 = 0.024$. Therefore R, which is $1 + r$, in our quarterly model is 1.024.

In some references, R is represented as $e^{r*(T-t)/n}$ where r is the continuously compounded equivalent of the risk-free rate of interest, e.g. in our example the continuously compounded equivalent of 10% p.a. is 9.53% and $R = e^{0.0953 * 0.25} = 1.024$.

In addition, the magnitudes of the potential upward and downward movements, u and d, which refer to the volatility of the underlying asset, must be determined from market information and be adjusted to reflect the number of binomial trials. Cox et al. (1979) have shown that the us and the ds are related to the standard deviation as follows:

$$u = e^{\sigma\sqrt{(T-t)/n}}$$

$$d = e^{-\sigma\sqrt{(T-t)/n}}$$

(8.32)

where $(T - t)$ is the life of the option in years (or fractions thereof) and n is the number of binomial trials to be calculated. In our example, the option has a life of one year and we are assuming four quarterly binomial trials, therefore $(T - t)/n = 0.25$.

Usually, it is also required that $u = 1/d$. This ensures that an up followed by a down movement gives the same result as a down followed by an up movement. The result is a recombining tree, more correctly known as a lattice.

Thus the multiplicative upward and downward movements are derived from the volatility variable, the standard deviation of the natural log of the asset price relative, i.e. the standard deviation of the continuously compounded return.

Returning now to our example, the values of u, d, R, p, and $1 - p$ are

$$u = e^{\sigma\sqrt{(T-t)/n}} = e^{0.2\sqrt{0.25}} = 1.10517$$

$$d = e^{-\sigma\sqrt{(T-t)/n}} = e^{-0.2\sqrt{0.25}} = 0.904837$$

$$R = (1+r)^{0.25} = 1.1^{0.25} = 1.024$$

$$(8.33)$$

$$p = \frac{(R-d)}{(u-d)} = 0.59539$$

$$(1-p) = \frac{(u-R)}{(u-d)} = 0.40461$$

Now, observe the binomial tree in Fig. 8.10. The life of the option has been divided into four quarterly time periods. We can calculate the various values of the underlying security after four periods, i.e. at the expiry of the option. The various values of the asset depend upon which path the asset price took. For example if the asset price rose in all four periods the appropriate node would be Su^4, with a value of 52.21. If on the other hand the asset price rose in two time periods and fell in two, the appropriate node would be Su^2d^2, with a value of 35.

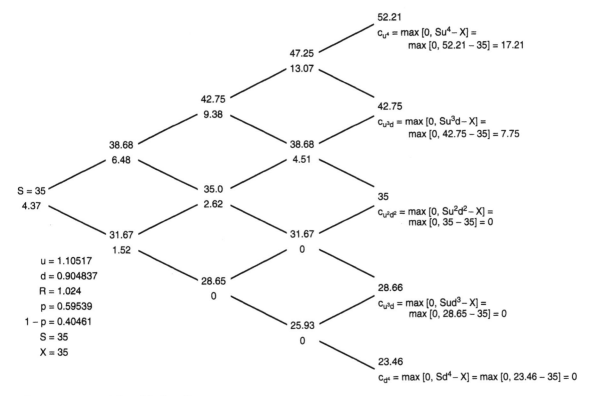

Figure 8.10 Binomial model of a call.

This tree also gives us the various potential values of the option at the end of the fourth period, i.e. at expiry. Being a call, the value at expiry must be max $[0, Su^4 - X]$, if, for example, the asset rises in price in each time period. In order to calculate the current value of the option, we start with these various boundary values and calculate the option's value at the end of the third period using the equation (8.31).

Next the potential values of the option at the end of the second period are derived from the values calculated at the end of the third period. Then the potential values of the option are calculated for the end of the first period from the values for the end of the second period. Finally, we calculate the current value of the option from the values calculated for the first period. Each of these stages will now be explained in detail with reference to the binomial tree in Fig. 8.10 above.

Stage 1: use expiration values to calculate values of third time period

The binomial tree above indicates that the possible values of the option at expiry are 17.21, 7.75, 0, 0 and 0. With this information we can calculate the option values **at the end of the third time period**.

First, recall that u, d and R have been changed to account for the quarterly nature of this model: $u = 1.10517$, $d = 0.904837$ and $R = 1.024$. As a consequence p becomes $(1.024 - 0.904837)/(1.10517 - 0.904837) = 0.59539$. Therefore, $1 - p$ becomes $(1.10517 - 1.024)/(1.10517 - 0.904837) = 0.40461$

$$
\begin{aligned}
c_{u^3} &= [pc_{u^4} + (1 - p)c_{u^3d}]/R \\
&= [0.59539 * 17.21 + 0.40461 * 7.75]/1.024 \\
&= [10.25 + 3.14]/1.024 = 13.07
\end{aligned}
$$

In the same manner we can calculate the option value for c_{u^2d} as

$$c_{u^2d} = [0.59539 * 7.75 + 0.40461 * 0]/1.024 = 4.51$$

We do not have to calculate any values for other nodes because the relevant values for the fourth time period were zero.

Stage 2: use third period values to calculate second period values

With the knowledge that $c_{u^3} = 13.07$ and $c_{u^2d} = 4.51$ we can calculate the values of the relevant nodes **at the end of the second time period** as follows:

$$
\begin{aligned}
c_{u^2} &= [pc_{u^3} + (1 - p)c_{u^2d}]/1.024 \\
&= [0.59539 * 13.07 + 0.40461 * 4.51]/1.024 = 9.38
\end{aligned}
$$

$$c_{ud} = [pc_{u^2d} + (1 - p)c_{ud^2}]/1.024$$

Remembering that c_{ud^2} is zero.

$$c_{ud} = [0.59539 * 4.51 + 0]/1.024 = 2.62$$

Stage 3: use second period values to calculate first period values

We can now calculate for cu and cd as follows:

$$c_u = [0.59539 * 9.38 + 0.40461 * 2.62]/1.024 = 6.48$$
$$c_d = [0.59539 * 2.62 + 0]/1.024 = 1.52$$

Stage 4: use first period values to calculate current option price

Finally we can calculate the current value of the option as:

$$c = [0.59539 * 6.48 + 0.40461 * 1.52]/1.024 = 4.37$$

In Chapter 4 we showed that the expected future value of a variable is the sum of various possible future values each multiplied by their corresponding probabilities. Thus we can apply the binomial probability equation from Chapter 4 to derive the value of a call by calculating the expected value of the asset in excess of the exercise price of the option at expiry of the option and discount that value back to the present. To value a put we would calculate the expected value of the asset below the exercise price of the option. Below is the general binomial formula for option valuation.

For a call:

$$C = \sum_{j=0}^{n} \left(\frac{n!}{(n-j)!\,j!} p^j (1-p)^{n-j} \text{MAX}[0, u^j d^{n-j} S - X] \right) / (1+r)^n \tag{8.34}$$

For a put:

$$P = \sum_{j=0}^{n} \left(\frac{n!}{(n-j)!\,j!} p^j (1-p)^{n-j} \text{MAX}[0, X - u^j d^{n-j} S] \right) / (1+r)^n \tag{8.35}$$

where p and $1 - p$ are as defined earlier, n is the total number of binomial trials, j is the number of up movements that occurred in reaching a given outcome and $n - j$ is the number of downward movements.

To illustrate this, recall that in the four-stage binomial tree given in Fig. 8.10 above, there are five outcomes. Su^4 is reached after four successes and the option would be worth 17.21 at expiry. Su^3d is reached after three successes and one failure. The option would have a value at expiry of 7.75. Su^2d^2 is reached after two successes, Sud^3 is reached after one success and Sd^4 is reached after no successes. In each of the last three situations the option value at expiry will be zero.

Thus there are three stages to determine the value of a European option under this process:

1. Derive the binomial probability of each of the outcomes at expiry.
2. Multiply the discounted value of the option at each of these outcomes by their corresponding probabilities.
3. Sum the discounted product of (1) and (2) above to get the value of the option.

Using the same information as in the four-stage binomial process above, the value of the call is given by the sum of the value of the following five equations.

Outcome after four upward movements, u^4

$$\frac{\dfrac{4!}{4!(4-4)}0.59539^4(0.40461)^{4-4}[17.21]}{(1.024)^4} = 1.967$$

Outcome after three upward movements and one downward movement, u^3d

$$\frac{\dfrac{4!}{3!(4-3)}0.59539^3(0.40461)^{4-3}[7.75]}{(1.024)^4} = 2.408$$

Outcome after two upward movements and two downward movements, u^2d^2

$$\frac{\dfrac{4!}{2!(4-2)}0.59539^2(0.40461)^{4-2}[0]}{(1.024)^4} = 0$$

Outcome after one upward movement and three downward movements, ud^3

$$\frac{\dfrac{4!}{1!(4-1)}0.59539^1(0.40461)^{4-1}[0]}{(1.024)^4} = 0$$

Outcome after four downward movements, d^4

$$\frac{\dfrac{4!}{0!(4-0)}0.59539^0(0.40461)^{4-0}[0]}{(1.024)^4} = 0$$

Summing the results gives a value of 4.37 to two decimal places, the same as that produced by the earlier method.

These examples have assumed that the life of the option is divided into only four discrete time periods. In reality, the time to expiry can be divided into infinitesimally small time periods. The greater the number of periods the greater the accuracy of the calculation. To illustrate this if the one-year period was divided into 6, 10, 20, 40 and 100 trials, the call prices would be 4.5256, 4.5723, 4.6081, 4.6262 and 4.6371. These values compare with the continuous time Black–Scholes model price of 4.6446. Thus we have an error in the 100 step binomial model of only 0.0075 units of option premium.

Thus we have an interesting point that as the time interval of each binomial trial gets smaller, i.e. as n gets larger, we move towards the continuous time stochastic process of asset returns to which the Black–Scholes model addresses itself. We will analyse the Black–Scholes model in Chapter 10, which covers continuous time mathematics in finance.

The trinomial equivalent of the binomial option pricing model

The binomial option pricing model has the advantages of being reasonably intuitive and very flexible regarding the options to which it can be applied. However, it has the considerable drawback of being much slower than models that have closed form solutions. In this section a trinomial equivalent to the binomial model is introduced. This model from Wilmot *et al.* (1993) has the same flexibility as the binomial model but is faster in calculating the option premium.

Under the trinomial process the underlying asset price can take one of three possible values at the end of each trinomial trial. It may rise by a multiplicative amount u to Su, remain at the current value Sq or fall to Sd, where $d = 1/u$.

In order to accommodate the assumptions that the options are valued in a risk-neutral environment, and that the asset follows a lognormal distribution, we must make some adjustments to the values of u and d (q is one by definition)

$$u = e^{2\sigma\sqrt{\delta t/2}}$$

$$d = e^{-2\sigma\sqrt{\delta t/2}} \tag{8.36}$$

$$q = 1$$

The one-period trinomial option pricing model is given as

$$C = \frac{p_u c_u + p_q c_q + p_d c_d}{R} \tag{8.37}$$

Like its binomial cousin this model can be applied for puts or calls given the different boundary conditions, i.e.

$$C = \max\,[S - X, 0]$$

$$P = \max\,[X - S, 0]$$

However, before we can apply this model we must calculate the values of p_u, p_q and p_d. The starting point is the value of p which is given as

$$p = \frac{e^{r\delta t/2} - e^{-\sigma\sqrt{\delta t/2}}}{e^{\sigma\sqrt{\delta t/2}} - e^{-\sigma\sqrt{\delta t/2}}} \tag{8.38}$$

We can then compute $p_u = p^2$, $p_q = 2p(1 - p)$ and $p_d = (1 - p)^2$.

To show the time-saving advantage of this model we will develop a two-trial tree and show that it has the same accuracy as the four-trial binomial tree.

Figure 8.11 shows the two-period tree of the underlying asset and the appropriate call value when $S = 35$, $X = 35$, $r = 0.09525$, vol $= 0.2$, $(T - t) = 1.0$ and $R = 1.048777$.

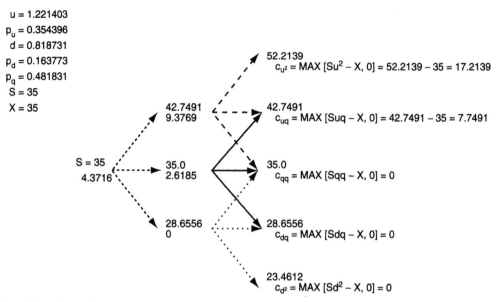

Figure 8.11 Trinomial option pricing model.

From the parameter values given in the figure and applying equation (8.37) terminal values of the underlying asset and the values of the option at the three nodes at time $t + 1$, c_u, c_q and c_d, are found in a manner analogous to the binomial approach. That is the value of the call at c_u is

$$c_u = \frac{0.354396 * 17.2139 + 0.481831 * 7.7491 + 0.163773 * 0}{1.048777} = 9.3769$$

The value of the call at c_q is

$$c_q = \frac{0.354396 * 7.7491 + 0.481831 * 0 + 0.163773 * 0}{1.048777} = 2.6815$$

The value at c_d is clearly zero.

Finally, applying equation (8.37) to the values of the options at nodes c_u, c_q and c_d will give the value at the current time period. That value is 4.3716. Thus we arrive after only two trinomial trials at the same price as the binomial model obtained after four trials. This economy of effort can be generalized to say that the trinomial model given above can deliver option values with the same degree of accuracy as the binomial model but in only half the number of trials.

Monte Carlo simulation

Monte Carlo simulation is the numerical method which enables us to model the future value of a variable by simulating its behaviour over time. Whilst much elegant mathematics has been developed to deal with stochastic processes of variables, it is the case that simple problems can lead to difficult mathematics, and that problems arise which are intractable under the analytic approach. The increased computing power available to today's analysts allows Monte Carlo simulation to be applied to solve many financial problems.

The probability function of a discrete random variable (or probability density function of a continuous random variable) provides information about the probability of the variable taking a particular value (or, in the continuous case, of it lying in a particular range). Even if the event which the random variable is modelling occurs only once, there is the perception that if it **were** to be repeated many times the random variable would take output values in proportion to those probabilities.

In simulating the process we simply imitate the behaviour of the underlying (input) random variables and follow through the consequences in the output variables in which we are interested. But the essential element is **repetition**. By repeating the process many, many times we produce a distribution for the output variable(s) from which we can construct a probability distribution. It is in allowing for this essential repetition that computing power is needed.

The five stages of a Monte Carlo simulation

The Monte Carlo simulation process can be divided into five stages as follows.

1. **Identify the stochastic nature** of the input variables. The identification will facilitate the choice of probability distribution which is to be used in the simulation.
2. **Imitate the movement of the input variables** by repeatedly drawing random numbers, which are adjusted to have the same probability distribution as the underlying variables. This entails converting random numbers from the uniform distribution, generated via a computer, into random variables with the same distributions as the variables to be imitated. These adjusted random variables become input variables.
3. **Simulate the underlying variable** by combining the input variables together according to the logic of the system. The logic of the system describes the way in which the input variables interact with each other to produce output variable(s). Thus from the repeated drawings of a random number, we generate future values of the underlying variable.
4. **Repeat this process** many (several thousand?) times. Take the mean of all the outcomes. This mean is the future (expected) value of the variable being simulated. Discount this future value by the appropriate discount rate to determine the present value of the variable being simulated.
5. **Apply the control variate technique** or other variance reduction process to increase the accuracy of the simulated results.

Stage 1: identify the probability distribution

The first stage of any Monte Carlo simulation is to identify the probability distribution(s) of the input variable(s). Most computer packages for Monte Carlo simulations will have a library of probability distributions built in. In addition they will provide a facility for constructing a probability distribution based on the researcher's own judgement. This is possible with modern computers because they are equipped with random number generators (actually pseudo random number generators). These provide the means to produce uniformly distributed numbers between zero and one. That is numbers between zero and one in which, say, a result between 0.1 and 0.2 has the same probability as one between 0.7 and 0.8, or any number between 0.3 and 0.5 will have the same probability as a number between 0.8 and 1.0.

Once we have generated our uniformly distributed random numbers, we must embark on the process of "converting" that uniform random variable to a random variable with the probability distribution of our choice.

The important consideration is the mechanism by which the uniformly distributed random variable is transformed into a random variable with a probability distribution which matches the empirical distribution. To do this, we need the **relative frequency distribution** of our random variable.

Take for example the relative frequency distribution of the returns to an asset depicted in Fig. 8.12. Recall that in the case of a relative frequency distribution, the height of each bar gives the percentage of total observations accounted for by the current value. In addition the sum of the percentages of each type of observation must sum to 100, or one if scaled appropriately.

This relative frequency distribution actually relates to the daily returns of the FTSE 100 index from 1984 to 1992. From this relative frequency distribution we construct a cumulative relative frequency distribution. This is given in Fig. 8.13.

Note that in the case of this cumulative frequency distribution the vertical axis has been scaled to sum to unity and thus represents the **empirical probability** associated with each corresponding value on the horizontal axis. The horizontal axis represents the level of returns.

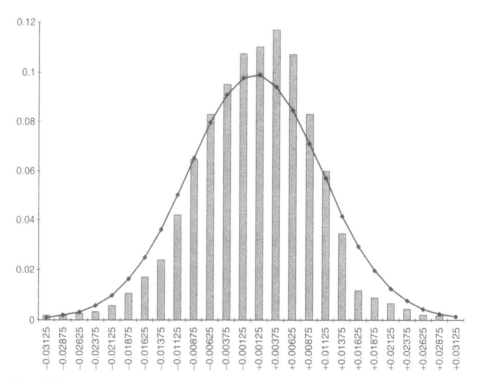

Figure 8.12

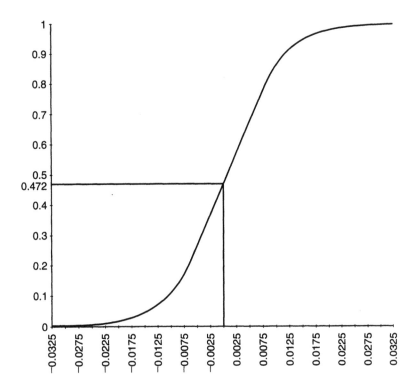

Figure 8.13

Stage 2: imitate the input variable(s)

The second stage is to simulate the behaviour of the underlying random variable. To do this we generate an appropriate large number of uniformly distributed random numbers ranging from zero to one. Each random number is then located on the vertical axis of the cumulative density function and the appropriate value of the random variable on the horizontal axis is read off. These are the **inputs to the simulation**. As the cumulative relative frequency distribution reflected both the magnitude of the underlying variable, the returns in our case, and the empirical probability distribution, the readings on the horizontal axis represent random observations of the underlying variable, with an empirical probability distribution that matches that of the original data.

To illustrate this, imagine generating a random number 0.472. The random number 0.472 is located on the vertical axis, then we move horizontally to the cumulative density function, drop perpendicularly to the axis and read off the randomly selected return to the variable. If our data relates to the daily returns of an asset, the result is a randomly generated one-day return, drawn from a probability distribution that matches the original empirical distribution. Of course with modern computer packages such manual processes are replaced with computerized processes, but the analogy is clear.

It is interesting to note that numbers from the random number generator between 0.4 and 0.6 produce outcomes between –0.25 and 0.25, whereas those between 0.7 and 0.9 produce outcomes between 0.52 and 1.28. Thus the steepness of the ogive in the middle produces "dense" outcomes underneath that steep portion, whereas the flatness in the tails produces sparse outcomes in those portions – exactly what the probability density function is describing.

Some readers will question why we go through such a process when we could use the assumption of the normal distribution to derive the same result. The answer is that the use of the normal distribution assumption is valid if the returns are independent. However, if there is any serial correlation in the returns data, that procedure will not give valid results. There is evidence that there is serial correlation in currency, interest rates and equity indices.

Stage 3: simulating the underlying variable

The third stage is to combine the inputs according to the logic of the system. This system describes the way in which each individual random input is combined to produce a single output which in our current example is one estimate of the future value of the FTSE 100 index. It is appropriate that we pause here and define what we mean by simulation. In the context of the Monte Carlo methodology we have the outcome of the overall simulation exercise, which we will call **the simulation**, which is the mean of the outcomes of many individual **simulation runs**.

To understand this, imagine that we wish to simulate the future price, S_T, of an asset. The simulated estimate, the outcome of **the simulation** of that asset, we will call $\hat{S}_T$. $\hat{S}_T$ will be the mean of many simulation runs which give individual estimates of S_T. We will call these individual estimates $\hat{S}_j$. Assume that we have data about the empirical distribution of the continuously compounded returns of our asset. We will call the individual returns observations r. The one-period ahead estimate, $\hat{S}_j$ will be

$$\hat{S}_j = S_0 e^{r_1} \tag{8.39}$$

If $\hat{S}_j$ is the n period ahead estimate, we make n successive drawings of the random returns observations, and derive $\hat{S}_j$ as

$$\hat{S}_j = S_0 e^{r_1} * e^{r_2} + \ldots + e^{r_n} \tag{8.40}$$

This is the same as

$$\hat{S}_j = S_0 e^{(r_1 + r_2 + \ldots + r_n)} \tag{8.41}$$

This process is the **simulation run**.

The simulation run that we have just described employs a new random variable for each sub-period over the time spanned by the simulation run. Thus, using daily returns data for example, a random path generated from 250 random observations of returns will give us the one-year simulated value, because there are 250 trading days in a year.

This process of generating random paths is important when the future value of the variable we are simulating depends not only on the final value but also how it arrived at that value. Options based on the average rate of the underlying asset, or on the minimum or maximum value of that asset, are examples.

Stage 4: produce the distribution of the future value of the asset

The fourth stage entails repeating many times the process in Stage 3 in order to produce the distribution of the future value of the asset. The whole process of repeatedly making, say, ten thousand, **simulation runs** is **the simulation**. The mean of the future values from the many simulation runs is the **simulated future value** of the random variable. The **simulated present value** is arrived at by **discounting** the simulated future value by an appropriate discount rate.

Thus the final simulated value of the random variable, i.e. the mean of the outputs of each simulation run, is given as

$$\hat{S}_T = \frac{1}{n}\sum_{j=1}^{n} \hat{S}_j \tag{8.42}$$

where $\hat{S}_T$ is the final simulated value of the random variable and $\hat{S}_j$ is the outcome of each simulated run.

To find the variance of $\hat{S}_T$, we first need to estimate the variance of $\hat{S}_j$. This is given by

$$\text{var}_{\hat{S}_j} = \frac{1}{n}\sum_{j=1}^{n}\left[\hat{S}_j - \hat{S}_T\right]^2 \tag{8.43}$$

The variance of $\hat{S}_T$ is given by the variance of $\hat{S}_j$ divided by n.

Thus the standard error is given as

$$\text{SE}_{\hat{S}_T} = \sqrt{\frac{\text{var}_{\hat{S}_j}}{n}} \tag{8.44}$$

Refer to Appendix 5.1 of Chapter 5.

Stage 5: apply variance reduction techniques to increase accuracy

The standard deviation of the estimate of S_T, i.e. $\hat{S}_T$, is given by

$$\sigma_{\hat{S}_T} = \sqrt{\frac{\text{var } S_T}{n}} \tag{8.45}$$

It follows that to reduce the standard deviation by a factor of ten, the number of simulations must be increased one-hundred fold (Boyle, 1977, p. 326). An alternative approach to reduce the required number of replications is to apply a variance reduction technique. Several techniques have been developed. We will discuss two:

1. the antithetic variate technique, and
2. the control variate technique.

The antithetic variate technique

Each time a random variable, r, is drawn, its complement, $1 - r$, is calculated and used to drive a parallel run of the simulation. Thus when the input variable is relatively large the parallel run will have a relatively small input variable. This tends to lead to output values from each run which are negatively correlated.

Since we are usually trying to estimate a mean, we can see the effect of this by considering $S = (S_1 + S_2)/2$, where S_1 and S_2 are the outputs from the parallel runs. The variance of S will be $(\text{var}(S_1) + \text{var}(S_2) + 2\text{cov}(S_1, S_2))/4$. Without the negative correlation we would have $\text{var}(S) = (\text{var}(S_1) + \text{var}(S_2))/4$. However, with a negative correlation, S_1 and S_2 will have a negative covariance which will reduce the variance.

The control variate technique

The idea behind the control variate technique is to find a variable similar to the random variable to be simulated and whose value is known. We will call this known value h.

The next step is to use the same random sampling used to simulate S_T to simulate h, giving the result $\hat{h}$. The point here is that the same random number is used to generate a sample for S and a sample of h. Although the same random numbers are used to derive both estimates, the values will not be the same because the logic of the processes of S_T and h will be different.

A new estimate of S, $S*$, can be calculated as

$$S* = h + (\hat{S} - \hat{h})\tag{8.46}$$

The variance of $S*$ is

$$\text{var } S* = \text{var } \hat{S} + \text{var } \hat{h} - 2\text{cov}(\hat{S}, \hat{h})\tag{8.47}$$

The standard deviation will be

$$\sigma S* = \sqrt{\text{var } \hat{S} + \text{var } \hat{h} - 2\,\text{cov}(\hat{S}, \hat{h})}\tag{8.48}$$

which will be less than $\sigma\hat{S}$ if

$$\rho_{\hat{S}, h} > \frac{\sigma\hat{h}}{2\sigma\hat{S}}\tag{8.49}$$

Thus the variance reduction capabilities of the control variate technique depend on finding a control variate that is highly correlated with the variable that is being simulated and at the same time has a similar probability distribution.

Application of Monte Carlo simulation to option pricing

We will now illustrate the use of Monte Carlo simulation to value a one-year call option on the asset that has the returns distribution depicted in Fig. 8.12. The current asset price is 1000, the exercise price of the option is also 1000, and the risk-free rate of interest is 6% p.a., continuously compounded. We are using the simulation technique because, as Fig. 8.12 shows, the empirical probability distribution is not normal.

The first stage is to identify the distribution. In our data we found a mean daily return of 0.000455 and a standard deviation of daily returns of 0.0100694. We need to convert a uniform random variable into a random variable with a distribution identical to the empirical distribution of the asset. The result is a series of random observations of daily returns.

In fact we do not use the observed mean daily return of r. We make an adjustment. Recall in the binomial option pricing model, the option was priced within a risk neutral framework because it was assumed that the written option position could be perfectly hedged. We make the same asumption in the Monte Carlo process. As such the relevant continuously compounded rate of return is the daily equivalent of the risk-free rate relevant to the life of the option. In this example we have assumed a rate of 6% p.a. Thus the daily continuously compounded rates need to be adjusted to reflect this. To do this we must remember that a normal distribution with a mean of μ and a standard deviation of σ gives rise to a lognormal distribution with a mean of $e^{\mu + \sigma^2/2}$. Thus with a lognormal distribution, to achieve an annual rate of 6% we need to adust the daily continuously compounded return, r, so that $e^{r + \sigma^2/2} = e^{0.06/250}$. Thus we require r such that $r + \sigma^2/2 = 0.06/250$. This gives for our data

$$r = \frac{0.06}{250} - \frac{0.0100694^2}{2} = 0.000189$$

(This will not be exactly correct for our empirical distribution. The correct value for r is $0.06/250 - \ln(\Sigma e^{r_i} \, prob(r_i))$.)

Consequently, the current asset price is compounded by the daily equivalent of 6% p.a but the probability distribution maintains its shape. Effectively the empirical distribution is shifted to the left, so the shape is unchanged but the mean is lower. This is depicted in Fig. 8.14.

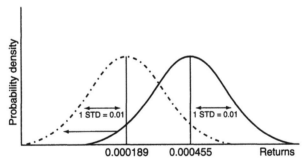

Figure 8.14 Recalibrating the returns distribution.

The second stage is to compound the current asset price by the random daily returns for each of the trading days during the life of the option. In our example we assume that there are 250 trading days in one year. As the empirical probability distribution relates to continuously compounded returns, we compound the current asset price as follows:

$$S_T = S_0 e^{r_1} * e^{r_2} * \ldots * e^{r_{250}}$$

We noted earlier this is the same as

$$S_T = S_0 e^{(r_1 + r_2 + \ldots r_{250})} \tag{8.50}$$

where r_i is a random observation of the one-day continuously compounded return drawn from the same empirical probability distribution as the underlying data. The combined effect of these 250 random observations is one simulation run. We need to repeat this many times in order to reduce the variability of our mean to be compatible with our accuracy requirements. In this example we have decided to require that our mean should have a standard error of less than 0.50. The annual volatility of the underlying data is 15.9%. Thus the annualized standard deviation of the asset price of 1000 is 159. Therefore, the standard error of our output mean will not be greater than $159/\sqrt{n}$, where n is the number of runs (the higher the strike price, the lower will be the actual standard error). We require an n of about 125 000 to reduce this to 0.50.

Note: the rule of thumb for determining the number of runs may be expressed as

$$n = \left(\frac{S * \sigma}{se} \right)^2$$

where se is the required standard error.

The process has therefore been repeated 125 000 times to give 125 000 alternative values of S_T.

We next have to apply the boundary conditions of the option. These are the same as we applied to the option priced under the binomial model. Thus the value of a call at time T will be $C_T = \max[S_T - X, 0]$. This boundary condition is applied to each of the alternative values of S_T, and the average of all 125 000 values of C_T taken. When discounted this average is the value of the option.

The result of our simulation of 125 000 runs is an option price of 93.52. A binomial model of 100 trials or the Black–Scholes model to be discussed in Chapter 10 gives a value of 95.03. We can see from the frequency distribution in Fig. 8.12 why our simulated option may have a lower price than indicated by the Black–Scholes equation. It will be noticed that the empirical distribution is very slightly skewed to the right, but also much more peaked just above the mean. As the option we are valuing is a call, this would have the effect of decreasing the value slightly.

The choice of 125 000 runs was motivated by the calculation of the standard error of the future asset price. However, we require the standard error of the option price. This will be less than the standard error of the asset price because applying the boundary conditions of the option, i.e. $\max(S - X, 0)$, will result in a narrower range of terminal values. The standard deviation of the 125 000 option prices was 125 giving

$$SE_{S_T} = \frac{125}{\sqrt{125\,000}} = \frac{125}{353.55} = 0.35$$

Thus we can say with 95% confidence that the real option price is 93.52 plus or minus 0.70, i.e. two standard errors.

Exercises

1. A bond matures in 2.5 years' time. It pays semi-annual coupons of 5, and will be redeemed at 100. Its current price is 102.

   Compute the gross redemption yield (GRY) for this bond using:

   (a) bisection
   (b) Newton–Raphson.

   Compare and contrast the two approaches.

2. The random variable X is distributed $N(6,9)$. Thus the probability density function for X is

$$f(x) = \frac{1}{3\sqrt{2\pi}} e^{-\frac{1}{2}\left(\frac{x-6}{3}\right)^2}$$

You are required to estimate $P(8 \leqslant X \leqslant 10)$,

(a) by using tables
(b) by using the trapezium rule with four intervals
(c) by using Simpson's rule with four intervals
(d) by using the polynomial approximation given.

3. Using a one-period binomial model, from the following data, satisfy yourself that the fair value of a call option is 7.95.

   Asset price = 50, exercise price = 50, rate of interest = 10% p.a., time to expiry one year. The asset price will move up or down by 25% over one year.

4. Prove to yourself that the hedge ratio, H, in the above example is 2.

5. From the data given below develop a six-period binomial model and satisfy yourself that the fair price of a call option is 4.52.

   Data: asset price = 35, exercise price = 35, volatility = 20%, short-term rate of interest = 10% p.a., time to expiry = one year.

6. Using the same data as in question 3 develop a three-stage trinomial and satisfy yourself that the fair price of the option is also 4.52.

7. Briefly describe the five stages of the Monte Carlo simulation process as applied to pricing an option on an underlying asset which is assumed to have a lognormal distribution.

8. Describe both the antithetic variate and the control variate techniques of variance reduction. Discuss the advantages and disadvantages of each when applying Monte Carlo simulation to option pricing.

Answers to selected questions

1. The GRY is given by x^2-1, where x is a solution to the equation:

   i.e. $102x^5 - 5x^4 - 5x^3 - 5x^2 - 5x - 105 = 0$

   (a) Output from an implementation of bisection:

   $(f(x) = 102x^5 - 5x^4 - 5x^3 - 5x^2 - 5x - 105)$

L	R	(L + R)/2	$f[(L + R)/2]$
1.01	1.505	1.2575	179.091
1.01	1.2575		58.424
1.01	1.1338	1.071	15.457
1.01	1.0719	1.040	−2.473
1.0409	1.0719	1.0564	6.217
1.0409	1.0564	1.0487	1.805
1.0409	1.0487	1.0448	−0.351
1.0448	1.0487	1.0467	0.723
1.0448	1.0467	1.0458	0.185
1.0448	1.0458	1.0453	−0.083
1.0453	1.0458	1.0455	0.051
1.0453	1.0455	1.0454	−0.016
1.0454	1.0454		

Root is 1.045, so GRY = 9.2%

(b) Newton–Raphson reorganization:
giving $x_1 = 1$
$x_2 = 1.05$
$x_3 = 1.045479935$
$x_4 = 1.045438601$
$x_5 = 1.045438598$
$x_6 = 1.045438598$
GRY = 9.3%

2. (a) 0.1613
   (b) and (c)

x	8	8.5	9	9.5	10
$f(x)$	0.10648267	0.09397063	0.08065691	0.0673329	0.05467002

(b) Trapezium: 0.16126389
(c) Simpson's: 0.16128010
(d) Polynomial: 0.908780897 − 0.74750848 = 0.1612848

References and further reading

Black, F. and Scholes, M.J. (1973) The pricing of options and corporate liabilities. *Journal of Political Economy*, **81**, 637–59.

Boyle, P.P. (1977) Options: a Monte Carlo approach. *Journal of Financial Economics*, **4**, 323–38.

Cox, J.C., Ross, S.A. and Rubinstein, M. (1979) Option pricing: a simplified approach. *Journal of Financial Economics*, **7**, 229–63.

Rendelman, R. and Barter, B. (1979) Two state option pricing. *Journal of Finance*, **34**, 1093–110.

Wilmot, P., Dewynne, J. and Howison, S. (1993) *Option Pricing: Mathematical Models and Computation*. Oxford Financial Press, Oxford.

Wilmot, P., Howison, S. and Dewynne, J. (1995) *The Mathematics of Financial Derivatives: A Student Introduction*. Cambridge University Press, Cambridge.

Winston, W.J. (1996) *Simulation Modeling Using @ Risk*. Wadsworth.

9

Optimization

Introduction

In this chapter we are concerned with optimization – in particular we are concerned with determining the structure of "optimal portfolios" where optimal is defined as having a minimum variance for a given level of return.

We learned in Chapter 3 that we could use calculus to find optimum points, and that these were either minimum points or maximum points of a function. Moreover we learned that we can find optimum points subject to constraints by using Lagrange multipliers. Finding an optimal point of any function is equivalent (if the function is differentiable, and subject to a few small difficulties such as stationary points of inflection) to solving the equation $f'(x) = 0$. Our optimization problem is to find a portfolio of minimum variance, where the portfolio variance is a function of the covariances and asset weightings. However, we also have a constraint in that we want the portfolio to achieve a minimum level of return.

In this chapter we will extend that analysis of optimum points to the multivariate example. We will also apply criteria known as the Kühn–Tucker conditions to allow the Lagrange multiplier technique to be applied when the constraints are inequalities rather than equalities.

Thus the structure of the chapter is as follows:

- the first section of this chapter will explain linear programming
- the second will explain the problem of selecting optimum portfolios of risky assets
- the third section will revisit Lagrangian multipliers and apply them to a multivariate example
- the fourth section will explain and apply the Kühn–Tucker conditions
- the fifth section will explain the Dantzig–Wolfe methodology which allows the linear programming technique to be used to solve a quadratic programming problem.

But first some definitions of some of the terms which will be used in this chapter.

Definitions

The objective function sets out the task which the optimization process is to achieve. For example, in this chapter we are concerned with **minimizing the risk** in a portfolio of assets. A typical objective function for a portfolio of risky assets would be

$$1 - \sum_{i=1}^{n-1} W_i \tag{9.1}$$

where Z is total risk and the W_is are the weightings of the assets in the portfolio. Recall from Chapter 2 that the risk of a portfolio of assets was a function of the variances and covariances of the assets, and that the variance of a single asset is the same as the covariance of an asset with itself. Recall also that the variance is a squared unit of measurement. Thus this objective function is non-linear; it is a **quadratic function**.

We will see later that some portfolio problems have linear objective functions.

Constraints. The objective function is usually set subject to a number of constraints. For example, it may be that some funds have to be invested in each asset. It may be subject to the constraint that all the funds must be fully invested. A third constraint may be that the minimum risk must be achieved subject to achieving a minimum level of return. These may be set out as below

1. $W_i > 0$ for all i

2. $\sum_{i=1}^{N} W_i = 1$ \hfill (9.2)

3. $R \geq 0.10$

Constraint 1 states that all assets must have a weighting in the portfolio that is greater than zero. This is a form of **inequality constraint**. Constraint 2 states that the sum of the weightings must equal one. This is an **equality constraint**. It simply says that all the funds must be invested in risky assets. Constraint 3 says that the portfolio must earn a rate of return that is equal to or greater than 10%. This is another inequality constraint.

Further constraints may be added such as the two given below. These are also inequality constraints and simply state the weighting of asset j in the portfolio must not exceed 10% whilst the weighting of asset k must not exceed 15%:

$$W_j \leqslant 0.10$$
$$W_k \leqslant 0.15 \tag{9.3}$$

These examples are not exhaustive; however, they do highlight the two forms of constraints, **equality constraints** and **inequality constraints**.

A **mathematical programming problem** is a problem in which a function of many variables (the objective function) is to be optimized subject to a number of constraints. The number of constraints is normally less than (usually substantially less than) the number of variables. The optimum is set in terms of maximization or minimization.

A **linear programming problem** is one in which objective function and constraints are linear.

A **quadratic programming problem** is one in which the objective function is a quadratic function of the variables, i.e. where some of the variables have squared values. However, the constraints are linear.

Linear programming

Constrained optimization problems in which the objective function is a linear function of the variables and in which the constraint functions are also linear are known as linear programming problems.

Linear programming problems in two variables can be solved by drawing graphs. Dantzig in the 1940s developed an algorithm, the **simplex algorithm**, which effectively translated the graphical approach into an algebraic method which could be coded for computing and in which any number of variables could be used. The simplex algorithm is an **iterative process** to find the optimum value of the objective function.

An application of linear programming (LP) to portfolio selection is the construction of portfolios within the framework of the Capital Asset Pricing Model (CAPM). A detailed analysis of this model and a discussion of the empirical evidence is to be found in Watsham (1993). Here we give a brief description.

The CAPM expresses the expected return of an asset as a linear function of the risk-free rate of return, the expected return on the market portfolio and the degree of systematic risk exhibited by the asset. The expected return to asset i is expressed as $E(r_i)$ $= r_f + \beta_i(E(r_M) - r_f)$, where $\beta_i(E(r_M) - r_f)$ is the expected risk premium for asset i and β_i is a measure of the systematic risk of that asset.

When we combine assets in a portfolio the returns of each asset combine in a linear form and the risk of the portfolio as represented by the portfolio β is also a linear combination, in this case a weighted average of the βs of the individual assets.

To illustrate this suppose that we wish to combine two assets in a portfolio in the proportions W_a and W_b ($W_a + W_b = 1$). Assume that the expected returns on these assets are $E(r_a)$ and $E(r_b)$, respectively, and that the βs of the assets are β_a and β_b, respectively.

Then the expected return to the portfolio will be $W_a E(r_a) + W_b E(r_b)$, i.e. the returns combine linearly. This is exactly the same as we noted in Chapter 2 when we looked at the returns of a portfolio of risky assets using mean-variance analysis.

Similarly, since in the CAPM model

$$E(r_a) = r_f + \beta_a(E(r_M) - r_f) \tag{9.4}$$

and

$$E(r_b) = r_f + \beta_b(E(r_M) - r_f) \tag{9.5}$$

it follows that the return to the portfolio will be given by

$$W_a[r_f + \beta_a(E(r_M) - r_f)] + W_b[r_f + \beta_b(E(r_M) - r_f)] \tag{9.6}$$

$$= r_f + [W_a\beta_a + WB\beta_b](E(r_M) - r_f) = r_f + \beta(E(r_M) - r_f) \tag{9.7}$$

where $\beta = W_a\beta_a + W_b\beta_b$.

Thus the β of a portfolio of assets is the weighted average β of the individual assets. Thus if the objective of the optimization exercise is to maximize portfolio return subject to a constraint on the maximum size of the portfolio β we have a problem where the objective function, the portfolio return, is linear and the constraints are linear. Thus we have a linear programming problem.

We will now illustrate the application of linear programming to maximize the return of a three-asset portfolio subject to a maximum level of portfolio β.

Selecting a three-asset portfolio – using linear programming to control systematic risk

Let us consider the problem of constructing a portfolio with the objective function of achieving maximum expected return, subject to the constraint that the β of the portfolio is to be no higher than 1.1. Assume that we have three assets to chose from, assets A, B and C. Their expected returns are 0.11, 0.15 and 0.08, respectively. The CAPM βs are 1, 1.2 and 0.9, respectively. The proportions of each asset in the portfolio are given by W_a, W_b and W_c. These weightings are at the discretion of the portfolio manager and are the variables that can be adjusted in order to achieve the objective. The expected returns and asset βs are fixed from the point of view of the portfolio manager because they are determined by the market. However, portfolio returns and β can be engineered by the portfolio manager by adjusting the proportions of each asset in the portfolio. The objective is to find that combination or weighting that maximizes the objective function subject to the constraints.

Thus the problem is to determine the optimum proportions (weights) of each asset which will give the maximum expected return subject to the maximum level of risk (β). This problem may be formulated mathematically as follows.

Objective function

Maximize (return)

$$0.11W_a + 0.15W_b + 0.08W_c \tag{9.8}$$

As the return to each asset is given, only the weights can be changed in order to achieve the objective function.

The constraints

We note above that the portfolio β must not exceed 1.1 – that is constraint 1. In this example there must be no short selling of any asset – that is constraint 2. And the fund must be fully invested – that is constraint 3. Thus the objective function and the constraints are given below.

Maximize (return) subject to

1. $W_a + 1.2W_b + 0.9W_c \leq 1.1$ (i.e. portfolio β must not exceed 1.1)   (9.9)
2. $0 \leq W_a \leq 1$
   $0 \leq W_b \leq 1$ (i.e. all assets must have non-negative weights)
   $0 \leq W_c \leq 1$ (9.10)
3. $W_a + W_b + W_c = 1$ (i.e. funds must be fully invested)   (9.11)

Note that all the constraints are linear (i.e. no squared or higher order values) and that there are both equality and inequality constraints.

We shall demonstrate the application of linear programming by showing how the above problem can be solved first graphically and then by the **simplex method**.

Graphical solution

To illustrate the solution graphically we shall use constraint 3 to eliminate W_c and thus produce a two-variable problem. This is possible because in any portfolio problem where the portfolio is to be fully invested, if we know $n - 1$ weightings we can determine the nth weighting. As all weights must sum to one, the nth weight must be

$$1 - \sum_{i=1}^{n-1} W_i$$

Note that this conversion from three to two variables is only to facilitate graphical demonstration of the technique. The problem as formulated would normally be submitted to an LP package in its original form – transformations reducing the number of variables would not usually be performed.

Using the constraint that $W_a + W_b + W_c = 1$ to replace W_c by $1 - W_a - W_b$ the objective function is restated as

Maximize

$$0.11W_a + 0.15W_B + 0.08(1 - W_a - W_b) = 0.03W_a + 0.07W_b + 0.08 \quad (9.12)$$

Subject to
$$1W_a + 1.2W_b + 0.9(1 - W_a - W_b) \leq 1.1$$
$$0 \leq W_a \leq 1$$
$$0 \leq W_b \leq 1$$
$$0 \leq (1 - W_a - W_b) \leq 1 \quad (9.13)$$

Given that we are treating the weighting of asset C as a residual, we can maximize the following two-asset objective function.

$$\text{Maximize} \quad 0.03W_a + 0.07W_b \tag{9.14}$$

When we have found the optimum weights for A and B from this function we can subtract the sum of these weights from one to derive the weighting for asset C.

Note that the "+0.08" in the objective is a constant. As such we do not have to include it in the function to be maximized, because if we find the values of W_a and W_b which maximize $0.03W_a + 0.07W_b$, then we will have found the values which maximize $0.03W_a + 0.07W_b + 0.08$. So once we have maximized $0.03W_a + 0.07b$ we must remember to add on the constant 0.08.

As we have substituted the expression for C into the objective function we must do the same with the constraints. Thus the constraints become

$$0.1W_a + 0.3W_b \leq 0.2$$
$$0 \leq W_a \leq 1$$
$$0 \leq W_b \leq 1$$
$$0 \leq W_a + W_b \leq 1 \tag{9.15}$$

When constructing a graphical solution (Fig. 9.1) we are aiming to define a "feasible region" between the two axes, which in this example are W_a and W_b.

By constraint 2 the maximum proportion that we can invest in A is 100% or one. This is shown as point "N". Similarly point "K" reflects 100% of the portfolio invested in asset B. As both W_a and W_b have to be less than or equal to one, these are plotted as vertical and horizontal lines from the appropriate axes. However, these constraints are overwhelmed by that which requires $W_a + W_b$ to be less than or equal to one. To understand the positioning of this constraint, assume that 100% is invested in A and nothing in B so the position of the constraint would be at point N. If 100% were invested in B and nothing in A the constraint would be positioned at point K. The points in between K and N represent combinations of A and B which sum to one.

Finally there is a constraint $0.1W_a + 0.3W_b \leq 0.2$. To illustrate this we draw the line $0.1W_a + 0.3W_b = 0.2$. This represents the boundary between the region of points satisfying $0.1W_a + 0.3W_b < 0.2$ and the region given by $0.1W_a + 0.3W_b > 0.2$. This inequality embodies the fact that the portfolio beta must not exceed 1.1. If we invest two-thirds of

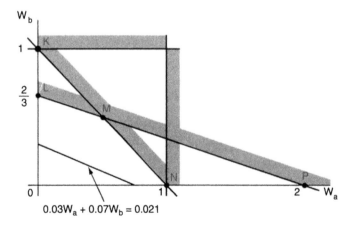

Figure 9.1

our money in asset B and nothing in A (and therefore, by implication invest one-third in C) then our portfolio will have a beta of 1.1. Thus $W_a = 0$, and $W_b = 2/3$ just satisfies the constraint, and this is represented by point L on the diagram. Similarly $W_a = 2$ and $W_b = 0$ just satisfies the same constraint, and this is marked as point P.

The points between K and M meet the constraint that $W_a + W_b = 1$, but they do not meet the constraint $0.1W_a + 0.3W_b \leqslant 0.2$. Similarly, the points between M and P meet the constraint $0.1W_a + 0.3W_b = 0.2$, but do not meet the constraint $W_a + W_b = 1$. Thus the feasible region in this example must be bounded by $0LMN$.

We now know the optimum combination of asset A and asset B lies within the area $0LMN$. However, we need to know which point within that area is the optimal point. To find that out we turn to the objective function, $0.03W_a + 0.07W_b$ and construct so called iso-profit lines. These are straight lines representing the objective function, where each point along the line represents the same value of the objective function. Let us consider the possibility of making the objective function equal to 0.021 (say). Thus the objective function $0.03W_a + 0.07W_b = 0.021$ gives us an iso-profit line of points representing portfolios (weighting of W_a, W_b and hence W_c) giving the objective of 0.021. This represents a portfolio return of 0.101 because we add back in the 0.08 relating to asset C.

This line is drawn on the diagram and it can be seen that some points on it are in the feasible region indicating that this return is achievable.

Can we do any better? For example can we achieve an objective value of 0.042 (i.e. a portfolio return of 12.2% when taking into account the constant which has to be added back)?

In both cases the lines contain points within the feasible region, so both returns are possible. We can extend this approach further from the origin until the iso-profit line is just touching a point on the boundary of the feasible region. The optimal combination will be represented by that point where the iso-profit line is furthest from the origin but still in contact with the feasible region. In this example this is at point M.

Since point M is on both constraint lines, we can find the coordinates of this point by solving simultaneously for both constraints

$$W_a + W_b = 1$$
$$0.1W_a + 0.3W_b = 0.2 \tag{9.16}$$

We will solve these simultaneous equations by the substitution method.

First we scale up the second equation by multiplying by 10 thus

$$W_a = W_b = 1$$
$$W_a + 3W_b = 2 \tag{9.17}$$

Subtracting the first from the second, we get

$$2W_b = 1 \tag{9.18}$$

therefore

$$W_b = 0.5 \tag{9.19}$$

Substituting this value back into the first equation gives $W_a = 0.5$.

Thus we get $W_a = 0.5$ and $W_b = 0.5$, and the objective value is $0.03 \times 0.5 + 0.07 \times 0.5 = 0.05$ and if we add back the constant of 0.08 we get 0.13 or 13%.

In reality it is not necessary to draw these "lines of equal profit". The fact that they exist, and that they are straight and parallel, demonstrates that the best solution is always at a vertex (the intersection of two lines) of the feasible region. Thus all that is needed is to check the vertices, so reducing the problem to a discrete problem.

Checking a vertex involves, as above, solving a system of simultaneous equations. In two dimensions the vertices are where two lines intersect and there are two equations in two unknowns. In three dimensions the vertices are where three planes intersect, giving three equations in three unknowns.

If we have 200 variables there would be 200 equations in 200 unknowns. This would represent quite a task. Therefore an efficient algorithm is needed to organize the search so that as few vertices as possible have to be checked and so that the checking can be done as simply as possible. This is achieved by the simplex algorithm. Computer realizations of this algorithm are extremely efficient and can deal routinely and efficiently with problems involving hundreds of variables and constraints.

The simplex method

The simplex method is an iterative process which starts at one "**initial solution**" and searches for superior solutions by moving round the boundary of the **feasible area** until it arrives at the optimum solution. To see how the simplex method actually works let us apply it to the same three-asset portfolio problem used earlier.

The simplex method assumes that all variables have non-negative values. This is not a stumbling block since a variable that needs to be allowed to take negative values can be replaced by the difference of two non-negative variables, for example X could take the value -3 by having $X_1 = 0$ and $X_2 = +3$ and replacing X by $X_1 - X_2$. Software packages incorporating the simplex algorithm are coded to perform such tricks automatically.

We will now restate our objective function and constraints as

Maximize

$$Z = + 0.03W_a + 0.07W_b$$

Subject to

$$0.1Wa + 0.3Wb \leq 0.2$$
$$W_a + W_b \leq 1 \tag{9.20}$$

The first step is to convert the inequalities into equalities by introducing extra variables known as **slack variables**. This problem has two inequalities, therefore we will need to apply two slack variables. Thus $0.1W_a + 0.3W_b \leq 0.2$ becomes $0.1W_a + 0.3W_b + s_1 = 0.2$, where s_1 is the slack variable. It represents the extent to which $0.1W_a + 0.3W_b$ is below 0.2. Similarly $W_a + W_b \leq 1$ becomes $W_a + W_b + s_2 = 1$.

Thus the inequality constraints translate into equality constraints as follows:

1. $0.1W_a + 0.3W_b \leq 0.2$ becomes $0.1W_a + 0.3W_b + s_1 = 0.2$
2. $W_a + W_b \leq 1$ becomes $W_a + W_b + s_2 = 1$

The objective is to maximize $0.03W_a + 0.07W_b$.

Note that we have ignored the non-negativity constraints because of the assumption of the simplex method that variables only take non-negative values.

One feasible, though not an optimal point is given by $W_a = 0$ and $W_b = 0$, i.e. the origin, because if nothing is invested in A or B, and we give s_1 a value of 0.2 and s_2 a value of one, the two equality constraints are met. We shall call this an initial feasible solution. It is the function of the simplex method to search for the optimal solution by repeatedly moving from one feasible solution to a better feasible solution.

It should be noted that it is often not easy to spot an initial feasible solution. When the inequality constraints are all of the "$\leqslant$" form, the origin will serve as an initial feasible solution. If there are any "$\geqslant$" constraints this will not be the case. Fortunately, by way of the sort of tortured inversion of procedures which appeals to mathematicians, the method may be turned on itself to produce its own initial feasible solution.

To summarize, the current state of the problem is as follows:

Maximize

$$Z = 0.03W_a + 0.07W_b$$

Subject to

$$0.1W_a + 0.3W_b + s_1 = 0.2$$
$$W_a + W_b + s_2 = 1 \qquad (9.21)$$

With $W_a = 0$, $W_b = 0$, $s_1 = 0.2$ and $s_2 = 1$, the solution is $Z = 0$. Here the solution represents the initial feasible solution. The method now moves through a sequence of improving feasible solutions until the best is found. To facilitate searching for a superior solution we will set out this initial situation in tableau form

	W_a	W_b	s_1	s_2	RHS
constraint 1	0.1	0.3	1	0	0.2
constraint 2	1	1	0	1	1
objective function	−0.03	−0.07	0	0	0

The bottom line represents the objective function. We have set up the objective in the form $Z - 0.03W_a - 0.07W_b + 0s_1 + 0s_2 = 0$. The first and second lines represent two constraints associated with this problem. The RHS represents the right-hand side of the equalities.

At this point we must define **basic** and **non-basic** variables. **Basic variables** are those where the column contains all zeros except for a single one. Thus in this initial solution s_1 and s_2 are basic variables. The current value of these variables may be seen by looking across from the "one" entry in their column, to the right-hand side. Thus the current value of S_1 is 0.2 and the current value of s_2 is one.

Other variables are known as **non-basic variables**. Non-basic variables have a current value of zero in this process.

In an iteration of the search process one basic variable is transformed into a non-basic variable. At the same time a non-basic variable is transformed into a basic variable (known as entering the solution).

At each stage of the procedure of searching for a superior solution, we must ensure that each constraint contains only one basic variable. The algorithm also requires the objective function to be expressed in terms of the non-basic variables. Our initial feasible solution satisfies these conditions.

Improving the solution

The first step in the process is to identify one entry in the objective row which has a negative sign. We will choose −0.03 (this is a free choice between −0.03 and −0.07). The column of figures in which −0.03 is located is known as the pivot column.

Recall that in the original form of the objective function, $Z = 0.03W_a + 0.07W_b$. In the tableau, Z is expressed in terms of the current non-basic variables, i.e. in the tableau this corresponds to $Z - 0.03W_a - 0.07W_b = 0$. Since in the initial solution, these non-basic variables have zero values, and since their coefficients are negative in the tableau, we can see that Z will increase if we increase the value either of W_a or of W_b, i.e. if we make either of them basic. We have a free choice – we have decided to increase W_a.

The second step is as follows: **using only the constraint rows**, divide each element in the right-hand side by the corresponding element in the pivot column. This is known as the ratio test. Find the quotient with the **minimum non-negative value**. The corresponding element in the pivot column identifies the pivot row and the element which corresponds to the intersection of the pivot column and the pivot row is known as the pivot.

Having carried out the ratio test we find that the one in constraint 2 is the pivot. We should transform this pivot into unity by dividing the whole pivot row by itself. However, in this case this is unnecessary because the pivot is already one.

The next step is to transform all the other elements in the pivot column to zero by adding or subtracting appropriate multiples of the pivot row (note: we use the new values of the pivot row, i.e. after the pivot has been transformed to unity). Thus to convert the 0.1 at the top of the pivot column to zero, we subtract 0.1 times the pivot row from each corresponding element of the top row. To convert the −0.03 to zero, we add 0.03 times the pivot row to each element in the bottom row. The resulting tableau is

	W_a	W_b	s_1	s_2	RHS
constraint 1	0	0.2	1	−0.1	0.1
constraint 2	1	1	0	1	1
objective function	0	−0.04	0	0.03	0.03

To interpret this solution note that those variables that are **basic** will have values given by the appropriate element on the right-hand side. All other variables will have values of zero. Thus the objective function (Z) has a value of 0.03 (ignoring the constant of 0.08), W_a has a value of one, s_1 has a value of 0.1 while W_b and s_2 have values of zero.

That we have not yet arrived at the optimal solution may be seen by examining the entries in the objective function row, where we note that there is still one entry with a negative sign.

So we repeat the process. Looking at the second tableau we see that the pivot column is that with −0.04 as the bottom element. Carrying out the ratio test we find the pivot to be 0.2 in the top constraint row which therefore becomes the pivot row. To convert the pivot to unity we divide all the elements in the pivot row by 0.2. Next we transform all other elements in the pivot column to zero, we subtract one times the newly transformed pivot row from the lower constraint, and we add 0.04 of the new pivot row to the objective function. The resulting tableau is

	W_a	W_b	s_1	s_2	RHS
constraint 1	0	1	5	−0.5	0.5
constraint 2	1	0	−5	1.5	0.5
objective function	0	0	0.2	0.01	0.05

Note that both W_a and W_b are now basic and that the two slack variables are non-basic.

Now consider the objective function. It says $Z = 0.05 - 0.2s_1 - 0.01s_2$. However, recall the constant of 0.08, so what the tableau really says is

Maximize
$$Z = 0.13 - 0.2s_1 - 0.01s_2$$

Subject to
$$W_b + 5s_1 - 0.5s_2 = 0.5$$
$$W_a - 5s_1 + 1.5s_2 = 0.5 \qquad (9.22)$$

However, the slack variables are non-basic and therefore have zero value.

We can now see, since $Z = 0.13 - 0.2s_1 - 0.01s_2$, and since we have a feasible point at which s_1 and s_2 both have value zero, that we can do no better. We have arrived at the optimal point.

The fact that $s_1 = 0$ and $s_2 = 0$ tell us that the constraints corresponding to those slacks/surpluses are "tight". In a bigger problem with more constraints, the values of other slack variables will be non-zero, and will give us the extent to which the corresponding constraints are "slack". Furthermore the coefficients of s_1 and s_2 in the objective function (0.2 and 0.01, respectively) give us the marginal worth of relaxing the corresponding constraints (which often involves employing additional resources).

However, recall that the original problem was to create a portfolio out of three assets A, B and C. The solution derived above shows, as did the graphical solution earlier, that the optimum portfolio contains only assets A and B.

It can be seen that the simplex method is encoding in algebra the process of chasing from vertex to vertex of the feasible region, where a move is always made in a direction which will increase the value of the objective function. (Note that there are exceptional cases in which the objective function value is allowed to remain constant for a step.) The advantage of switching from geometry to algebra is, of course, that the algebra is just as valid in 200 dimensions as in two – the pictures are just more difficult to imagine!

Constructing portfolios to minimize total variance

The CAPM assumes that only the exposure to systematic risk in each individual asset is relevant to portfolio construction. However, the model originally developed by Markowitz (1952), and still the most widely applied, uses the total risk of each individual asset. Consequently, the covariances between each pair of potential assets for the portfolio must be considered when constructing portfolios and determining the overall risk of the portfolio.

In Chapter 4 it was explained that when the returns to a risky asset are a random variable, the returns to a portfolio are a value-weighted average of the returns to the individual assets, i.e.

$$R_p = \sum_{i=1}^{n} W_i r_i \qquad (9.23)$$

However, the standard deviation of a portfolio is not a value-weighted average of the standard deviations of the individual securities because the covariance between each pair of assets has to be taken into account. To illustrate this, the standard deviation of a portfolio of two assets is

$$\sigma_p = \sqrt{W_a^2\sigma_a^2 + W_b^2\sigma_b^2 + 2W_aW_b(\rho_{ab}\sigma_a\sigma_b)} \qquad (9.24)$$

where

$$\sigma_p = \text{standard deviation of the portfolio}$$
$$W_a \text{ and } W_b = \text{weights of } a \text{ and } b \text{ in portfolio}$$
$$\sigma_a^2 \text{ and } \sigma_b^2 = \text{variances of returns of } a \text{ and } b$$
$$\rho_{ab} = \text{correlation of returns of } a \text{ and } b$$
$$\sigma_a \text{ and } \sigma_b = \text{standard deviations of the returns of } a \text{ and } b$$
$$(\rho_{ab}\,\sigma_a\,\sigma_b) = \text{the } \textbf{covariance} \text{ of the returns of } a \text{ and } b.$$

Equation (9.25) can be generalized to

$$\sigma_p^2 = \sum_{i=1}^n W_i\sigma_i^2 + \sum_{i=1}^n \sum_{j=1,i\neq j}^n W_iW_j\sigma_{ij} \qquad (9.25)$$

where σ_{ij} is the covariance between pairs of assets.

For a portfolio with assets a to n, this can be written in matrix format as

$$\sigma_p^2 = [W_aW_bW_c\ldots W_n] \begin{bmatrix} \sigma_a^2 & \sigma_{ab} & \cdots & \sigma_{an} \\ \sigma_{ba} & \sigma_b^2 & \cdots & \sigma_{bn} \\ \cdot & \cdot & \cdots & \cdot \\ \cdot & \cdot & \cdots & \cdot \\ \sigma_{na} & \sigma_{nb} & \cdots & \sigma_n^2 \end{bmatrix} \begin{bmatrix} W_a \\ W_b \\ W_c \\ \cdot \\ \cdot \\ W_n \end{bmatrix} \qquad (9.26)$$

Each variance element in the variance/covariance matrix is multiplied by its respective asset weighting twice, thus the weightings relating to variances have a squared influence, hence the W_i^2. Each covariance is multiplied once by the weighting for each asset in the pair of assets, and there are two covariances for each possible pair, hence the 2 cov W_iW_j.

The efficient frontier

We showed in Chapter 2 that if the correlation coefficient between pairs of assets is less than 1.0, diversification can improve the relationship between expected portfolio risk and expected portfolio return. This is because while the return variable is a linear function of average returns the risk factor is a quadratic function of the variance of security returns. The degree of improvement depends on the weightings each asset has in the portfolio and the correlation between those assets.

The clearest way to demonstrate this is to look at a two-asset example. Consider the data in Table 9.1 which represents the various standard deviations of a portfolio constructed from two risky assets under the assumptions that the correlation (Cor) is 0.6 or 0.9 and where the proportions of each asset held in the portfolios is changed in 10% blocks. Figure 9.2 is a diagram of the efficient frontiers corresponding to the portfolios constructed upon the assumption of Cor = 0.60 and Cor = 0.90. Asset A has an expected return of 10% with a standard deviation of 14%, and asset B has an expected return of 12% with a standard deviation of 15%.

Table 9.1

Weight W_b	Weight W_a	Return R_p	Standard deviation (Cor = 0.6)	Standard deviation (Cor = 0.9)
0	1	10.0	14.0	14.0
0.1	0.9	10.2	13.55	13.965
0.2	0.8	10.4	13.22	13.961
0.3	0.7	10.6	13.01	13.988
0.4	0.6	10.8	12.92	14.045
0.5	0.5	11.0	12.97	14.133
0.6	0.4	11.2	13.15	14.251
0.7	0.3	11.4	13.45	14.397
0.8	0.2	11.6	13.86	14.571
0.9	0.1	11.8	14.38	14.773
1.0	0	12.0	15.0	15.0

For an assumed degree of correlation, the standard deviation is calculated for some of the various portfolios that can be constructed from those two assets and plotted in Fig. 9.2.

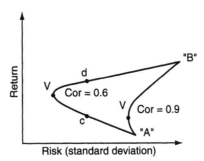

Figure 9.2 Efficient frontiers.

First look at the data in the column of Table 9.1 headed Cor = 0.6 and the plot in Fig. 9.2 labelled Cor = 0.6. Both show the benefits of diversification when assets are moderately well correlated. The data and plot labelled Cor = 0.9 show that diversification has a beneficial effect on the risk-return trade-off even when assets are highly, but imperfectly, correlated. Note that in both cases the efficient frontier is concave. The greater the degree of concavity, the greater the benefits of diversification. Note, however, that not all points on the frontier are efficient, only the upper part of each concave frontier (marked V to B in Fig. 9.2) is efficient.

The top part of each of the lines AB represents an efficient frontier of possible portfolios in that on the frontier it is not possible to achieve more return without bearing more risk. Above the line is a space of unattainable combinations of risk and return because of the limited characteristics of securities A and B. Below the line are inferior combinations of risk and return which could be improved upon by simply moving to any point on the line VB. This is achieved by selling existing assets and buying A and/or B. For example, portfolio c is on the lower part of the frontier marked Cor = 0.6. The investor can increase his or her utility by selling that portfolio and buying a combination of A and B represented by any of the points on the efficient frontier. For example, by moving to point d, the investor bears the same level of risk but gets a higher return than at c.

It will be noted that there is no unique best portfolio. The thick lines represent many "efficient portfolios". The frontier is efficient because it is not possible to increase return without increasing risk or to reduce risk without reducing return. The optimal combination of risk and return will depend upon the objective function (utility function of the investor).

However, given the generally less than perfectly correlated returns to individual assets the theory would suggest that the most diversified portfolio and therefore the portfolio that gives the best return per unit of risk will be a portfolio that contains all risky assets. This is worth some consideration by investment managers because their portfolios are usually constrained to contain only cash, bonds and equities.

The portfolio optimization problem

Now that we understand the relationship between risk and return and the influence of the covariance, we can specify the portfolio optimization problem. **The portfolio optimization problem** is to determine what proportion of the portfolio should be allocated to each investment so that the amount of expected return and the level of risk **optimally** meet the investor's objectives. In this section we will assume that the investor's objective is to minimize the risk of the portfolio, where risk is measured by the portfolio variance.

In practice it is usual for the investor to impose constraints on the way in which the portfolio may be constructed. For example, the objective function may be to minimize risk, but subject to a minimum level of return. There may also be constraints on the minimum and maximum proportions which may be invested in each asset. We will explain how to handle these constraints later.

We will illustrate the portfolio problem by considering constrained optimization within the context of a three-asset portfolio.

The risk and return of a three-asset portfolio, with assets a, b and c and the weights being depicted as Ws, are as follows.

The expected return

$$E(r_p) = W_a r_a + W_b r_b + W_c r_c \tag{9.27}$$

The portfolio variance is

$$\sigma_p^2 = W_a^2 \sigma_a^2 + W_b^2 \sigma_b^2 + W_c^2 \sigma_c^2$$
$$+ 2W_a W_b(\text{cov}_{ab})$$
$$+ 2W_a W_c(\text{cov}_{ac})$$
$$+ 2W_b W_c(\text{cov}_{bc}) \tag{9.28}$$

As explained earlier, this portfolio variance can be expressed using vectors of weights and a matrix of variances and covariances. Expressing the portfolio variance as Z we have

$$Z = [W_a W_b W_c] \begin{bmatrix} \text{var}_a & \text{cov}_{ab} & \text{cov}_{ac} \\ \text{cov}_{ba} & \text{var}_b & \text{cov}_{bc} \\ \text{cov}_{ca} & \text{cov}_{cb} & \text{var}_c \end{bmatrix} \begin{bmatrix} W_a \\ W_b \\ W_c \end{bmatrix} \tag{9.29}$$

We will now proceed to illustrate constrained optimization.

Constrained optimization

Investor requirements usually constrain the selection process. For example, the investor may require the risk to be minimized subject to the expected return being greater than or equal to a given level.

The portfolio problem is thus to minimize the portfolio variance, subject to a minimum level of return. As we saw above the portfolio variance (Z) can be can be expressed as the product of the W vector transposed, $\underline{W}^T$, the variance/covariance matrix Ω and the W vector $\underline{W}$. Thus the problem is a quadratic programming problem and may be formally set out as

Minimize $\quad\quad\quad\quad\quad\quad Z = \underline{W}^T \Omega \underline{W}$

Subject to $\quad\quad\quad\quad\quad W_a + W_b + W_c = 1$

$\quad\quad\quad\quad\quad\quad\quad\quad\quad W_a E(r_a) + W_b E(r_b) + W_c E(r_c) \geqslant R$ $\quad\quad\quad$ (9.30)

where R is the minimum acceptable level of return.

Note: in matrix notation the bolding and underlining of expressions such as $\underline{W}$ indicate that the expressions represent vectors.

We will illustrate this optimization process with the aid of a three-asset example.

The application of quadratic programming to a three-asset portfolio selection problem – finding the optimal (minimum variance) portfolio

Suppose that we have three assets, A, B and C with expected returns of 0.11, 0.15 and 0.08, respectively. The variance/covariance matrix, which we will designate as Ω is

$$
\begin{array}{ccc}
 & \text{A} & \text{B} & \text{C} \\
\text{A} & & & \\
\text{B} \quad \Omega = & \begin{bmatrix} 0.00015 & 0.00005 & -0.00007 \\ 0.00005 & 0.00025 & -0.00003 \\ -0.00007 & -0.00003 & 0.00010 \end{bmatrix} \\
\text{C} & & &
\end{array}
$$

We wish to find the proportions, the Ws, to invest in each asset so as to obtain a required return of 11% with minimum variance, i.e. we wish to find $\underline{W}$ (i.e. $[W_a W_b W_c]^T$) to solve the following problem

Minimize $\quad\quad\quad\quad\quad\quad Z = \underline{W}^T \Omega \underline{W}$

Subject to $\quad\quad\quad\quad\quad W_a + W_b + W_c = 1$

$\quad\quad\quad\quad\quad\quad\quad\quad\quad 0.11W_a + 0.15W_b + 0.08W_c = 0.11$

Note: this is an equality constrained problem. In this example we assume that negative positions in each asset are not possible.

The variance $Z = \underline{W}^T \Omega \underline{W}$, multiplies out to give

$$
Z = \begin{bmatrix} W_a & W_b & W_c \end{bmatrix} \begin{bmatrix} 0.00015 & 0.00005 & -0.00007 \\ 0.00005 & 0.00025 & -0.00003 \\ -0.00007 & -0.00003 & 0.00010 \end{bmatrix} \begin{bmatrix} W_a \\ W_b \\ W_c \end{bmatrix}
$$

$$(= W_a^2\,\sigma_a^2 + W_b^2\sigma_b^2 + W_c^2\sigma_c^2 + 2W_aW_b\mathrm{cov}_{ab} + 2W_aW_c\mathrm{cov}_{ac} + 2W_bW_c\mathrm{cov}_{bc})$$

$$= 0.00015W_a^2 + 0.00025W_b^2 + 0.00010W_c^2 + 0.00010W_aW_b -$$
$$0.00014W_aW_c - 0.00006W_bW_c.$$

Thus in "long hand" our optimization problem is to

Minimize

$$Z = 0.00015W_a^2 + 0.00025W_b^2 + 0.00010W_c^2 + 0.00010W_aW_b -$$
$$0.00014W_aW_c - 0.00006W_bW_c.$$

Subject to

$$W_a + W_b + W_c = 1$$
$$0.11W_a + 0.15W_b + 0.08W_c = 0.11$$

Optimization under equality constraints: using Lagrange multipliers

As our problem as specified above only has equality constraints, the problem can be solved by using Lagrange multipliers, one for each constraint. The use of Lagrange multipliers in optimizing just one variable has been covered in Chapter 3.

Recall that when implementing Lagrange multipliers in the single variable case, we first set the constraint to equal zero and then combined it with the function to be optimized (refer to Chapter 3):

Minimize $Z = f(\underline{W})$ (The underlining of the bold W indicates that W is in fact a vector of, say, n assets)

Subject to: $\underline{g}(\underline{W}) = \underline{0}$ (The fact that g is expressed as a vector indicates that there are a number, m say, of inequalities. Note that $m < n$.)

In our particular examples

$$\underline{g} = \begin{bmatrix} g_1(\underline{W}) \\ g_2(\underline{W}) \end{bmatrix} = \begin{bmatrix} 0 \\ 0 \end{bmatrix} \tag{9.31}$$

where

$$g_1(\underline{W}) = W_a + W_b + W_c - 1$$
$$g_2(\underline{W}) = 0.11W_a + 0.15W_b + 0.08W_c - 0.11$$

We construct the Lagrangian: $L(\underline{W}, \underline{\lambda})$ where $L(\underline{W}, \underline{\lambda}) =$

$$L(\underline{W}, \underline{\lambda}) = Z - \sum_{i=1}^{m} \lambda_i g_i(\underline{W}) \tag{9.32}$$

The λ_i are the Lagrange multipliers.

Thus the Lagrangian is constructed by subtracting from the original objective function all the individual constraint functions which have been multiplied by their respective Lagrange multipliers.

Note that we now have a function of $n + m$ variables. We construct the n partial derivatives with respect to the x_i variables (note: we do not construct m the partial derivatives for the Lagrange multipliers because partial differentiating with respect to the λs only returns the equality constraints which we already have). Next we try to find values of the x_is and λs which make these partial derivatives zero and which satisfy the equality constraints simultaneously. Thus we have a system of $n + m$ equations in $n + m$ unknowns. As we have the same number of equations as we have unknowns, we can find our solution by establishing the problem as a series of simultaneous equations, and by solving accordingly.

However, whether or not a solution does exist depends on whether or not the constraints are contradictory. If we were subject to one constraint that $x_{20} < 50$ (say) and another that $x_{20} > 100$, then clearly there could be no solution, but in a large and complex problem similar inconsistencies might be more difficult to spot. Depending on the complexity of the function L, we can solve this analytically or by a numerical technique – provided that a solution exists.

Applying this to our problem

Minimize

$$Z = 0.00015W_a^2 + 0.00025W_b^2 + 0.00010W_c^2 + 0.00010W_aW_b - 0.00014W_aW_c - 0.00006W_bW_c$$

Subject to

$$W_a + W_b + W_c = 1$$
$$0.11W_a + 0.15W_b + 0.08W_c = 0.11$$

Thus the Lagrangian is

$$L(\underline{W}, \underline{\lambda}) = 0.00015W_a^2 + 0.00025W_b^2 + 0.00010W_c^2 + 0.00010W_aW_b$$
$$- 0.00014W_aW_c - 0.0006W_bW_c - \lambda_1(W_a + W_b + W_c - 1)$$
$$- \lambda_2(0.11W_a + 0.15W_b + 0.08W_c - 0.11)$$

Partially differentiating and setting equal to 0 (note that $\underline{W}^T\Omega\underline{W}$ differentiates to $2\Omega\underline{W}$ – or it can be done term by term. We will proceed term by term)

$$\partial L/\partial W_a = 0.00030W_a + 0.00010W_b - 0.00014W_c - \lambda_1 - 0.11\lambda_2 = 0$$
$$\partial L/\partial W_b = 0.00050W_b + 0.00010W_a - 0.00006W_c - \lambda_1 - 0.15\lambda_2 = 0$$
$$\partial L/\partial W_c = 0.00020W_c - 0.00014W_a - 0.00006W_b - \lambda_1 - 0.08\lambda_2 = 0$$

and remembering the constraints

$$W_a + W_b + W_c = 1$$

$$0.11W_a + 0.15W_b + 0.08W_c = 0.11$$

Thus we have five equations in five unknowns. We will solve this system of equations using the same matrix algebra that we used in Chapter 6 to solve the simultaneous equations in the multiple regression model. However, the problem here is slightly more straightforward because the matrix is square (5×5).

First the partial derivatives and constraints are set out in matrix format

$$
\begin{array}{l}
0.00030W_a + 0.00010W_b - 0.00014W_c - 1 \; - 0.11\lambda_2 = 0 \\
0.00010W_a + 0.00050W_b - 0.00006W_c - \lambda_1 - 0.15\lambda_2 = 0 \\
-0.00014W_a - 0.00006W_b + 0.00020W_c - \lambda_1 - 0.08\lambda_2 = 0 \\
1W_a \qquad\qquad 1W_b \qquad\qquad 1W_c \qquad\qquad\qquad\quad = 1 \\
0.11W_a \qquad\quad 0.15W_b \qquad 0.08W_c \qquad\qquad\quad\; = 0.11
\end{array}
$$

Next we place the left-hand side elements into a 5 × 5 matrix, followed by a 5 × 1 vector of the variables including the λs. On the right-hand side we place an identity matrix and a 1 × 5 vector of the right-hand side of differentiation

$$
\begin{bmatrix}
0.00030 & 0.00010 & -0.00014 & -1 & -0.11 \\
0.00010 & 0.00050 & -0.00006 & -1 & -0.15 \\
-0.00014 & -0.00006 & 0.00020 & -1 & -0.08 \\
1 & 1 & 1 & 0 & 0 \\
0.11 & 0.15 & 0.08 & 0 & 0
\end{bmatrix}
\begin{bmatrix} W_a \\ W_b \\ W_c \\ \lambda_1 \\ \lambda_2 \end{bmatrix}
=
\begin{bmatrix}
1 & 0 & 0 & 0 & 0 \\
0 & 1 & 0 & 0 & 0 \\
0 & 0 & 1 & 0 & 0 \\
0 & 0 & 0 & 1 & 0 \\
0 & 0 & 0 & 0 & 1
\end{bmatrix}
\begin{bmatrix} 0 \\ 0 \\ 0 \\ 1 \\ 0.11 \end{bmatrix}
$$

We will begin by scaling both the first row of both matrices by 3333.33 in order to bring the element in the top left-hand corner to 1.

Next we subtract 0.10 times the new first row from the second row of both sides. Then we add 0.00014 times the new first row to the third row of both sides, then subtract 1 times the new first row from the fourth row and finally subtract 0.11 times times the new first row from the fifth row of both sides. The result is that the first column of the left-hand matrix has 1 in the top element and 0 for all other elements thus

$$
\begin{bmatrix}
1 & 0.3333 & -0.46666 & -333.33 & -366.66 \\
0 & 0.000466 & -0.000013 & -0.66666 & -0.11333 \\
0 & -0.000013 & 0.00020 & -1.4666 & -0.131333 \\
0 & 0.9999 & 1.46666 & 3333.33 & 366.666 \\
0 & 0.11333 & 0.13133 & 366.666 & 40.3333
\end{bmatrix}
\begin{bmatrix} W_a \\ W_b \\ W_c \\ \lambda_1 \\ \lambda_2 \end{bmatrix}
=
\begin{bmatrix}
3333.33 & 0 & 0 & 0 & 0 \\
-0.33333 & 1 & 0 & 0 & 0 \\
-0.46666 & 0 & 1 & 0 & 0 \\
-0.3333.33 & 0 & 0 & 1 & 0 \\
0366.666 & 0 & 0 & 0 & 1
\end{bmatrix}
\begin{bmatrix} 0 \\ 0 \\ 0 \\ 1 \\ 0.11 \end{bmatrix}
$$

We now proceed similarly, first scaling the second rows so that the second entry in the second column of the matrix on the left-hand side (0.000466) becomes one, and then subtracting or adding scaled copies of the row so as to make the remaining entries in the second column all zero. Repeating this process for the third, fourth and fifth columns of the left-hand matrix has the effect of transforming it to the identity matrix, whilst performing operations on the right-hand matrix produces the multiplicative inverse. We finally end up with

$$
\begin{bmatrix}
1 & 0 & 0 & 0 & 0 \\
0 & 1 & 0 & 0 & 0 \\
0 & 0 & 1 & 0 & 0 \\
0 & 0 & 0 & 1 & 0 \\
0 & 0 & 0 & 0 & 1
\end{bmatrix}
\begin{bmatrix} W_a \\ W_b \\ W_c \\ \lambda_1 \\ \lambda_2 \end{bmatrix}
=
\begin{bmatrix}
1991.87 & -853.659 & -1138.21 & 0.79350 & -4.22764 \\
-853.659 & 365.854 & 487.805 & -1.48293 & 16.0976 \\
-1138.21 & 487.805 & 650.407 & 1.68943 & -11.8699 \\
-0.79350 & 1.48293 & -1.68943 & 0.00155 & 0.01542 \\
4.22764 & -16.0976 & 11.8699 & -0.01542 & 0.15837
\end{bmatrix}
\begin{bmatrix} 0 \\ 0 \\ 0 \\ 1 \\ 0.11 \end{bmatrix}
=
\begin{bmatrix} 0.328 \\ 0.288 \\ 0.384 \\ -0.000157 \\ 0.002 \end{bmatrix}
$$

giving the result $W_a = 0.328$, $W_b = 0.288$ and $W_c = 0.384$ (with $\lambda_1 = -0.000157$ and $\lambda_2 = 0.002$).

Of course, the computations as described above are particularly tedious and virtually impossible to do by hand without error (even using a calculator). There are many efficient computing routines available to do the task, or we could use a spreadsheet. However, it is worth noting that is also possible to use a linear programming package. This is possible if the objective function is simply set to some artificial form and the constraints set out as below

$$0.00030W_a + 0.00010W_b - 0.00014W_c - \lambda_1 - 0.11\lambda_2 = 0$$
$$0.00010W_a + 0.00050W_b - 0.00006W_c - \lambda_1 - 0.15\lambda_2 = 0$$
$$-0.00014W_a - 0.00006W_b + 0.00020W_c - \lambda_1 - 0.08\lambda_2 = 0$$
$$W_a + W_b + W_c \; 0\lambda_1 \quad 0\lambda_2 = 1$$
$$0.11W_a + 0.15W_b + 0.08W_c \; 0\lambda_1 \quad 0\lambda_2 = 0.11$$

Effectively the linear programming package is simply being used to find a feasible solution to the system of equations. Because there are as many equations as there are unknowns there will in general be a unique feasible solution and therefore that must give the optimal solution to the optimization problem.

Quadratic programming with inequalities

In practical applications portfolio selection problems may often include a number of inequality constraints, often putting bounds on the amounts to be invested in particular areas. Suppose, for example, that we impose the extra constraint $W_c \leqslant 0.25$ on the above problem.

The problem here is that we do not know in advance whether or not our extra constraint will be binding. Of course we can solve this dilemma simply by solving two problems, one in which we have no constraint on x_3 and, if the answer gives $W_c > 0.25$, one in which we insist that $W_c = 0.25$. However, with two such constraints there would be four possibilities to check – neither constraint tight; the one tight but not the other; the other tight but not the one; both tight. With 10 such constraints there are $2^{10} = 1024$ possibilities. Whilst it is not possible to avoid this combinatorial explosion it would be helpful to be able to automate the search in some way. The Kühn–Tucker conditions provide the mechanism for this.

Note that we cannot now guarantee to be on the efficient frontier, so we must also relax our equality constraints on the total invested and/or on the required return – we may end up with a more risky portfolio but we may not need to use all of our money and/or we may end up with a larger return!

Kühn–Tucker conditions

The first step in this analysis is to manipulate the inequalities so that the problem is presented in **standard form**

Minimize $Z = f(\underline{W})$
Subject to $g_i(\underline{W}) = 0$ for i in some set E (i.e. the set of equality constraints)
                   $g_i(\underline{W}) \geqslant 0$ for i in some set I (the inequality constraints)

So, for our problem (note that there are now no equality constraints)

Minimize $\quad Z = 0.00015W_a^2 + 0.00025W_b^2 + 0.00010W_c^2 + 0.00010W_aW_b$
$\qquad\qquad\qquad - 0.00014W_aW_c - 0.00006W_bW_c$

Subject to $\quad 1 - W_a - W_b - W_c \geqslant 0$
$\qquad\qquad 0.11W_a + 0.15W_b + 0.08W_c - 0.11 \geqslant 0$
$\qquad\qquad 0.25 - W_c \geqslant 0$

Note that it is necessary to present the inequality $W_a + W_b + W_c \leqslant 1$ in the **standard form**, i.e. $1 - W_a + W_b + W_c \geqslant 0$.

Again the Lagrangian is constructed and analytical methods to find the unconstrained minimum of the Lagrangian are considered. However, there is the added complication that not all of the inequality constraints need be active. The analysis copes with this by allowing the corresponding Lagrange multipliers to be zero (there are exceptional circumstances in which Lagrange multipliers for active equations can also be zero).

This is expressed formally by searching for Kühn–Tucker points, that is points satisfying the Kühn–Tucker conditions

$$\frac{\delta L(\underline{W}, \lambda)}{\delta W_i} = 0 \text{ for each value of } i \tag{9.33}$$

and

$$g_i(\underline{W}) = 0 \text{ for } i \text{ in } E$$
$$g_i(\underline{W}) \geqslant 0 \text{ for } i \text{ in } I$$
$$\lambda_i \geqslant 0 \text{ for all } i$$
$$\lambda_i g_i(\underline{W}) = 0 \text{ for all } i$$

The last condition is called the complementarity condition, and is saying that inactive constraints have a zero multiplier.

Again, for our problem (noting the third Lagrange multiplier corresponding to $0.25 - W_c \geqslant 0$)

$$0.00030W_a + 0.00010W_b - 0.00014W_c + \lambda_1 - 0.11\lambda_2 = 0$$
$$0.00010W_a + 0.00050W_b - 0.00006W_c + \lambda_1 - 0.15\lambda_2 = 0$$
$$- 0.00014W_a - 0.00006W_b + 0.00020W_c + \lambda_1 - 0.08\lambda_2 + \lambda_3 = 0$$
$$1 - W_a - W_b - W_c \geqslant 0$$
$$0.11W_a + 0.15W_b + 0.08W_c - 0.11 \geqslant 0$$
$$0.25 - W_c \geqslant 0$$
$$\lambda_1 \geqslant 0$$
$$\lambda_2 \geqslant 0$$
$$\lambda_3 \geqslant 0$$
$$\lambda_i g_i(\underline{W}) = 0 \text{ for all } i.$$

Kühn–Tucker analysis lays the basis for an adaptation of LP to solve the problem.

Dantzig–Wolfe

Having established the Kühn–Tucker conditions we need to find a Kühn–Tucker point, i.e. a point satisfying each of these conditions. Note that we have nine linear constraints. Thus, apart from complementarity and the lack of an objective function, we have an LP. Our task is to find a feasible point as against an optimal point.

Finding a first feasible point is in fact the first step in the simplex algorithm. It involves creating an artificial solution in which all of the "real" variables have value zero, and in which artificial variables, $a_1, \ldots, a_9$ are added, one to each constraint, with values equal to the corresponding right-hand sides. The objective is then to minimize the sum of these a_is. When a solution is found with this sum equal to zero we will have a feasible solution to our problem.

So all that is required is to take that initial part of the simplex algorithm and add in the complementarity conditions. In our example there are three such conditions. They are

at least one of λ_1 and $(W_a + W_b + W_c - 1)$ must be zero
at least one of λ_2 and $(0.11W_a + 0.15W_b + 0.08W_c - 0.11)$ must be zero
at least one of λ_3 and $(0.25 - W_c)$ must be zero

These conditions are easy to check. Thus the use of the hitherto purely theoretical Kühn–Tucker conditions enables us to convert our particular quadratic programming problem into a modified linear programming problem.

The result produced for our problem is $W_a = 0.436$, $W_b = 0.280$ and $W_c = 0.25$. This is the minimum variance solution, but note that not all of our money is used, $W_a + W_b + W_c = 0.966$. This shouldn't be too surprising since our extra constraint is preventing us from using as much as we would like of the least risky asset. The solution is telling us that we can achieve our target return by using less than we can afford of the riskier assets.

If we insist on investing all of our cash then we can return to the LP again, changing the $W_a + W_b + W_c \leqslant 1$ back to $W_a + W_b + W_c = 1$. This produces the result $W_a = 0.533$, $W_b = 0.217$ and $W_c = 0.25$. The return is now 0.111, which is 0.1% more than we needed.

A summary of results is given in Table 9.2.

Table 9.2

	Mean/variance efficient portfolio	Min variance portfolio subject to additional constraint	Best investment subject to additional constraint and with all funds invested
Proportion invested in security no.1	0.328	0.436	0.533
Proportion invested in security no.2	0.288	0.280	0.217
Proportion invested in security no.3	0.384	0.25	0.25
Total proportion invested	1	0.966	1
Return	0.11	0.11	0.111
Risk (variance)	0.0000368	0.0000472	0.0000503
(standard deviation)	0.00607	0.00687	0.00709

A Cook's tour of hill-climbing methods

Increasingly computer packages, spreadsheets and so on have provided in-built optimization packages. There are many robust algorithms available, and no particular technique will cover all that may be met. However, we will now attempt to impart a flavour of them by indicating how the Markowitz problem might be solved by more general methods.

The Dantzig–Wolfe method is specific to the Markowitz portfolio management problem. There are many circumstances in which different quadratic programming problems arise, so more general approaches are needed. Increased computing power has allowed the development of quadratic programming packages using these more general, and thus less powerful, approaches. Indeed the Markowitz portfolio problems are now more likely to be solved by such packages.

The mathematical programming approach seen above involves moving from point to point around the boundary of the feasible region, always increasing the value of the objective function (for a maximization problem). Thus the focus is on the constraints, the objective providing directionality for the iteration. The alternative approach to optimization is to move from point to point within the feasible region, focusing on the objective and going as far as possible on each iteration within the constraints.

The general philosophy of the approach is to use the characteristics of the objective function at the current point to determine the direction of movement. A move is then made in that direction, within the feasible region, to a new point, and the process is repeated. The term "hill-climbing" is sometimes used to describe this approach. The choice of direction can be made by studying or estimating the gradient of the function – hence the term "gradient methods" – or by other means – "direct search methods".

One such approach is now used extensively to solve quadratic programming problems. There are various embodiments of the approach, and they are known as **active set methods**.

Active set methods for quadratic programming problems

Active set methods are search methods in which, at each iteration, a subset of the inequality constraints, together with all of the equality constraints, are *active*. The iteration then consists of attempting to move from the current point, which satisfies all active constraints as equalities, to the **best** point which satisfies all active constraints as equalities.

Finding that best point, i.e. solving an equality constrained quadratic programming problem, is relatively easy. It can, for instance, be done by using the equalities to eliminate variables or by the Lagrange multiplier approach above. But moving to that point may not be possible, since it may lead to an infringement of one or more of the inactive constraints. Accordingly a move is made as far as possible in the direction from the current point to the best point (for the current active set).

If the move stops short of the best point then that is because another constraint has become active. That constraint is then added to the active set and the process is ready to begin the next iteration.

If the best point (for the current active set) is reached, i.e. if it is feasible, then the Lagrange multipliers for the active constraints are computed. If these are all non-negative then a solution has been found. If any one is negative then this implies that the corresponding inequality may be allowed to become non-active. It is thus removed from the active set, preparing the way for a fresh iteration.

Exercises

1. What do you understand by the terms: objective function, constraints, mathematical programming, linear programming and quadratic programming?

2. Provide a graphical solution to the following linear programming problem.

   Maximize (return): $0.10W_a + 0.18W_b + 0.07W_c$

   Subject to:
   (a) $0.8W_a + 1.3W_b + 1.1W_c \leqslant 1.1$
   (b) $0 \leqslant W_a \leqslant 1$
       $0 \leqslant W_b \leqslant 1$
       $0 \leqslant W_c \leqslant 1$
   (c) $W_a + W_b + W_c = 1$

3. What do you understand by the terms "basic" and "non-basic" in relation to linear programming?

4. Use the simplex method to solve the problem set in question 2.

5. Consider three assets A, B and C, with expected returns of 8%, 18% and 7%, respectively. The variance–covariance matrix is

$$\Omega = \begin{bmatrix} 0.00013 & 0.00004 & -0.00008 \\ 0.00004 & 0.00022 & -0.00002 \\ -0.00008 & -0.00002 & 0.00009 \end{bmatrix}$$

   Set up the optimization problem which minimizes the variance of the portfolio subject to achieving a minimum return of 12%. Note: you do not have to solve the problem, just set out the objective function and the constraints.

6. Set out the Lagrangian applicable to the above problem.

7. What do you understand by the term "Kühn–Tucker conditions". Set out the Kühn–Tucker conditions appropriate to the problems in questions 5 and 6.

Answers to selected exercises

2. Use $W_a + W_b + W_c = 1$ to express W_c as $1 - W_a - W_b$.

   Substituting gives:

   Maximize $0.10W_a + 0.18W_b + 0.07(1 - W_a - W_b) = 0.03W_a + 0.11W_b + 0.07$

   Subject to $0.8W_a + 1.3W_b + 1.1(1 - W_a - W_b) \leqslant 1$, which simplifies to $W_b \leqslant 1.5W_a$
                $0 \leqslant W_a \leqslant 1$
                $0 \leqslant W_b \leqslant 1$
                $0 \leqslant W_a + W_b \leqslant 1$
   Solution: $W_a = 0.4; W_b = 0.6; W_c = 0;$ Return $= 0.148$

4. Using the transformed, two-variable problem, as in question 2 (note that the second row states that $-3W_a + 2W_b \leq 0$, which is the same as $W_b \leq 1.5W_a$):

Ret	W_a	W_b	s_1	s_2	s_3	s_4	RHS
1	−0.03	−0.11	0	0	0	0	0.07
0	−3	2	1	0	0	0	0
0	1	0	0	1	0	0	1
0	0	1	0	0	1	0	1
0	1	1	0	0	0	1	1
1	−0.195	0	0.055	0	0	0	0.07
0	−1.5	1	0.5	0	0	0	0
0	1	0	0	1	0	0	1
0	1.5	0	−0.5	0	1	0	1
0	2.5	0	−0.5	0	0	1	1
1	0	0	0.016	0	0	0.078	0.148
0	0	1	0.2	0	0	0.6	0.6
0	0	0	0.2	1	0	−0.4	0.6
0	0	0	−0.2	0	1	−0.6	0.4
0	1	0	−0.2	0	0	0.4	0.4

5. Let W_a, W_b and W_c be the weights of each asset in the portfolio.

Minimize $0.00013W_a^2 + 0.00022W_b^2 + 0.00009W_c^2$
$\qquad\qquad + 0.00008W_aW_b - 0.00016W_aW_c - 0.00004W_bW_c$
Subject to $0.08W_a + 0.18W_b + 0.07W_c \geq 0.12$
$\qquad\qquad 0 \leq W_a \leq 1$
$\qquad\qquad 0 \leq W_b \leq 1$
$\qquad\qquad 0 \leq W_c \leq 1$
$\qquad\qquad W_a + W_b + W_c \leq 1$

6. The problem is simplified considerably by insisting that $W_a + W_b + W_c = 1$, i.e. that all of the money is invested. Furthermore, if short selling is allowed then the $0 \leq \ldots \leq 1$ inequalities are not needed. The Lagrangian for this simplified problem is

$L(W_a, W_b, W_c; \lambda, \mu) = 0.00013W_a^2 + 0.00022W_b^2 + 0.00009W_c^2$
$\qquad\qquad + 0.00008W_aW_b - 0.00016W_aW_c - 0.00004W_bW_c$
$\qquad\qquad - \lambda(0.12 - 0.08W_a - 0.18W_b - 0.07W_c)$
$\qquad\qquad - \mu(W_a + W_b + W_c - 1)$

7. Ignoring the $0 \leqslant \ldots \leqslant 1$ inequalities

$$0.00026W_a + 0.00008W_b - 0.00016W_c + 0.08\lambda - \mu = 0$$
$$0.00044W_b + 0.00008W_a - 0.00004W_c + 0.18\lambda - \mu = 0$$
$$0.00018W_c - 0.00016W_a - 0.00004W_b + 0.07\lambda - \mu = 0$$
$$0.08W_a + 0.18W_b + 0.07W_c \geqslant 0.12$$
$$W_a + W_b + W_c \leqslant 1$$
$$\lambda \geqslant 0$$
$$\mu \geqslant 0$$
$$\lambda(0.12 - 0.08W_a - 0.18W_b - 0.07W_c) = 0$$
$$\mu(W_a + W_b + W_c - 1) = 0$$

References and further reading

Fletcher, R. (1987) *Practical Methods of Optimization*, 2nd edn. John Wiley, New York.
Wilkes, F.M. (1994) *Mathematics for Business Finance and Economics*. Routledge, London.
Markowitz, H. (1952) Portfolio selection. *Journal of Finance*, pp. 77–99.

10

Continuous time mathematics in finance: asset prices as a stochastic process

Introduction

A **stochastic process** is a process that describes the change in one or more variables, where the changes are subject to uncertainty. It is particularly appropriate to apply these processes to analysing the future changes in asset prices because these changes are indeed uncertain.

In finance we are particularly concerned with two broad groups of stochastic processes. Discrete time/discrete variable processes are those that allow discrete variables to change at discrete intervals of time. We have already met these stochastic processes in the form of the binomial and trinomial models in Chapter 8.

In this chapter we are concerned with continuous time/continuous variable stochastic process; processes that allow continuous variables, asset prices and asset returns in this chapter, to change continuously in time. The mathematics of such processes is known as continuous time mathematics and the processes lead to applications of stochastic calculus.

This chapter will begin with an explanation of continuous time stochastic processes of asset prices and returns. It will then explain the application of stochastic calculus to the pricing of derivatives. Finally, it will introduce the concept of risk neutrality and then go on to explain in detail the pricing of derivative securities within a risk neutral framework, applying the model developed by Black and Scholes (1973).

The stochastic process of asset prices

Asset prices cannot be negative but can be infinitely positive. As a result price relatives i.e. P_1/P_0, cannot be negative but can be infinitely positive. This empirical fact is the basis for suggesting that security price relatives are lognormally distributed, and that continuously compounded security returns, i.e. $\ln P_1/P_0$, are normally distributed. When the individual observations of the random variable are ordered, as they are when the observations occur over time, the random variable, security price returns in our example, may follow a random walk, a special case of a martingale. A continuous time random walk is referred to as a diffusion process. The characteristics of martingales, and the particular example of a random walk, were studied in Chapter 7.

There are a group of random variables where the displacement is dependent on the immediately previous state of that variable. Such a stochastic process assumes that only the current value of the stochastic (random) variable is of any importance in predicting future values of that variable. Such stochastic processes are known as **Markov process**, after the Russian mathematician of that name. However, the important feature of all the processes that interest us is that the stochastic change or innovation of the variable is independent and identically distributed (IID).

A body of financial theory, known as the efficient markets hypothesis (EMH), suggests that asset prices reflect all historical information regarding the asset, and that markets respond immediately to new information regarding that asset. This response is manifested by a change in price. If markets do indeed respond to new information immediately and each piece of new information is independent of previous information, changes in asset prices will follow a Markov process.

Although the history of the movements of the random variable over time is of no use in predicting future movements, the statistical characteristics of the past movements may be useful in predicting the probability distribution of future movements. For example, the mean and standard deviation (the volatility) of past movements may be useful in predicting future movements in a probabilistic sense.

There is a whole family of Markov processes; some, such as the binomial and trinomial models, we have already met. In this chapter, we will be interested in three continuous time/continuous variable processes, the basic Wiener process, the generalized Wiener process and the Ito process.

The Wiener process – also known as Brownian motion

A particular type of Markov process that is used as a starting point to specify the stochastic process of asset prices is the basic **Wiener process** or **Brownian motion.** It describes the stochastic process of a variable that is subjected to numerous small shocks. This would seem, at first glance, to be a good place to start the description of the process of asset prices. Asset prices are often hypothesized to change randomly through time because of the cumulative effects of numerous independent random shocks caused by the receipt of pieces of information.

The basic Wiener process

In order to understand the relevance of Wiener process to the movement of asset prices, we will begin by explaining the basic Wiener process.

Consider S to be any random variable, and t to be time. Over a small interval of time, Δt, that random variable, S, can change by ΔS. If S follows a Wiener process, i.e. Brownian motion, the change in S over a small interval of time, i.e. ΔS, will be related to Δt by the equation

$$\Delta S = \epsilon \sqrt{\Delta t} \tag{10.1}$$

where ϵ is a random sampling from a standardized normal variable, with a mean of zero and a standard deviation of one.

At the limit this is written as

$$dS = \epsilon \sqrt{dt} \tag{10.2}$$

Since ϵ is a standard normal variable, it follows that ΔS must be normally distributed with a mean of zero, a variance of Δt, and a standard deviation of $\sqrt{\Delta t}$.

Thus in effect we have a variable S, that is changing randomly by ΔS, because it is dependent on another random variable, $\epsilon \sqrt{\Delta t}$ (maybe the effect of randomly arriving new information), that has an expected value of zero, a variance of Δt and a standard deviation of $\sqrt{\Delta t}$.

It is an important property of the Wiener process that the variables are independent. It then follows that if the separate values of ΔS are independent and as the ϵs come from a normal distribution, ΔS is IID ($\Delta S \sim N(0, \sqrt{\Delta t})$). As the variance of each individual ΔS is Δt, the variance over a longer time period will be $\Sigma \Delta t$.

To understand this, consider independent observations of ΔS over two, very small, but consecutive time periods, t_1 and t_2. As the observations of S are independent, the variance of the observations during the combined time period T ($T = t_1 + t_2$), will be the sum of the variances in each of the shorter time periods $\text{Var}_T = \text{Var } t_1 + \text{Var } t_2$.

Consequently, if the random variable follows a basic Wiener process, the change over a long period of time ($T = \Sigma \Delta t$), $\Delta S(T)$ will have a mean (expected value) of zero, but a variance of T and a standard deviation of $\sqrt{T}$. Thus the change in the random variable will have an expected value of zero and a standard deviation equal to the square root of the future time period in question.

Figure 10.1 illustrates a Wiener process. Note how the degree of variability of the level tends to increase as the time interval Δt increases, whereas the degree of variability of the innovations is constant. This is further demonstrated in Fig. 10.2 where we have three instants of the same process.

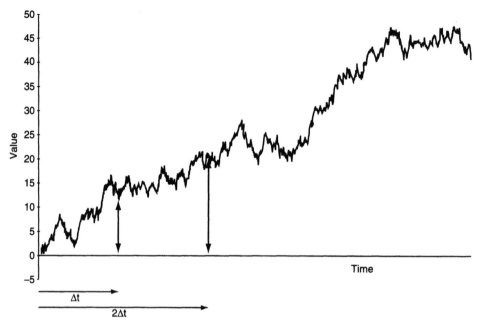

Figure 10.1 Brownian motion.

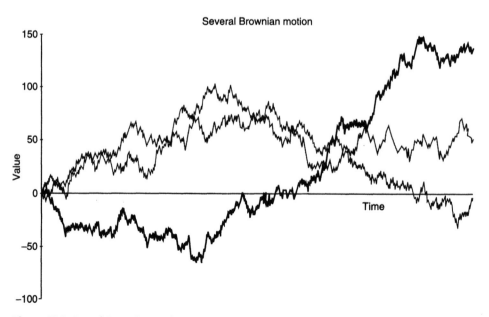

Figure 10.2 Several Brownian motions.

Can we apply this basic Wiener process to describe the stochastic process of asset prices? Unfortunately, not in its current form. Applying the basic Wiener process to asset prices is not valid in three counts:

1. Assets have differing degrees of volatility. In the process described above, the volatility was one.
2. Risky assets have positive expected return on average. In the above process, the mean value of ΔS was assumed to be zero. Thus on average the future price will be the current price.
3. The Wiener process assumes that the absolute change in price, ΔS, is independent of the magnitude of S. However, in reality we would not expect this to be the case. We would expect the absolute price changes of high priced assets to be larger on average than the absolute changes of low priced assets. It is the proportional changes in asset prices, $\Delta S/S$, that we would expect to be independent of S; a proportional or percentage change could be the same irrespective of the price of the assets. Moreover, as an indicator of the importance of a given change in the asset price, it is more appropriate to measure the relative return $\Delta S/S$ rather than ΔS.

Problem 1: different assets have different degrees of randomness or volatility

Firstly, we know that different securities have different degrees of volatility, thus the impact of ϵ will differ between securities as reflected by the volatility of the security.

This problem is easily addressed by scaling ϵ by σ, the annualized standard deviation of ΔS. Thus $\sigma\epsilon$ still has an expected value of zero, but now has a standard deviation of $\sigma * 1 = \sigma$.

Thus equation (10.2) becomes

$$\Delta S = \sigma\epsilon\sqrt{\Delta t} \tag{10.3}$$

And again at the limit this becomes

$$dS = \sigma\epsilon\sqrt{dt} \tag{10.4}$$

ΔS has an expected value of zero, a standard deviation of $\sigma\sqrt{\Delta t}$ and a variance of $\sigma^2\Delta t$. The right-hand side of equation 10.4, the $\epsilon\sqrt{dt}$, is often abbreviated to dz.

Problem 2: risky assets have a positive expected return

The second problem arises because the expected value of the random variable, ϵ, is zero and thus the expected change in S, ΔS, is also zero. Yet the expected return from holding risky assets must be positive, on average, in order to reward investors for bearing risk. We must therefore adapt the basic Wiener process to a generalized Wiener process to account for positive expected return.

A **generalized Wiener** process is a basic Wiener process with a "drift" parameter added.

With regard to stochastic processes, the term drift is used to represent the positive or negative trend in the time series of the stochastic variable. When that variable is a financial asset it is appropriate to assume a **positive drift** because, as noted above, risky assets should offer a positive return to compensate investors for the risk that they bear. Thus drift is analogous to expected return.

The drift parameter α, represents the change in S, per small unit of time, dt. If we were only considering the drift, the dS resulting from an expected return of α per small unit of time, would be αdt. However, combined with the Wiener process, the stochastic process of a random variable that has both a drift rate and basic Wiener process. That random variable now has an expected change for two reasons. The first reason is our expectation of return over a small interval of time, αdt. The second reason is the random change $\sigma \epsilon \sqrt{dt}$ described by the basic Wiener process discussed earlier. Thus a small change in the asset price over a small interval in time can be modelled by a **stochastic differential equation** as follows:

$$dS = \alpha dt + \sigma \epsilon \sqrt{dt} \tag{10.5}$$

This can also be written in discrete form as

$$\Delta S = \alpha \Delta t + \sigma \epsilon \sqrt{\Delta t} \tag{10.6}$$

As ΔS is normally distributed, the mean (expected value) is $\alpha \Delta t$, the standard deviation is $\sigma \sqrt{\Delta t}$ and the variance of ΔS is $\sigma^2 \Delta t$. Thus σ^2 becomes the variance per unit of time; σ is known as volatility.

Therefore, the generalized Wiener process incorporates a basic Wiener process and a drift element. The drift element is deterministic, i.e. non-random, and the basic Wiener process is the stochastic element. Figure 10.3 illustrates such a process. The effect of the positive drift is clear to see. A generalized Wiener process can be considered to be the continuous time equivalent of a sub-martingale.

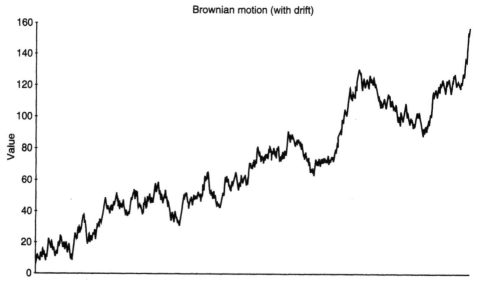

Figure 10.3 Brownian motion (with drift).

Problem 3: magnitude of changes in the asset price should be independent of the magnitude of the asset price

To understand this problem, recall that in the above argument, α is the **absolute** return per unit of time, and is constant, but is not independent of the asset price. However, investors require a **percentage rate** of return dependent upon the risks involved and therefore which is independent of the level of the asset price. Thus if investors need 8% return on a given asset, they require 8% whether the asset price is £1 or £10.

Thus the desired stochastic process of asset prices should incorporate an absolute return which is a function of the asset price but a rate of return which is independent of the asset price.

Thus to satisfy our needs for a model of asset prices, it is necessary for the general Wiener process developed so far, to be replaced by a more general type of stochastic process known as an **Ito process**.

An Ito process is a generalized Wiener process in which the parameters α, the expected return and, σ^2, the variance, are both functions of the underlying variables. In a generalized form the Ito process is $dx = \alpha(x, t)dt + \sigma(x, t)\epsilon\sqrt{dt}$. For us the underlying variables are the asset price (S) and time (t). Therefore our Ito process is written as $dS = \alpha(S, t)dt + \sigma(S, t)\epsilon\sqrt{dt}$. As the underlying variables change, so the absolute rate of drift changes and so does the volatility.

To convert our Wiener process into an Ito process, we will specify μ as the expected **rate return** expressed in decimal form, so μS is the absolute return. Thus for a small period of time, Δt, the expected absolute return will be $\Delta S = \mu S \Delta t$, which at the limit is $dS = \mu S dt$. Dividing both sides by S, gives the rate of return of $dS/S = \mu dt$.

So we have an absolute change in the asset price ΔS which is a function of the asset price, i.e. $\mu S \Delta t$, and a rate of return, dS/S, which is independent of the asset price, i.e. $= \mu dt$.

Although we may specify the absolute expected return to be a function of the asset price, the degree of uncertainty regarding that expected return, over a small period of time, will be independent of the asset price. That is, that the investor is equally uncertain about future outcomes if the asset price is £1 or £10.

Thus dS/S is a function of μdt and $\sigma\epsilon\sqrt{dt}$. Combining these we have

$$\frac{dS}{S} = \mu dt + \sigma\epsilon\sqrt{dt}$$

$$\text{(10.7)}$$

$$\text{or } dS = \mu S dt + \sigma S \epsilon\sqrt{dt}$$

which is an Ito process. Whereas Wiener processes are often referred to as Brownian motion, our particular Ito process is sometimes referred to as geometric Brownian motion.

Figure 10.4 shows an example of an Ito process (geometric Brownian motion). Notice the exponential-like growth which is a consequence of the rate of growth being proportional to the level of the process, S.

We have already noted that although the degree of uncertainty will be independent of the asset price, the absolute return will be larger the higher the asset price. Thus the actual dispersion of the expected asset prices will depend on the magnitude of the current asset price. Consequently, the standard deviation of the absolute change will be larger, the higher the asset price.

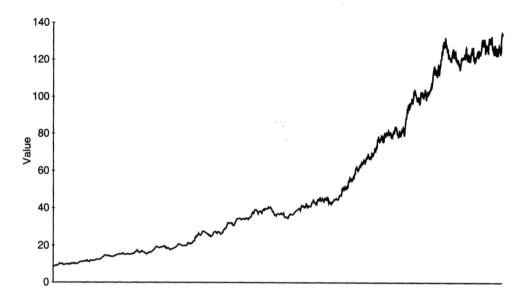

Figure 10.4 Ito process.

The variance of the actual change in the asset price, which depends on the magnitude of the asset price, and is given as

$$\sigma^2 S^2 \Delta t \tag{10.8}$$

and the resulting instantaneous variance becomes

$$\sigma^2 S^2 \tag{10.9}$$

These results arise because, if we specify the standard deviation to be proportional to the asset price, the variance will be proportional to the square of the asset price.

Expressing variability in terms of standard deviations, we now have the expected (deterministic) element of return expressed as a function of the asset price and time, and we also have a stochastic element of the variability of the asset price as a function of the asset price and time. Such a model of the movement in the price of an asset can therefore be represented in the form

$$dS = \mu S dt + \sigma S \epsilon \sqrt{dt} \tag{10.10}$$

which is an Ito process because the asset prices and the variability of those prices are a function of the underlying variable, the asset price, S, and of time t.

The first element, the mean, $\mu S dt$, is fixed, i.e. non-stochastic. The second element is the stochastic component, $\sigma S \epsilon \sqrt{dt}$, which causes the whole function to be stochastic. Moreover as ϵ is drawn from a standard normal distribution, dS is normally distributed, but with a mean of $\mu S dt$, and a standard deviation of $\sigma S \sqrt{dt}$. In addition, dS/S is normally distributed, with a mean of μdt, and a standard deviation of $\sigma \sqrt{dt}$.

Applying Ito's lemma to derivative pricing

We explained above how the asset prices follow an Ito process

$$dS = \mu S dt + \sigma S \epsilon \sqrt{dt} \qquad (10.11)$$

We can use knowledge of this Ito process to value derivative securities such as forwards, futures and options. Ito (1951) showed that any variable that is a function of another variable which follows an Ito process, will itself follow an Ito process of the form

$$dW = \left(\frac{\partial W}{\partial X} \alpha + \frac{\partial W}{\partial t} + \frac{1}{2} \frac{\partial^2 W}{\partial X^2} \sigma^2 \right) dt + \frac{\partial W}{\partial X} \sigma \epsilon \sqrt{dt} \qquad (10.12)$$

where W is a function (derivative) of X, and X itself follows the Ito process

$$dX = \alpha dt + \sigma \epsilon \sqrt{dt} \qquad (10.13)$$

Thus W also follows an Ito process. The parameters, α, σ^2 and σ, in equation (10.12) are the drift rate (expected return), instantaneous variance and volatility respectively, of X, the original variable that follows an Ito process. Note that there is a basic Wiener process ($\epsilon \sqrt{\delta t}$) within each of these Ito processes.

The price of a derivative security, W, is a function of the price of the underlying security X and time. The derivative variable, W, has a drift rate of

$$\frac{\partial W}{\partial X} \alpha + \frac{\partial W}{\partial t} + \frac{1}{2} \frac{\partial^2 W}{\partial X^2} \sigma^2 \qquad (10.14)$$

and a volatility rate of

$$\left(\frac{\partial W}{\partial X} \right) \sigma \qquad (10.15)$$

Applying Ito's lemma to derivatives, recall that S is the price of the underlying asset, the expected return, variance and standard deviation of the underlying asset price were given as μS, $\sigma^2 S^2$ and σS. So a derivative variable W, which is a function of S and t, will follow the following Ito process

$$dW = \left(\frac{\partial W}{\partial S} \mu S + \frac{\partial W}{\partial t} + \frac{1}{2} \frac{\partial^2 W}{\partial S^2} \sigma^2 S^2 \right) dt + \frac{\partial W}{\partial S} \sigma S \epsilon \sqrt{dt} \qquad (10.16)$$

We shall see below how this result can be used in valuing derivatives.

Pricing derivatives in a risk-neutral framework

Risk neutrality is an artificial state of the world where investors are assumed to be indifferent towards risk. As such they neither demand a risk premium for bearing risk, nor do they pay others for bearing risk. Consequently in a risk-neutral world all risky assets earn only the risk-free rate of return. Such a world is not the world which we generally recognize as representing our financial markets. However, this intellectual artefact is very useful in enabling us to overcome three interrelated problems associated with the pricing of derivative securities.

Firstly, we need to know the investor's expectation of return for the particular asset in question. This will provide us with the drift rate, μ.

Secondly, we need to know the investor's expectation of risk, the spread of the probability distribution of future returns of the asset. This is proxied by the standard deviation.

Thirdly, we need to know the return required by investor's. This will give us the discount rate to be used in the derivatives valuation. The required return will be related to investors expectation of risk, and their degree of risk aversion. Yet we have no way of knowing what those risk preferences are.

All three variables, expected return, expected risk and the degree of risk aversion, will differ between investors. However, there is a clever solution to two of our problems. Recall that there is a stochastic process $\sigma S \epsilon \sqrt{dt}$ found in the Ito process of the underlying variable and in the Ito process of the derivative, but applied in different proportions. Therefore, it is possible to go long the underlying variable and short the derivative variable so that the two stochastic processes will cancel out. This is possible in the asset markets, for example, by buying the underlying asset and selling some number of the derivative in question, say a call option. Recall that this is the way in which the one period binomial model was developed in Chapter 8.

Remember that it is the stochastic process $\sigma S \epsilon \sqrt{dt}$ which makes both the underlying variable, and the derivative, stochastic and thus risky. Consequently, a portfolio that is long the underlying variable and short the derivative variable, in such a way that the two stochastic processes cancel out, must be riskless. As such, in efficient financial markets, that combination (portfolio) should earn only the risk-free rate of interest. Consequently, only the risk-free rate of interest need be used to discount the future value to its present value.

To see the reasoning behind this, recall that the Ito process of the underlying asset S is

$$dS = \mu S dt + \sigma S \epsilon \sqrt{dt} \tag{10.17}$$

and that the Ito process of W, which is a derivative of S is

$$dW = \left(\frac{\partial W}{\partial S} \mu S + \frac{\partial W}{\partial t} + \frac{1}{2} \frac{\partial^2 W}{\partial S^2} \sigma^2 S^2 \right) dt + \frac{\partial W}{\partial S} \sigma S \epsilon \sqrt{dt} \tag{10.18}$$

Now representing the asset S and its derivative W by their respective Ito process, if we construct a portfolio, Π, that is short one unit of the derivative security and long $\partial W/\partial S$ units of the underlying security **the change** in the value of the portfolio, $\delta\Pi$, will be

$$\delta\Pi = -\left[\left(\frac{\partial W}{\partial S} \mu S + \frac{\partial W}{\partial t} + \frac{1}{2} \frac{\partial^2 W}{\partial S^2} \sigma^2 S^2 \right) \delta t + \frac{\partial W}{\partial S} \sigma S \epsilon \sqrt{\delta t} \right]$$

$$+ \frac{\partial W}{\partial S} \left(\mu S \delta t + \sigma S \epsilon \sqrt{\delta t} \right) \tag{10.19}$$

$$= \left(-\frac{\partial W}{\partial t} - \frac{1}{2} \frac{\partial^2 W}{\partial S^2} \sigma^2 S^2 \right) \delta t$$

To show how the final result is arrived at, we have expanded the second line of equation (10.19) and shown the cross-cancelling

$$\delta\Pi = -\left[\frac{\partial W}{\partial S}\mu S\delta t + \frac{\partial W}{\partial t}\delta t + \frac{1}{2}\frac{\partial^2 W}{\partial S^2}\sigma^2 S^2\delta t + \frac{\partial W}{\partial S}\sigma S\epsilon\sqrt{\delta t}\right]$$

$$+\frac{\partial W}{\partial S}\mu S\delta t + \frac{\partial W}{\partial S}\sigma S\epsilon\sqrt{\delta t}$$

$$= -\frac{\partial W}{\partial t}\delta t - \frac{1}{2}\frac{\partial^2 W}{\partial S^2}\sigma^2 S^2\delta t$$

$$= \left(-\frac{\partial W}{\partial t} - \frac{1}{2}\frac{\partial^2 W}{\partial S^2}\sigma^2 S^2\right)\delta t$$

(10.20)

Recall that our portfolio, Π, was short one unit of derivative W, and long $\partial W/\partial S$ of the underlying asset S. As our portfolio is instantaneously riskless, it should only earn the instantaneous risk-free rate of interest. Thus any change in Π over a small interval of time, should equal the risk-free rate of return times the value of the risk-free portfolio over the same small interval of time. Thus

$$\delta\Pi = \left(-\frac{\partial W}{\partial t} - \frac{1}{2}\frac{\partial^2 W}{\partial S^2}\sigma^2 S^2\right)\delta t = r\left(-W + \frac{\partial W}{\partial S}S\right)\delta t$$

(10.21)

Rearranging gives us

$$\frac{\partial W}{\partial t} + rS\frac{\partial W}{\partial S} + \frac{1}{2}\sigma^2 S^2\frac{\partial^2 W}{\partial S^2} = rW$$

(10.22)

This is a particularly useful finding because it means that the expected return or drift rate, μ, can be replaced by the risk-free rate of interest, r. The consequences are that we do not need to know the expected rate of return of the underlying asset when pricing the derivative, so problem one is solved. In addition as the portfolio is risk free, at least for the instant in time, dt, only the risk-free rate of interest need be used to discount the future value of the derivative to its present value, or current price.

However, note that our second problem, the need to know the standard deviation, or volatility, of the underlying variable, is not solved! In effect we still need to know the probability distribution of the underlying asset.

The Black–Scholes partial differential equation and option pricing

Equation (10.22) is the partial differential equation developed by Black and Scholes (1973) to value any derivative. There are different solutions to the equation for different types of derivatives, each solution is dependent on the boundary conditions of the type of derivative being priced. In the case of European options, the boundary conditions are the same as those explained in Chapter 8, i.e.

$$C = [\max 0, S - X]$$

$$P = [\max 0, X - S]$$

Boundary conditions play a particularly important role in the solving of partial differential equations in that solving such equations is analogous to integration in Chapter 3. The partial differential equation describes rates of change of a variable, it does not describe that level of that variable. We need other information to fix the level, and that information is provided by the boundary conditions.

Assumptions – the Ito process and lognormality

Amongst the various assumptions made by Black and Scholes, two are particularly important here. The first is that security prices follow an Ito process. The second is that security prices are lognormally distributed and thus the continuously compounded returns are normally distributed. These two assumptions are inextricably linked as will now be shown.

The Ito process

If it is assumed that a security price, S follows an Ito process

$$dS = \mu S dt + \sigma S \epsilon \sqrt{dt} \tag{10.23}$$

the natural log of that variable will be normally distributed

$$\ln S_T \sim N\left[\ln S_t + \left(\mu - \frac{\sigma^2}{2}\right)(T-t), \sigma^2(T-t)\right] \tag{10.24}$$

where t is the current time, T is the future point in time and $T - t$ is thus the period of time over which the security is being analysed.

To understand this, let $W = \ln S$. Since W is a function of S it follows that if S follows an Ito process so W will follow an Ito process

$$W = \ln S$$

$$\frac{\partial \ln S}{\partial S} = \frac{1}{S}$$

$$\frac{\partial^2 \ln S}{\partial S^2} = -\frac{1}{S^2} \tag{10.25}$$

$$\frac{\partial \ln S}{\partial t} = 0$$

These derivatives can be confirmed by recalling pages 88–92 in Chapter 3.

Recalling from equation (10.18) above that the Ito process for the derivative W was given as

$$dW = \left(\frac{\partial W}{\partial S}\mu S + \frac{\partial W}{\partial t} + \frac{1}{2}\frac{\partial^2 W}{\partial S^2}\sigma^2 S^2\right)dt + \frac{\partial W}{\partial S}\sigma S \epsilon \sqrt{dt} \qquad (10.26)$$

and substituting the values in equation (10.25) into equation (10.26) we get

$$d\ln S = \left(\frac{1}{S}\mu S + 0 + \left(-\frac{1}{2}\frac{1}{S^2}\sigma^2 S^2\right)dt + \frac{1}{S}\sigma S \epsilon \sqrt{dt}\right) \qquad (10.27)$$

Cancelling out and recalling that ϵ is assumed to have a value of one, gives us

$$d\ln S = \left(\mu - \frac{\sigma^2}{2}\right)dt + \sigma \epsilon \sqrt{dt} \qquad (10.28)$$

Thus $\ln S$ follows a Weiner process.

The lognormal distribution

It follows from the appendix of Chapter 4 we showed that a variable is lognormally distributed if its natural logarithm is normally distributed. Moreover, the lognormal distribution is an attractive one to assume in relation to asset prices because it has a range from 0 to $+\infty$. This is exactly the theoretical range of asset prices because they cannot be negative but could attain very high positive values.

Moreover, we showed in Chapter 4 that if the current asset price, S_t, is indeed lognormally distributed, the price at future time T, $\ln S_T$ will be normally distributed

$$\ln S_T \sim N\left[\ln S_t + \left(\mu - \frac{\sigma^2}{2}\right)(T-t), \sigma^2(T-t)\right] \qquad (10.29)$$

It is known that $\ln S_T - \ln S_t = \ln S_T/S_t$. Thus the difference between two logs of a variable is the continuously compounded rate of return for the period $T - t$. This will be normally distributed with a mean and standard deviation as follows:

$$\text{Mean} = \left(\mu - \frac{\sigma^2}{2}\right)(T-t)$$
$$\qquad (10.30)$$
$$SD = \sigma\sqrt{T-t}$$

If we rewrite equation (10.30) in terms of the very small intervals of time underlying continuous time mathematics we get

$$\text{Mean} = \left(\mu - \frac{\sigma^2}{2}\right)\delta t$$
$$\qquad (10.31)$$
$$SD = \sigma\sqrt{\delta t}$$

which is exactly what we obtained from assuming that asset prices followed an Ito process. Thus the assumption that an Ito process underlies asset prices and that asset prices are lognormally distributed are compatible.

Black and Scholes recognized the partial differential equation that they had developed was analogous to the equation that describes the diffusion of heat over time through a solid. The solution to the so-called "heat equation" was already explained by Churchill (1963). Thus with the given boundary conditions and the assumptions that r and σ are constant they were able to develop an exact or "closed-form solution" to the value of a European option on an asset which makes no cash distribution such as dividends during the life of the option.

That exact solution for a European call is

$$c = SN(d_1) - Xe^{-r(T-t)}N(d_2) \qquad (10.32)$$

where

$$d_1 = \frac{\ln(S/X) + (r + \sigma^2/2)(T-t)}{\sigma\sqrt{(T-t)}} \qquad (10.33)$$

$$d_2 = d_1 - \sigma\sqrt{T-t} \qquad (10.34)$$

and

$$
\begin{aligned}
c &= \text{call option premium} \\
S &= \text{current asset price} \\
X &= \text{exercise price} \\
T-t &= \text{time to expiry in decimals of a year} \\
\sigma &= \text{the standard deviation of the asset price in decimals} \\
\ln &= \text{natural logarithm} \\
N(\cdot) &= \text{cumulative standard normal probability distribution} \\
d_1 \text{ and } d_2 &= \text{standardized normal variables} \\
r &= \text{risk-free rate of interest in decimals}
\end{aligned}
$$

To give an example assume that the current asset price is 35.0, the exercise price is 35.0, the risk-free rate of interest is 10%, the volatility is 20% and the time to expiry is one year. Thus $S = 35$, $X = 35$, $(T-t) = 1.0$, $r = 0.1$, and $\sigma = 0.2$.

First we calculate d_1, then d_2 and the present value of the exercise price $Xe^{-r(T-t)}$

$$d_1 = \frac{\ln(35/35) + (0.1 + 0.2^2/2)*1.0}{0.2\sqrt{1.0}} = 0.60$$

$$d_2 = d_1 - 0.2\sqrt{1.0} = 0.40$$

and

$$Xe^{-r(T-t)} = 35e^{-(0.1*1.0)} = 31.66934$$

The equation for the call then looks like this

$$c = 35N(0.6) - 31.6693N(0.4)$$

The next step is to look up the values of a **cumulative standardized normal probability distribution** at points 0.6 and 0.4 from tables available for the purpose. An alternative procedure for calculating the values of the cumulative standardized normal probability distribution is to fit a polynomial as explained in Chapter 8.

Given that d_1 is a standardized normal random variable, and remembering that $N(d_1)$ is a cumulative standardized normal probability distribution, $N(d_1)$ represents the area under the standardized normal curve from $z = -\infty$ to $z = d_1$.

These values are 0.7257 and 0.6554 and are included in the equation as follows:

$$c = 35(0.7257) - 31.6693(0.6554) = 4.6434$$

Thus the price of the call option in question is 4.6434. The reader may wish to compare this price with those derived from the binomial and the trinomial models discussed in Chapter 8.

Exercises

1. What do you understand by the term "Markov process"?

2. Explain what a basic Wiener process is. Why is it that if a variable follows a basic Wiener process, mean change will be 0, and the standard deviation of the changes will be $\sqrt{T}$?

3. What do you understand to be a generalized Wiener process?

4. What is an Ito process? Why is it more applicable to describing the stochastic process of asset prices than either the basic Wiener process or the generalized Wiener process?

5. Explain, with the use of an Ito process, how it is possible to value derivative instruments in a risk-neutral framework.

6. Explain how two of the assumptions of the Black–Scholes option pricing model, viz. (1) that security prices are lognormally distributed; (2) that security prices follow an Ito process; are compatible.

7. Using the Black–Scholes option model for a European option on an equity that does not pay dividends, derive the value of a three-month **call** option, where the asset price and the exercise price are both 35, the short-term rate of interest is 10% p.a., the volatility is 20%.

8. Compare your result for questions 7 with that for question 3 in Chapter 8. Explain any differences.

9. Explain how finite difference methods can be applied to solve partial differential equations. Explain the difference between the working of the explict finite difference method and the implicit finite difference method.

10. Explain how to manage problems regarding convergence and stability in the explicit finite difference method.

11. From the data in question 7 value the call option using the explicit finite difference method and a six-period grid. Compare your results with those of question 3 in Chapter 8.

Answers to selected questions

7. 1.85.

11.

X	delta t	r	delta S	vol
35	0.04	0.1	2.5	0.2
		0.0066667		

S							
50							15
47.5						12.645228	12.5
45					10.289854	10.145228	10
42.5				7.9338793	7.7898538	7.6452282	7.5
40			5.603448	5.443196	5.2898538	5.1452282	5
37.5		3.4588128	3.2569941	3.0500355	2.8418287	2.6452282	2.5
35	1.7827782	1.5772272	1.3558335	1.1109992	0.8284516	0.4792531	0
32.5		0.4649837	0.3272028	0.1958625	0.0801407	0	0
30			0.0370406	0.0115722	0	0	0
27.5				0	0	0	0
25					0	0	0
22.5						0	0
20							0

References

Black, F. and Scholes, M. (1973) The pricing of options and corporate liabilities. *Journal of Political Economy*, **81**, (May–June), pp. 637–59.

Churchill, R.V. (1963) *Fourier Series and Boundary Value Problems*, 2nd edn. McGraw-Hill, New York.

Hull, J.C. (1993) *Options, Futures and other Derivatives*, 2nd edn. Prentice-Hall, Englewood Cliffs, NJ.

Ito, K. (1951) On stochastic differential equations. *Memoirs, American Mathematical Society*, **4**, pp. 1–51.

Watsham, T.J. (1992) *Options and Futures in International Portfolio Management*. Chapman & Hall, London.

Appendix 10.1: finite difference methods applied to the Black–Scholes partial differential equation

In Chapter 10 we showed that combining the assumption of risk neutrality with the assumption that asset prices follow an Ito process allows us to develop a partial differential equation that must be satisfied by all derivative securities. This Black–Scholes partial is repeated below

$$\frac{\partial W}{\partial t} + rS\frac{\partial W}{\partial S} + \frac{1}{2}\sigma^2 S^2 \frac{\partial^2 W}{\partial S^2} = rW \tag{A.10.1}$$

Such equations are called partial differential equations because they contain partial derivatives of one or more variables. Here we have a partial differential equation incorporating the first derivative of the price of an option with respect to time, $\partial W/\partial t$, the first derivative of the option price with respect to the price of the underlying security, $\partial W/\partial S$, and the second derivative of the option with respect to the price of the underlying security, $\partial^2 W/\partial S^2$.

It is only possible to obtain closed-form solutions to differential equations like the one studied earlier in this chapter under restrictive conditions. In the Black–Scholes case, we have a European option that does not pay dividends, giving rise to the boundary conditions stated earlier. Other types of option, for example, the so-called "exotic options" have more complex boundary conditions, and closed-form solutions are often not possible. When closed-form solutions are not possible, it is necessary to resort to numerical methods. Fortunately, there are numerical approaches which are easy to understand and apply. The finite difference method is one such approach. There are actually two methods, the explicit finite difference and the implicit finite difference methods.

The approach substitutes **finite difference approximations** in place of the partial derivatives in the partial differential equation. In the explicit method the approximations are

$$\frac{\partial W}{\partial t} \approx \frac{W(t + \Delta t, S) - W(t, S)}{\Delta t}$$

$$\frac{\partial W}{\partial S} \approx \frac{W(t + \Delta t, S + \Delta S) - W(t + \Delta t, S - \Delta S)}{2\Delta S} \qquad (A.10.2)$$

$$\frac{\partial^2 W}{\partial S^2} \approx \frac{W(t + \Delta t, S + \Delta S) - 2W(t + \Delta t, S) + W(t + \Delta t, S - \Delta S)}{\Delta S^2}$$

Substituting the finite difference approximations into the Black–Scholes differential equation gives

$$\frac{W(t + \Delta t, S) - W(t, S)}{\Delta t} + rS \frac{W(t + \Delta t, S + \Delta S) - W(t + \Delta t, S - \Delta S)}{2\Delta S}$$

$$+ \frac{1}{2}\sigma^2 S^2 \frac{W(t + \Delta t, S + \Delta S) - 2W(t + \Delta t, S) + W(t + \Delta t, S - \Delta S)}{\Delta S^2} = rW(t, S) \qquad (A.10.3)$$

Manipulating the algebra gives

$$W(t, S) = \frac{1}{1 + r\Delta t}\left[\frac{S}{2\Delta S}\Delta t\left(\frac{S}{\Delta S}\sigma^2 + r \right)W(t + \Delta t, S + \Delta S) + \right.$$

$$\left. \left(1 - \left(\frac{S}{\Delta S}\right)^2 \sigma^2 \Delta t \right)W(t + \Delta t, S) + \frac{S}{2\Delta S}\Delta t\left(\frac{S}{\Delta S}\sigma^2 - r \right)W(t + \Delta t, S - \Delta S) \right] \qquad (A.10.4)$$

We will illustrate this by applying the technique to the same option to which we applied the Black–Scholes model earlier.

The process begins by creating a grid of equally-spaced prices of the underlying asset on the vertical axis, and equally-spaced time steps on the horizontal axis. The difference between each equally-spaced asset price will constitute ΔS, and the equally-spaced time steps will constitute Δt. From the boundary conditions of the option, all the possible values of the option at expiry (at time T) will be known and constitute the right-hand values of the grid. These will be similar to the values at expiry that were found at the right-hand edge of the binomial and trinomial trees.

In our example, a grid with asset prices (S) ranging from 10 to 60 in steps (ΔS) of 2.5, and time steps (Δt) of one tenth of a year 0.1 is given in Fig. 10.5 below.

	X	delta t	r	delta S	vol							
	35	0.10	0.1	2.5	0.2							
60												25
57.5											22.84653	22.5
55										20.68964	20.34653	20
52.5		0.016							18.52934	18.18964	17.84653	17.5
50								16.36569	16.02934	15.68964	15.34653	15
47.5							14.1987	13.86569	13.52934	13.18964	12.84653	12.5
45						12.05241	11.73283	11.36569	11.02934	10.68964	10.34653	10
42.5					9.975935	9.658943	9.26024	8.927471	8.529345	8.189638	7.846535	7.5
40				8.015471	7.687618	7.285476	6.939675	6.511311	6.155919	5.689638	5.346535	5
37.5			6.210491	5.877035	5.49624	5.141786	4.728753	4.352501	3.876839	3.485565	2.846535	2.5
35	4.600594	4.281332	3.938053	3.600486	3.232397	2.871997	2.459606	2.070626	1.546642	1.143564	0	0
32.5		2.647781	2.355079	2.051135	1.74779	1.427594	1.113621	0.763501	0.456294	0	0	0
30			1.187463	0.963614	0.742513	0.532283	0.329068	0.157218	0	0	0	0
27.5				0.337589	0.220145	0.120385	0.046231	0	0	0	0	0
25					0.036596	0.011443	0	0	0	0	0	0
22.5						0	0	0	0	0	0	0
20							0	0	0	0	0	0
17.5								0	0	0	0	0
15									0	0	0	0
12.5										0	0	0
10											0	0

Figure 10.5

As the option being valued has an exercise price of 35, the right-hand boundary of the grid represents the boundary conditions for the option, i.e. $c = \max [S_T - 35, 0]$
Now consider Fig. 10.6 which is an enlarged detail of Fig. 10.5.

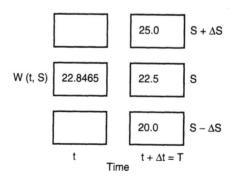

Figure 10.6

At time T, which is the far-right column, the option prices refer to the boundary conditions, in this case relating to a call with an exercise price of 35. Now consider the top three values at time T. The box with 22.50 in it is the value of the option at point $W(t + \Delta t, S)$. The value of 25.00 in the top box is $W(t + \Delta t, S + \Delta S)$, and the value of 20.00 is $W(t + \Delta t, S - \Delta S)$.

Next, these top three terminal values are taken from the grid and incorporated into equation (A10.4) along with the values for S, ΔS, Δt, volatility and the risk-free rate of interest. The value of S is that value on the left-hand edge of the grid that corresponds to the middle of the terminal values of the option at time T, i.e. 57.5 for our first group of three numbers.

This is illustrated below for the above option. $\Delta t = 0.1$, $r = 0.10$, $\Delta S = 2.50$ and $\sigma = 0.20$. Taking S to be 57.50, $S + \Delta S = 60.0$ and $S - \Delta S = 55.0$ Given the exercise price of the option is 35, the corresponding option prices at time T are 25, 22.50 and 20. Putting these values into equation (A.10.4) we get

$$
W(t, S) = \frac{1}{1 + (0.1 * 0.1)} \left[\frac{\left(\frac{57.5}{2.5}\right) 0.1}{2} \left(\frac{57.5}{2.5} 0.2^2 + 0.1 \right) 25 + \right.
$$

$$
\left. \left(1 - \left(\frac{57.5}{2.5}\right)^2 0.2^2 0.1 \right) 22.5 + \frac{\frac{57.5}{2.5} 0.1}{2} \left(\frac{57.5}{2.5} 0.2^2 - 0.1 \right) 20 \right] = 22.8465
$$

Thus the value call at point $W(S, t)$ on the grid is 22.8465. This process is repeated by shifting down one grid point and using the three corresponding values at $t + \Delta t$. In our example they are 22.5, 20, and 17.5. When these are incorporated into equation (A.10.4) we get a value for the option of 20.3465. Thus we work from the boundary condition back through the grid until we come to the current time period and an option price of 4.6006. The analogy between this method and the binomial and trinomial models discussed in Chapter 8 will be clear.

Note that it may also be necessary to use the following boundary conditions

$$W(t, 0) = 0 \text{ for a call, or } X \text{ for a put } (X \text{ being the strike price})$$

For large S, $W(t, S) = S - X$ for a call, or 0 for a put.

For an American call option at each stage in the computation the computed value of $W(t, S)$ is compared with $S - X$ ($X - S$ for a put), and the latter is substituted if it is larger. This represents early exercise. This illustrates the flexibility that the numerical approach has in being able to cope with other than plain vanilla options.

Convergence and stability

The explicit finite difference approach offers the prospect of constructing option prices for all combinations of time to expiry and security price. However, the numerical procedures introduce problems concerning the proliferation of rounding errors which can lead to wild oscillations in neighbouring values and/or negative values, i.e. **instability**.

The value of 4.6006 compares with 4.57 for a ten-step binomial model, and 4.64 for the Black–Scholes model. Consequently we will wish to improve the accuracy of our approximation. Like the binomial model we could consider shortening the time steps on the grid and reduce the interval between S values (i.e. the vertical intervals on the above graphs, known as the mesh size), in order to reduce the approximation errors and **converge** to the true result.

Fortunately it can be proved that these two objectives – achieving convergence and avoiding instability – may both be achieved provided that, as we make our reductions in time interval and mesh size, we ensure that $\Delta t/\Delta S^2$ remains between 0 and 0.5. This, however, creates its own problems since it indicates that halving the mesh size will need to be compensated by reducing the time step by a factor of four. Thus the computation time will increase by a factor of eight.

The implicit finite difference method

In practice large numbers of accurate results are required, so computational efficiency is important. An alternative procedure, the implicit finite difference method, uses the following more obvious and more accurate finite difference approximations to the partial derivatives

$$\frac{\partial W}{\partial t} \approx \frac{W(t + \Delta t, S) - W(t, S)}{\Delta t}$$

$$\frac{\partial W}{\partial S} \approx \frac{W(t, S + \Delta S) - W(t, S - \Delta S)}{2\Delta S} \tag{A.10.5}$$

$$\frac{\partial^2 W}{\partial S^2} \approx \frac{W(t, S + \Delta S) - 2W(t, S) + W(t, S - \Delta S)}{\Delta S^2}$$

The explicit finite difference method furnishes an **explicit** formula for $W(t, S)$ in terms of $W(t + \Delta t, S + \Delta S)$, $W(t + \Delta t, S)$ and $W(t + \Delta t, S - \Delta S)$. In the **implicit** version we have to solve a system of simultaneous equations to find $W(t, S)$.

Iterating backwards from the known values at time $t + \Delta t$ will involve solving a system of simultaneous equations. The number of equations to be solved is a function of the size of the grid.

Note that it may also be necessary to use the upper and lower boundary conditions as above

$$W(t, 0) = 0 \text{ for a call, or } X \text{ for a put } (X \text{ being the strike price})$$

$$\text{For large } S, W(t, S) = S - X \text{ for a call, or } 0 \text{ for a put.}$$

Appendix 10.2: derivation of Black–Scholes using expectations

We can calculate the value of an option as an expectation – a probabilistically weighted sum. We shall produce the result for a call option, and quote the result for a put. An elementary approach is taken, which makes the algebra unavoidably tedious.

Recall that the value of a European call option at maturity is max $(0, S_T - X)$, where X is the exercise price.

$$E \text{ (value at time } T) = \int_X^\infty (S_T - X)f(S_T) \, dS_T$$

$$= \int_X^\infty (S - X)f(S) \, dS \quad \text{(this is known as a stochastic integral)}$$

$$= \int_X^\infty Sf(S)dS - \int_X^\infty Xf(S) \, dS$$

where f is the probability density function (pdf) of S_T.

We deal with these integrals separately. The second is the easier

$$\int_X^\infty Xf(S) \, dS = X \int_X^\infty f(S) \, dS$$

$$= X\left(1 - N\left(\frac{\ln X - \left(\ln S(t) + \left(\mu - \frac{\sigma^2}{2}\right)(T - t)\right)}{\sigma\sqrt{(T - t)}}\right)\right)$$

since S is lognormally distributed, and where N gives the left-tail probability of the standard normal distribution (see Fig. 10.8).

$$= X\left(1 - N\left(\frac{-\ln\frac{S}{X} - \left(\mu - \frac{\sigma^2}{2}\right)(T - t)}{\sigma\sqrt{(T - t)}}\right)\right) = XN\left(\frac{\ln\frac{S}{X} + \left(\mu - \frac{\sigma^2}{2}\right)(T - t)}{\sigma\sqrt{(T - t)}}\right)$$

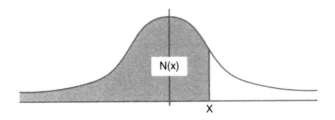

$N(x)$

X

Figure 10.8

The first integral is harder – it needs the substitution $w = \ln S$, which, in turn will require us to know the pdf of w (see Fig. 10.9).

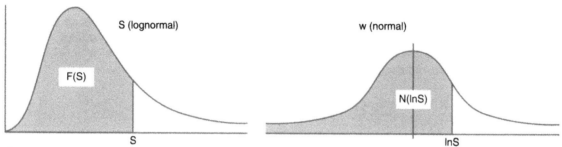

S (lognormal)   w (normal)

F(S)   N(lnS)

S   lnS

Figure 10.9

$F(S) = N(\ln S)$, so, differentiating

$$f(S) = n(\ln S) \times \frac{1}{S} = \frac{n(w)}{e^w}$$

where n is the pdf for the normal distribution.

$$\therefore \int_X^\infty Sf(S)\,dS = \int_{\ln X}^\infty e^w \frac{n(w)}{e^w} e^w dw$$

$$= \int_{\ln X}^\infty e^w \frac{1}{b\sqrt{2\pi}} \exp\left(\frac{-(w-a)^2}{2b^2}\right) dw \qquad \text{where } a = \ln S + \left(\mu - \frac{\sigma^2}{2}\right) \text{ and } b = \sigma\sqrt{T-t}$$

$$= \int_{\ln X}^\infty \frac{1}{b\sqrt{2\pi}} \exp\left(\frac{-(w^2 - 2aw + a^2 - 2wb^2)}{2b^2}\right) dw$$

$$= \int_{\ln X}^\infty \frac{1}{b\sqrt{2\pi}} \exp\left(\frac{-[(w-(a+b^2))^2 - 2ab^2 - b^4]}{2b^2}\right) dw$$

$$= \int_{\ln X}^\infty \frac{1}{b\sqrt{2\pi}} \exp\left(\frac{-(w-(a+b^2))^2}{2b^2}\right) \exp\left(a + \frac{b^2}{2}\right) dw$$

$$= \exp\left(a + \frac{b^2}{2}\right) \int_{\ln X}^\infty \frac{1}{b\sqrt{2\pi}} \exp\left(\frac{-(w-(a+b^2))^2}{2b^2}\right) dw$$

$$= \exp\left(a + \frac{b^2}{2}\right)\left(1 - N\left(\frac{\ln X - (a+b^2)}{2b^2}\right)\right)$$

$$= \exp\left(\ln S + \left(\mu - \frac{\sigma^2}{2}\right)(T-t) + \frac{\sigma^2(T-t)}{2}\right)\left(1 - N\left(\frac{\ln X - \ln S - \left(\mu - \frac{\sigma^2}{2}\right)(T-t) - \sigma^2(T-t)}{\sigma\sqrt{(T-t)}}\right)\right)$$

$$= e^{\mu(T-t)}S\left(1 - N\left(\frac{-\ln\frac{S}{X} - \left(\mu + \frac{\sigma^2}{2}\right)(T-t)}{\sigma\sqrt{(T-t)}}\right)\right) = e^{\mu(T-t)}SN\left(\frac{\ln\frac{S}{X} + \left(\mu + \frac{\sigma^2}{2}\right)(T-t)}{\sigma\sqrt{(T-t)}}\right)$$

Thus the expected value of the option at time T is $e^{\mu(T-t)}S(t)N(d_1) - XN(d_2)$, where

$$d_1 = \frac{\ln\frac{S}{X} + \left(\mu + \frac{\sigma^2}{2}\right)(T-t)}{\sigma\sqrt{(T-t)}} \quad \text{and } d_2 = \frac{\ln\frac{S}{X} + \left(\mu - \frac{\sigma^2}{2}\right)(T-t)}{\sigma\sqrt{(T-t)}}$$

Finally, we must discount back to the present time, time t. There is an important subtlety involved in this – what should we use as the discount rate?

If the drift rate of the underlying security, μ, is greater than the risk-free rate, r, then the security will offer a higher return, but at higher risk and return. We will be able to compute the expected future value of an option on the security – but we will not then be able to discount to give a current value since we shall not know what discount rate to use.

The fact that the Black–Scholes differential equation for the price of a derivative security does not contain μ implies that the solution (the value of the derivative security) must be independent of μ. We can thus sidestep our difficulty by moving into an imaginary risk-neutral world in which arbitrage will ensure that all securities earn the risk-free rate. This will not affect the price of an option on the security.

We can thus value the option by computing its expected value, imagining that the security has a drift rate of r rather than μ, and then discounting at the risk-free rate, r.

Thus the present value of a call option on the security is

$$e^{-r(T-t)}(e^{r(T-t)}S(t)N(d_1) - XN(d_2)) = S(t)N(d_1) - Xe^{-r(T-t)}N(d_2)$$

This is the Black–Scholes formula.

Multivariate analysis: principal components analysis and factor analysis

Introduction

Principal components analysis and factor analysis are techniques for analysing the structure of data within a multivariate framework. Along with multiple regression, which was covered in Chapter 6, and multivariate cointegration, which was covered in Chapter 7, they are among the most widely used techniques for multivariate analysis. They differ from multiple regression in that in the regression case the objective is to identify the relationship between an exogenous variable, and multiple endogenous variables. In the case of principal components analysis and factor analysis, the structure of the relationship between the endogenous variables only is investigated. Unlike cointegration the relationships between the endogenous variables need not be stationary.

In finance, the social sciences, or in many other fields of interest, is it often necessary to identify the characteristics of multivariate structures. Two such characteristics are

1. the volatility in a multivariate structure
2. the correlation or collinearity between the variables.

Generally, some variables will have a major influence on the overall changes in the structure, while others will only have a minor or even insignificant influence.

One difficulty is to determine which of the variables to include in the structure and therefore which variables to measure. For example, if two variables were to turn out to be perfectly correlated then one could be dispensed with – the second adds no further information. This is analogous to the multicollinearity problem in multiple regression.

In general, the issue of what variables to include and which to exclude will not be clear cut, and a mechanism will be needed to enable us to identify variables, or combinations of variables, which impart the available information in the most efficient and effective way.

Principal components analysis (PCA) is the technique which is used when the volatility of a multivariate structure is being analysed. Factor analysis (FA) is used when the correlation between the variables of a multivariate structure is being analysed. Both rely on analysing the variance/covariance matrix, since it is this which contains all the information on the extent to which the variables in use vary with each other, i.e. the extent to which they duplicate or complement each other. In this chapter we will label the variance/covariance matrix as C.

Although both PCA and FA utilize the variance–covariance matrix, they differ from the mean-variance analysis met in Chapters 4 and 9 in that mean-variance analysis measures the total collective variability of a group of variables, without specifically identifying which sub-groups (linear combinations of variables) contribute to that variability. PCA identifies and ranks the linear combinations and their contribution to that total variability. Each linear combination is a "principal component" and is identified by the strength of the covariance between its constituents. The contribution of each principal component to total variability is ranked according to the collective variance of the linear combination.

In applying PCA the total variability within the data is measured by the sum of the eigenvalues (which will be equal to the sum of the elements on the leading diagonal of C, known as its trace). Components (linear combinations of variables) are then chosen, in order of decreasing eigenvalues, until a sufficiently large proportion of the total variability has been accounted for. In this way the dimensionality of the system is reduced and the most important components (directions) are identified.

It will be instructive to refresh our understanding of matrix algebra that we introduced in the Appendix to Chapter 6, by showing how it is used to calculate the total variance of a portfolio.

Recall from Chapter 2 that the variance of a portfolio was the weighted sum of the covariances between each pair of assets in the portfolio, where the variance is considered to be the covariance of the asset with itself.

Imagine a portfolio of two assets A and B. The variance of the returns to asset A is 0.00015, the variance of returns to asset B is 0.00025, and the covariance between A and B is 0.00005. The variance–covariance matrix, C, is

$$C = \begin{bmatrix} 0.00015 & 0.00005 \\ 0.00005 & 0.00025 \end{bmatrix}$$

If we assume an equally weighted portfolio, the portfolio variance is derived by multiplying the C matrix by a horizontal $1 * n$ vector of weights, and then multiplying the product matrix by an $n * 1$ vector of weights. Thus

$$[0.5\ 0.5] \begin{bmatrix} 0.00015 & 0.00005 \\ 0.00005 & 0.00025 \end{bmatrix} \begin{bmatrix} 0.5 \\ 0.5 \end{bmatrix} = 0.000125 \tag{11.1}$$

Thus the total variance of this portfolio is 0.000125.

Principal components analysis

The application of PCA has two objectives. The first is to reduce the dimensionality of the data from one of many variables to one of a few variables. This is achieved by identifying groups of the original variables that have the qualities that the group members are correlated with each other, but the group as a whole is linearly independent of other groups or individual variables. These linearly independent groups of variables are the principal components.

The second objective, which is facilitated by the first, is interpretation of the data. This is made possible because PCA identifies the linear combinations of variables, and orders them according to their contribution to the total variance of the original data. Thus, the first principal component is the linear combination of variables which has the highest variance, the second principal component is that linear combination that has the second highest variance, and so on. We would like to be able to explain a very large proportion of the total variance with as few principal components as possible.

There are many situations in practical finance when it is desirable to identify the major variables, or linearly independent combinations of variables, that are the major contributors to risk. In our simple example using a two-asset portfolio, there will only be two components. However, in a large n variable portfolio, there will be up to n major contributing components. Some of those components may be highly correlated with others, such that the sub-group as a whole contributes to risk in a manner independent of the contribution made by other variables or linear combinations of variables. PCA enables us to identify those independent linear combinations of variables and their contribution to total variance. We thus have a much richer understanding of what influences the risk of our portfolios and we are therefore more able to manage that risk.

PCA is often a means to an end rather than an end in itself. For example, the identification of the principal components may serve to construct inputs into regression equations, such that the dependent variable is not regressed on the raw independent variables but upon the principal components. Later in this chapter we will see how identifying the principal components in the movement of interest rates will enable us better to measure the interest rate risk inherent in bond portfolios.

PCA will first be explained with the aid of a simple two-asset example using the same hypothetical data as in equation (11.1) above. Then our understanding will be enhanced by illustrating with a four-asset example using actual asset returns data. Finally we will explain how the technique can be applied to determining the riskiness of a bond.

The hypothetical two-asset example

The application of PCA enables us to extract from the variance–covariance matrix a number of linear combinations of asset variances and covariances which explain the covariability of the assets, each combination being independent of the other combinations. This is possible because the symmetric structure of the variance–covariance matrix ensures that it is possible to subject it to a process known as diagonalization. This process involves three stages. These are:

1. find the eigenvectors and their respective eigenvalues
2. construct three matrices Q, D and Q^{-1}
3. identify the linear combinations from the eigenvectors, and rank these combinations in order of highest to lowest according to eigenvalue.

Stage 1: finding the eigenvectors and eigenvalues

The first stage is to find the **eigenvectors** and the associated **eigenvalues** of the variance–covariance matrix, C. We must find the **eigenvectors** because they give us the linearly independent combinations of variables, the principal components, that contribute to the total variance. We must find the **eigenvalues** because they give us the proportion of total risk accounted for by each principal component.

Mathematically, the eigenvectors are vectors, x_i, which each have an associated scalar λ_i, known as its eigenvalue, such that where the variance–covariance matrix C is used to multiply the vector x_i, it simply produces a scaling of the vector i.e. $Cx = \lambda x$. The symmetry of the C matrix ensures that there are n of these vectors (provided that C is a non-singular matrix, i.e. C has an inverse) and that they are orthogonal (perpendicular to each other).

An example is given below. The left-hand 2×2 matrix is the variance–covariance matrix we used earlier. That matrix can be multiplied by a vector such that the product is simply a scaled version of the eigenvector, e.g.

$$\begin{bmatrix} 0.00015 & 0.00005 \\ 0.00005 & 0.00025 \end{bmatrix} \begin{bmatrix} 1 \\ m \end{bmatrix} = \lambda \begin{bmatrix} 1 \\ m \end{bmatrix} \tag{11.2}$$

where

$$\begin{bmatrix} 1 \\ m \end{bmatrix} \tag{11.3}$$

is an eigenvector and λ is the associated eigenvalue.

Expanding this we get

$$0.00015 + 0.00005m = \lambda$$
$$0.00005 + 0.00025m = \lambda m \tag{11.4}$$

Therefore, multiplying both by m and cancelling out the λ, we get

$$0.00015m + 0.00005m^2 = 0.00005 + 0.00025m$$

$$0.00005m^2 - 0.00010m - 0.00005 = 0$$

$$5m^2 - 10m - 5 = 0$$

$$m^2 - 2m - 1 = 0$$

$$m = \frac{+2 \pm \sqrt{4+4}}{2} =$$

$$\frac{2 \pm \sqrt{8}}{2} = 1 \pm \sqrt{2}$$

Thus, the eigenvectors are

$$\begin{bmatrix} 1 \\ 1 + \sqrt{2} \end{bmatrix} \text{ and } \begin{bmatrix} 1 \\ 1 - \sqrt{2} \end{bmatrix} \tag{11.5}$$

We now have to normalize these so that the vectors have a length of one. This entails dividing each component of the eigenvector by the square root of the sum of the squares of each component. The resulting normalized vectors are the eigenvectors

$$\begin{bmatrix} \dfrac{1}{\sqrt{1^2 + \left(1+\sqrt{2}\right)^2}} \\[2em] \dfrac{1+\sqrt{2}}{\sqrt{1^2 + \left(1+\sqrt{2}\right)^2}} \end{bmatrix} = \begin{bmatrix} 0.383 \\ 0.924 \end{bmatrix}, \quad \begin{bmatrix} \dfrac{1}{\sqrt{1^2 + \left(1-\sqrt{2}\right)^2}} \\[2em] \dfrac{1-\sqrt{2}}{\sqrt{1^2 + \left(1-\sqrt{2}\right)^2}} \end{bmatrix} = \begin{bmatrix} 0.924 \\ -0.383 \end{bmatrix}$$

Remember, there will be as many eigenvectors as there are variables in the variance–covariance matrix. Thus in a 2×2 matrix there will be two eigenvectors, and in an $n \times n$ matrix there will be n eigenvectors.

We can prove that the above vectors are eigenvectors because eigenvectors are vectors which when multiplied by the matrix C simply give scalar multiples of the vector. For example

$$\begin{bmatrix} 0.00015 & 0.00005 \\ 0.00005 & 0.00025 \end{bmatrix} \begin{bmatrix} 0.383 \\ 0.924 \end{bmatrix} = \begin{bmatrix} 0.000104 \\ 0.000250 \end{bmatrix} = 0.000271 \begin{bmatrix} 0.383 \\ 0.924 \end{bmatrix}$$

Here we have multiplied matrix C by the vector to arrive at a new vector. This vector also happens to be the product of 0.000271 and the original vector. Thus we see that the left-hand vector is a eigenvector, and that the scalar, 0.000271, is an eigenvalue.

The same process applied to the right-hand vector shows that it too is an eigenvector

$$\begin{bmatrix} 0.00015 & 0.00005 \\ 0.00005 & 0.00025 \end{bmatrix} \begin{bmatrix} 0.924 \\ -0.383 \end{bmatrix} = \begin{bmatrix} 0.000119 \\ -0.000050 \end{bmatrix} = 0.000129 \begin{bmatrix} 0.924 \\ -0.383 \end{bmatrix}$$

Here we see the eigenvalue for this vector is 0.000129.

Stage 2: construct matrices Q, D and Q^{-1}

The next step is to construct three matrices: Q, D and Q^{-1}. Matrix Q is constructed from the eigenvectors by writing them as the columns of the matrix, ranking them in the order of their respective eigenvalues. Thus the Q matrix applicable to the above example is

$$Q = \begin{bmatrix} 0.383 & 0.924 \\ 0924 & -0.383 \end{bmatrix} \tag{11.6}$$

Matrix D is a diagonal matrix (zeros everywhere except on the main diagonal), the diagonal elements being the eigenvalues written in the same order as the eigenvectors were written when constructing Q. Thus the D matrix relating to the above example is

$$D = \begin{bmatrix} 0.000271 & 0 \\ 0 & 0.000129 \end{bmatrix} \tag{11.7}$$

We shall then have $CQ = QD$. Again provided that C is non-singular, so will be Q, and then we can then write $C = QDQ^{-1}$. However, provided that the eigenvectors are of unit length, i.e. the squares of the components of the eigenvector sum to one, the inverse of Q is Q itself so we can write $C = QDQ$

$$
\underset{C}{\begin{bmatrix} 0.00015 & 0.00005 \\ 0.00005 & 0.00025 \end{bmatrix}} = \underset{Q}{\begin{bmatrix} 0.383 & 0.924 \\ 0.924 & -0.383 \end{bmatrix}} \underset{D}{\begin{bmatrix} 0.000271 & 0.0 \\ 0.0 & 0.000129 \end{bmatrix}} \underset{Q^{-1}}{\begin{bmatrix} 0.383 & 0.924 \\ 0.924 & -0.383 \end{bmatrix}} \tag{11.8}
$$

Stage 3: identify the linear combination of variables and rank the combinations according to their eigenvalues

Now consider the variance/covariance matrix, C. We will assume that it relates to two assets X and Y. The variance of this two-asset portfolio can be written as

$$
[X\ Y]\begin{bmatrix} 0.00015 & 0.00005 \\ 0.00005 & 0.00025 \end{bmatrix}\begin{bmatrix} X \\ Y \end{bmatrix} \tag{11.9}
$$

But as $C = QDQ^{-1}$ this may be written as

$$
[X\ Y]\begin{bmatrix} 0.383 & 0.924 \\ 0.924 & -0.383 \end{bmatrix}\begin{bmatrix} 0.000271 & 0.0 \\ 0.0 & 0.000129 \end{bmatrix}\begin{bmatrix} 0.383 & 0.924 \\ 0.924 & -0.383 \end{bmatrix}\begin{bmatrix} X \\ Y \end{bmatrix} \tag{11.10}
$$

which becomes

$$
[0.383X + 0.924Y \quad 0.924X - 0.383Y]\begin{bmatrix} 0.000271 & 0 \\ 0 & 0.000129 \end{bmatrix}\begin{bmatrix} 0.383X + 0.924Y \\ 0.924X - 0.383Y \end{bmatrix}
$$

From the above example of diagonalization we can identify the linear combinations of variables, X and Y in this example, that independently contribute to the variance of the overall portfolio. These combinations are identified by the eigenvectors. Thus, in the above example, $0.383X + 0.924Y$ constitute one such linearly independent combination (i.e. one principal component), and $0.924X - 0.383Y$ represents the other.

We need to be able to rank the eigenvectors according to their importance in contributing to the overall variance of the portfolio. This ranking is made through the eigenvalues. The eigenvector with the highest eigenvalue, contributes most to the overall variance, the eigenvector that makes the second largest contribution to total variance is that with the second largest eigenvalue, and so on.

The proportion of total variance contributed by each combination of variables is given by the proportion that each eigenvalue represents in the sum of eigenvalues. The sum of the eigenvalues is given by the sum of the diagonal elements in the D matrix. For example, the sum of the eigenvalues in the D matrix above is 0.000400. Thus the eigenvector with an eigenvalue of 0.000271 contributes 0.000271/0.000400 or 67.75% of the total variance.

An example using the FTSE 100, gilt, S&P 500 and US$ currency returns

Consider an equally weighted portfolio of four asset classes, the FTSE 100 index, the S&P 500 index, the UK gilt index and the £/$ currency exchange rate. Using monthly data of returns for the period from September 1989 to December 1993, the variance–covariance matrix is as given below

	Gilt	FTSE	S&P	$/£
Gilt	5.5884	1.6749	0.2233	−1.0528
FTSE	1.6749	22.3660	10.0886	−5.5765
S&P	0.2233	10.0886	10.7730	−0.5839
£/$	−1.0528	−5.5765	−0.5839	16.2161

The four eigenvalues and corresponding eigenvectors are

eigenvalues: $\lambda_1 = 30.2572$, $\lambda_2 = 14.8258$, $\lambda_3 = 5.61280$, $\lambda_4 = 4.2477$

eigenvectors:

$$
\begin{bmatrix} 0.074904 \\ 0.824166 \\ 0.438093 \\ -0.351135 \end{bmatrix}
\begin{bmatrix} -0.058582 \\ 0.199571 \\ 0.362687 \\ 0.908404 \end{bmatrix}
\begin{bmatrix} 0.921223 \\ -0.153459 \\ -0.322417 \\ 0.154422 \end{bmatrix}
\begin{bmatrix} 0.377235 \\ -0.507399 \\ 0.756690 \\ -0.166314 \end{bmatrix}
$$

The first eigenvector has a corresponding eigenvalue of 30.2572 and accounts for 55.07% of the total variance. The second eigenvector has a corresponding eigenvalue of 14.8258 and accounts for 26.98% of the total variance. Cumulatively the first two eigenvectors account for 82.05% of the total variance. The third and fourth eigenvectors have eigenvalues of 5.6128 and 4.2477, respectively, with corresponding proportions of total variance being 10.22% and 7.73%.

Interpretation of the eigenvectors will be carried out in relation to the standardized form which is analysed below.

An example using standardized variables

One problem that arises with PCA is to do with different data being of different orders of magnitude. Readers will recall that this is a problem with covariances generally. That is that the magnitude of the covariance is a function of the magnitude of the data as well as the difference between X and $\bar{X}$. Readers will also recall that the suggested solution was to calculate the correlation coefficient, which in effect is a standardized covariance.

A similar problem can occur with PCA based on a covariance matrix derived from unstandardized data. For this reason when the data for separate variables exhibit different orders of magnitude, or when the units of measurement differ between the variables, it is usual to standardize the data before applying the principal components technique.

In the following example, the raw returns data have been standardized because the data have considerable differences in the variances and covariances between variables. The standardization was made as follows:

$$z_i = \frac{(x_i - \bar{x})}{\sigma_x}$$

	Gilt	FTSE	S&P	$/£
Gilt	1.0	0.1498	0.0288	−0.1106
FTSE	0.1498	1.0	0.6499	−0.2928
S&P	0.0288	0.6499	1.0	−0.0442
£/$	−0.1106	−0.2928	−0.0442	1.0

The four eigenvalues and corresponding eigenvectors are

eigenvalues: $\lambda_1 = 1.76345$, $\lambda_2 = 1.05032$, $\lambda_3 = 0.88860$, $\lambda_4 = 0.29764$

eigenvectors:

$$\begin{bmatrix} 0.205868 \\ 0.688288 \\ 0.612746 \\ -0.329272 \end{bmatrix}' \begin{bmatrix} -0.698552 \\ 0.102372 \\ 0.424376 \\ 0.566966 \end{bmatrix}' \begin{bmatrix} 0.679614 \\ -0.010147 \\ 0.166887 \\ 0.714262 \end{bmatrix}' \begin{bmatrix} -0.088133 \\ 0.718106 \\ -0.645443 \\ 0.244866 \end{bmatrix}$$

As the variables were standardized before the analysis, the sum of the eigenvalues will be four, because there are four variables. With this in mind, the percentages of total variance accounted for by each eigenvector are

$$\lambda_1 = 44.09, \quad \lambda_2 = 26.26, \quad \lambda_3 = 22.22, \quad \lambda_4 = 7.44$$

It should be noted that the process of standardization is not inconsequential, as the magnitude of the eigenvectors and eigenvalues will be different if derived from the standardized data compared with the magnitudes derived from the unstandardized data. Take the example given above. When using unstandardized data the first principal component accounts for 55% of the total variance and the third principal component accounts for just 10%.

However, using standardized data, the first principal component accounts for only 44% of total variance and the third principal component accounts for 22%, with the second and fourth principal components remaining basically unchanged.

Interpretation of principal components

Interpretation of the principal components is somewhat subjective. For example, consider the eigenvectors relating to the standardized variables above. The eigenvectors, reflecting the principal components, can be used to construct measures of the exposure of a portfolio to the elements of the stochastic variability corresponding to these principal components. This gives the portfolio manager the opportunity to adjust the weights of the portfolio components so as to reduce those elements of risk.

We demonstrate this in a different context – the management of risk in the bond markets.

An application to bond markets

PCA has been applied to the development of risk models of bond markets. Recall from Chapter 3 that duration and convexity are widely used by bond market practitioners to summarize the interest rate sensitivity of individual bonds and bond portfolios. However, both duration and convexity rely on the assumptions that the yield curve is flat and only shifts in parallel. Casual empiricism indicates that these assumptions are not met in practice. In particular the individual spot rates are not perfectly correlated, thus the term structure and therefore the yield curve does not shift in parallel.

PCA has been used by Kahn (1989), Kahn and Gulrajani (1993) and Karki and Reyes (1994) to identify the major themes of common movement among the many spot rates that constitute the term structure of interest rates. Then this knowledge of co-movement can be combined with knowledge of the bond's exposure to the individual spot rates within the term structure to derive a summary measure of risk associated with each form of co-movement.

Changes in the value of a bond arise from three causes:

1. the life of the bond shortens
2. the term structure changes
3. changes in the market's assessment of the characteristic of the bond, e.g. default, embedded option values, etc.

The shortening maturity of the bond is known and thus the effect on the changing value of the bond is not uncertain. However, the changes in the value of a bond due to changes in the term structure and the market's assessment of the bond characteristics are uncertain and therefore the cause of risk.

Recalling that the term structure is a series of spot rates of interest relating to the maturity spectrum of the bond market, a bond will be exposed to a movement in the term structure depending on how its expected cash flows are distributed along the term structure, i.e. how each cash flow is associated with each spot rate. For example, consider two zero coupon bonds. One has a maturity of 10 years the other has a maturity of, say, three years. If the term structure were to shift up or down in parallel, both bonds would be affected in the same way, the magnitude of the effects being indicated by the duration and convexity. However, consider the case where long rates rise and short rates fall, with the pivot point at five years. In such circumstances, the 10-year bond will fall in price but the three-year bond will actually rise in price.

Now consider a coupon paying bond. This is simply a portfolio of zero coupon bonds. Each of the coupons and the final redemption payment is discounted at the appropriate spot rate. The value of the bond is simply the sum of the present values; the variance of the bond is simply the weighted sum of the covariances of each of those discounted cash flows.

Thus the first task in applying PCA to bond risk analysis is to specify the maturity profile of the term structure of interest rates. We will refer to each point on this maturity profile as a vertex. Typically these will be at monthly intervals for the first three months, then quarterly for the next nine months and then half yearly thereafter for a sufficient number of years to accommodate all the bonds at issue.

Recall that PCA relates to linear combinations of variables, and from Chapter 1 recall that bond prices are linear combinations of present values. Each present value is the result of applying an appropriate spot rate to a particular cashflow. We can use PCA to identify linear combinations of the changes in those present values resulting from changes in spot rates of interest. From the linear combinations of the changes in present values it is a simple matter to relate to the combinations of changes in the spot rates which caused changes in the present values.

As the return on a bond is the result of a non-stochastic element (that due to the maturity shortening) as well as the changes in interest rates, it is necessary to remove the non-stochastic element. To do this we work, not in total return, but in **excess return** and define risk as the variance of the excess returns. Excess returns are defined as the returns to a zero coupon bond minus the risk-free return. The risk-free return is that element of return on that particular zero coupon bond that is due only to the shortening of maturity.

We must therefore begin by defining the bond return and the risk-free return. To understand the concept of the bond return consider a fixed holding period of one month. We will designate this holding period as Δt. The current time is t. The residual maturity (i.e. the remaining maturity of the bond at the **end of the holding period**) is w_i. Thus, the life of the bond at time t is $\Delta t + w_i$. Now consider investing £1 at time t, in a bond maturing at time $t + \Delta t + w_i$ and selling it at time $t + \Delta t$. The change in value of that investment will be the return over the holding period. This return will be due to the bond life shortening as well as any changes in the term structure and assessment of the characteristics.

Now consider investing £1 at time t in a pure discount government bond maturing $t + \Delta t$. The return to that investment will be certain, and thus is the risk-free return. Subtract this risk-free holding period return from the holding period return in the bond maturing at time $t + \Delta t + w_i$ and the result is the excess return to a cash flow maturing at vertex $t + \Delta t + w_i$.

By subtracting the return on the risk-free investment from the return on the bond maturing at time $t + \Delta t + w_i$ any change in value of that bond due only to the shortening of the maturity has been removed. Thus the resulting excess return is a purely stochastic element of return.

If we collect time series of these holding period returns for each vertex of the term structure, we can construct a variance–covariance matrix of the excess returns.

From the variance–covariance matrix of these excess returns, we derive the eigen-vectors and associated eigenvalues. In government bond markets, three principal components account for 99% of the term structure risk. In the studies by Kahn, by Kahn and Gulrajani and by Karki and Reyes cited earlier, the first principal component can be interpreted as a change in the general level of the term structure, analogous to a parallel shift. The second principal component can be interpreted as a change in the slope of the term structure. The third principal component can be interpreted as a change in the curvature of the term structure.

To explain how such a model can be used to identify the risks in a coupon paying bond consider first how we would use matrix algebra to derive the total variance of a bond with five annual cash flows of CF_1, CF_2, CF_3, CF_4 and CF_5, respectively. The total variance V, would be

$$[CF_1\ CF_2\ CF_3\ CF_4\ CF_5]\begin{bmatrix} var_1 & cov_{12} & cov_{13} & cov_{14} & cov_{15} \\ cov_{21} & var_2 & cov_{23} & cov_{24} & cov_{25} \\ cov_{31} & cov_{32} & var_3 & cov_{34} & cov_{35} \\ cov_{41} & cov_{42} & cov_{43} & var_4 & cov_{45} \\ cov_{51} & cov_{52} & cov_{53} & cov_{54} & var_5 \end{bmatrix}\begin{bmatrix} CF_1 \\ CF_2 \\ CF_3 \\ CF_4 \\ CF_5 \end{bmatrix} = V \tag{11.11}$$

Now using PCA to identify the themes of co-movement that contribute to this variance, the variance–covariance matrix is transformed into a group of three matrices Q, D and Q^{-1} and the total variance of the bond can be derived by multiplying the $1 \times n$ vector weights by $n \times n$ matrices, Q, D and Q^{-1}, and then the resulting product by the $n \times 1$ vector of weights as follows:

$$[CF_1\ CF_2\ CF_3\ CF_4\ CF_5]Q\ D\ Q^{-1}\begin{bmatrix} CF_1 \\ CF_2 \\ CF_3 \\ CF_4 \\ CF_5 \end{bmatrix} \tag{11.12}$$

It will be recalled that the **Q** matrix is the matrix of **eigenvectors** ranked from left to right in the order of the size of their respective **eigenvalues**. Thus in equation (11.13) below, $EV_{1(1)}$ to $EV_{1(n)}$ are the 1 to n elements of the first eigenvector. Thus, multiplying the vector of the size of the cash flows by the **Q** matrix will give a vector of exposures of the individual bond cash flows to the principal components as follows:

$$[CF_1\ CF_2\ CF_3\ CF_4\ CF_5\ 0\ 0\ ...\ 0_n]\begin{bmatrix} EV_{1(1)} & \cdots & EV_{n(1)} \\ EV_{1(2)} & \cdots & EV_{n(2)} \\ \cdot & \cdots & \cdot \\ \cdot & \cdots & \cdot \\ EV_{1(n)} & \cdots & EV_{n(n)} \end{bmatrix} \tag{11.13}$$

Note that the vector of cash flows includes the actual cash flows of the bond in question, together with a series of zeros. These zeros relate to those vertices of the term structure to which the bond in question has no cash flows exposed.

Thus, the product of the cash flow vector and the first eigenvector gives a single measure (a summary statistic) of the exposure of the bond to the first principal component. The product of the cash flow vector and the second eigenvector is the measure of exposure of the bond to the second principal component and so on. The resulting row vector of exposures would be

$$[exp_1\ exp_2\ exp_3\ exp_4\ exp_5\ 0\ 0\ 0\ ...\ 0_n] \tag{11.14}$$

Note that because, as explained on page 362 above, $Q^{-1} = Q$ itself, Q^{-1} multiplied by the column vector of cash flows gives the transpose of 11.14 which is a column vector of exposures.

Now recall that the D matrix is a diagonal matrix where the diagonal elements are the eigenvalues (λ) written in the same order as the eigenvectors they relate to. Pre-multiplying the diagonal matrix D by the row vector of exposures, and post-multiplying the product by the column vector of exposures, will give the total variance in terms of principal components as follows:

$$[\exp_1 \exp_1 \exp_1 \exp_1 \exp_1 \, 0 \, 0 \ldots 0_n] \begin{bmatrix} \lambda & 0 & \ldots & 0_n^1 \\ 0_1^2 & \lambda_2 & \ldots & 0_n^2 \\ \cdot & \cdot & \ldots & \cdot \\ \cdot & \cdot & \ldots & \cdot \\ 0_1^n & \cdot & \ldots & \lambda_n \end{bmatrix} \begin{bmatrix} \exp_1 \\ \exp_2 \\ \exp_3 \\ \exp_4 \\ \exp_5 \\ 0 \\ \cdot \\ 0_n \end{bmatrix} = v \qquad (11.15)$$

This variance can be expanded into a sum of products of eigenvalues and respective exposures, as follows:

$$V = (\exp_1^2 * \lambda_1) + (\exp_2^2 * \lambda_2) + \ldots + (\exp_n^2 * \lambda_n) \qquad (11.16)$$

Thus the riskiness of a bond is a function of the eigenvalues of each principal component and the exposure of cash flows to each principal component.

Factor analysis

Factor analysis (FA) represents another approach to elucidating the structure of a variance–covariance matrix. To understand the use of FA, we must begin by looking more closely at the concept of variance. Refer back to the Markowitz model of portfolio risk. That model disaggregated total variance into systematic and unsystematic variance. The systematic variance was that which could not be removed by judicious diversification, whereas the unsystematic risk could be removed. In effect the systematic risk is common to all the securities in a portfolio, whereas the unsystematic risk is unique to each security.

The distinction between systematic risk and unsystematic risk is at the heart of FA. With PCA we were concerned with explaining total variance. With FA we are concerned with determining the amount of systematic risk (known as communality in FA parlance) within a covariance structure.

In PCA we extract linear combinations of the underlying variables, so that, at each stage, the component which is extracted explains the greatest possible amount of the remaining variability. In FA we are effectively splitting the total variability of the data into two parts – that which the variables share (communality) and that which is specific to individual variables. The analysis uses estimates of the communality to construct explanatory factors.

To illustrate the application of FA, let us suppose that we have run a PCA which has identified m significant underlying components.

An attempt is then made to express each of the original data variables as a linear combination of that number of factors. The linear combinations which are selected are those which do the best job – they produce modelled variable values which explain as much of the systematic variance of the data as possible.

Consider the standardized variance–covariance matrix of the four financial asset classes given on page 364 and repeated here

	Gilt	FTSE	S&P	$/£
Gilt	1.0	0.1498	0.0288	−0.1106
FTSE	0.1498	1.0	0.6499	−0.2928
S&P	0.0288	0.6499	1.0	−0.0442
£/$	−0.1106	−0.2928	−0.0442	1.0

Recall that this matrix results from 51 monthly returns observations from September 1989 to December 1993. Recall also that the first three principal components accounted for 93% of total variance. Thus we have assumed that there are three significant factors for the purposes of FA.

Each standardized variable is hypothesized to be a linear combination of the three factors of the form

$$\text{GILT} = t_1 = \lambda_{11}f_1 + \lambda_{12}f_2 + \lambda_{13}f_3 + e_1$$

$$\text{FTSE} = t_2 = \lambda_{21}f_1 + \lambda_{22}f_2 + \lambda_{23}f_3 + e_2$$

$$\text{S\&P} = t_3 = \lambda_{31}f_1 + \lambda_{32}f_2 + \lambda_{33}f_3 + e_3$$

$$\text{\$/£} = t_4 = \lambda_{41}f_1 + \lambda_{42}f_2 + \lambda_{43}f_3 + e_4 \tag{11.17}$$

The result is a 4×1 vector of variables, a 4×3 matrix of λ_3^s, a 3×1 vector of factors and a 4×1 vector of error terms.

Now consider n observations of each of our variables. So we have a $4 \times n$ matrix of t values

$$T = \begin{bmatrix} t_{11} & t_{12} & \cdots & t_{1n} \\ t_{21} & t_{22} & \cdots & t_{2n} \\ t_{31} & t_{32} & \cdots & t_{3n} \\ t_{41} & t_{42} & \cdots & t_{4n} \end{bmatrix} \tag{11.18}$$

Similarly we have a $3 \times n$ matrix of F values

$$F = \begin{bmatrix} f_{11} & f_{12} & \cdots & f_{1n} \\ f_{21} & f_{22} & \cdots & f_{2n} \\ f_{31} & f_{32} & \cdots & f_{3n} \end{bmatrix} \tag{11.19}$$

and a $4 \times n$ matrix of e values

$$E = \begin{bmatrix} e_{11} & e_{12} & \cdots & e_{1n} \\ e_{21} & e_{22} & \cdots & e_{2n} \\ e_{31} & e_{32} & \cdots & e_{3n} \\ e_{41} & e_{42} & \cdots & e_{4n} \end{bmatrix} \tag{11.20}$$

Our model now becomes

$$T = \lambda F + E \tag{11.21}$$

where T is $4 \times n$, λ is 4×3, F is $3 \times n$ and E is $4 \times n$.

Now we need some extra assumptions:

- the f_i values are independent
- the e_i values are independent
- f_i and e_i values are uncorrelated.

Now $C = TT^T$ which equals

$$(\lambda F + E)(\lambda F + E)^T \tag{11.22}$$

This is equal to

$$\lambda F(\lambda F)^T + \lambda FE^T + E(\lambda F)^T + EE^T$$
$$= \lambda FF^T \lambda^T + \lambda FE^T + EF^T \lambda^T + EE^T \qquad (11.23)$$

Because the f and e values are assumed to be uncorrelated the FE^T matrix is the zero matrix, and the EF^T is also [0]. Thus equation (11.23) reduces to

$$\lambda FF^T \lambda^T + [0] + [0] + EE^T \qquad (11.24)$$

As we are using standardized variables, and because we assumed that the f values are independent, FF^T becomes an identity matrix, because the correlation of F with itself is one by definition and the covariances are assumed to be zero.

The EE^T matrix is a diagonal matrix. Thus the correlation of the group of assets is the product of the lambda and lambda transpose matrices added to the EE^T matrix

$$COR = \lambda\lambda^T + EE^T \qquad (11.25)$$

where the λ is a 4×3 matrix and λ^T is a 3×4 matrix. The result is a 4×4 matrix which has the communalities along the main diagonal. The EE^T matrix represents that proportion of the risk which is unsystematic.

The objective in factor analysis is to solve equation 11.25 in such a way that the largest possible proportion of total variance is accounted for by the $\lambda\lambda^T$ and as little as possible is accounted for by the diagonal matrix.

So **if** the three-factor model holds for our four variables, then we will be able to write the correlation matrix in the form $\lambda\lambda^T$ and a diagonal matrix.

An iterative process is used to attempt this decomposition. We will simply demonstrate this process by showing the computer output.

The λ matrix is

	Factor 1	Factor 2	Factor 3
GILT	0.27338	−0.71591	0.64064
FTSE	0.91401	0.10492	−0.00956
S & P	0.81370	0.43492	0.15732
£/$	−0.43726	0.58105	0.67333

$\lambda =$ (for the above matrix)

The λ^T matrix is

$$\begin{bmatrix} 0.27338 & 0.91401 & 0.81370 & -0.43726 \\ -0.71591 & 0.10492 & 0.43492 & 0.58105 \\ 0.64064 & -0.00956 & 0.15732 & 0.67330 \end{bmatrix}$$

The $\lambda\lambda^T$ matrix, known as the matrix of communalities, is

$$\lambda\lambda^T = \begin{bmatrix} 0.997688 & 0.168637 & 0.011869 & -0.104177 \\ 0.168637 & 0.846516 & 0.787853 & -0.345136 \\ 0.011869 & 0.787853 & 0.876006 & 0.002840 \\ -0.104177 & -0.345136 & 0.002840 & 0.982154 \end{bmatrix}$$

Under this initial specification, the variance accounted for by each factor is

$$\text{Factor 1} = 1.763450$$

$$\text{Factor 2} = 1.050317$$

$$\text{Factor 3} = 0.888597$$

given by $\lambda^T\lambda$. This totals 3.702364, so as the total variance sums to 4 (because we are using standardized variables), the three factors account for 92.56% of the total variance of the portfolio of assets concerned. As a consequence, the unsystematic risk only accounts for 7.44%.

Note that the decomposition leaves the residual matrix

$$\begin{bmatrix} 0.002312 & -0.018837 & 0.016931 & -0.006423 \\ -0.018837 & 0.153484 & -0.137953 & 0.052336 \\ 0.016931 & -0.137953 & 0.123994 & -0.047040 \\ -0.006423 & 0.052336 & -0.047040 & 0.017846 \end{bmatrix}$$

This is not diagonal. Whether it is sufficiently close to diagonal to allow us to regard the model as useful is a matter of judgement.

While FA attempts to identify underlying factors which account for our observations, the factors themselves are not directly observable (or have not been at the point of the analysis). Thus the emphasis is more on understanding than on use. Nevertheless, the identification of useful factors can be instrumental in guiding the direction of subsequent work.

To further complicate FA it may be shown that, given a set of factors, any orthogonal transformation (i.e. rotation) of those factors will achieve exactly the same effect. We are therefore free to choose whichever of these yields more meaningful results. A procedure known as the varimax procedure was used to choose factors so that some of the factor loadings are large and others are small. This associates the variables with fewer and more different factors.

We can illustrate this by applying the varimax procedure to the above results.

First we give the "orthogonal transformation matrix" which is

$$\begin{array}{c c c c} & 1 & 2 & 3 \\ 1 & \begin{bmatrix} 0.91024 & -0.35424 & 0.21441 \\ 2 & 0.39490 & 0.58692 & -0.70681 \\ 3 & 0.12454 & 0.72804 & 0.67413 \end{bmatrix} \end{array}$$

We multiply this 3×3 matrix by the original $4 \times 3\lambda$ matrix. The result is the new λ matrix resulting from the rotation process as follows:

$$\lambda = \begin{array}{c} \text{Gilt} \\ \text{FTSE} \\ \text{S\&P} \\ \text{£/\$} \end{array} \begin{array}{c c c} \text{Factor 1} & \text{Factor 2} & \text{Factor 3} \\ \begin{bmatrix} 0.04591 & -0.05062 & -0.99650 \\ 0.87221 & -0.26917 & 0.11537 \\ 0.93200 & 0.08155 & -0.02689 \\ -0.08470 & 0.98612 & -0.05055 \end{bmatrix} \end{array}$$

These factors have the following interpretation. Factor 1 has virtually no gilts, but large combinations of both equity indices; it also only has a very small, negative, element of currency. Factor 2 has as its largest element the currency with virtually no gilts or S&P and a small, negative, element of the FTSE. Factor 3 has a large, negative, element of gilts, and all other elements are very small.

The new λ^T matrix is

$$\lambda^T = \begin{bmatrix} 0.04591 & 0.87221 & 0.93200 & -0.08470 \\ -0.05062 & -0.2917 & 0.08155 & 0.98612 \\ 0.99650 & 0.11537 & -0.02689 & -0.05055 \end{bmatrix}$$

The EE^T matrix is unchanged by this process.
The variance explained by each factor is now

$$\text{Factor } 1 = 1.638670$$

$$\text{Factor } 2 = 1.054089$$

$$\text{Factor } 3 = 1.009606$$

(again given by $\lambda^T\lambda$). These figures sum to 3.702365, thus these three factors still account for 92.56% of total variance.

The arbitrage pricing theory

A classical application of this process was in the early testing of the arbitrage pricing theory (APT). The theory assumes that in markets where arbitrage is possible, all assets with similar characteristics will trade at similar prices because the activities of arbitragers will remove any pricing differences.

APT assumes that because investors are free to hold a widely diversified portfolio, only systematic risk is priced in the market. However, that systematic risk is a function of several factors. Those factors are unspecified by the theory but must be identified through empirical research. APT postulates the existence of a multiplicity of underlying "risk factors" which together account for the observed variability of asset returns. Given this assumption, it follows that the expected return of an asset (effectively its change in price) is a linear combination of its factor loadings, which measure the exposure of the asset to each risk factor.

These risk factors are not observable – only the historical returns are – but we can use FA to estimate the factor loadings and hence the risk premia associated with each risk factor.

For example, consider a market where there are three such risk factors: F_1, F_2, F_3. A three-factor model of asset returns would relate the expected return on a particular security to a limited number of linearly related economic influences or factors and some idiosyncratic elements

$$R_{it} = rf + b_{i1}F_{1t} + b_{i2}F_{2t} + b_{i3}F_{3t} + e_{it} \tag{11.26}$$

where

rf = the risk-free return
R_{it} = the returns on asset i at time t
b_{it} = the factor loadings
e_i = the idiosyncratic elements effecting security i
F_{it} = the risk premium of factor i.

There are two important statistical assumptions related to this model. The first is that the e values are independent of each other. The other assumption is that the F values are also independent.

The risk associated with holding a particular security comes from two sources. The first source of risk is the macroeconomic factors that affect all securities. These factors provide systematic risk, and their influence pervades the whole asset market and cannot be diversified away. The second source of risk is the idiosyncratic element. This element is unique to each security and in a broadly diversified portfolio can be diversified away. Thus, an efficient market will only reward the risks associated with the systematic (macroeconomic) factors.

As these risk factors affect all assets, the return to each of these risk factors is the market price of bearing that particular risk. Consequently, the return to each factor is the same for each and every asset. In other words the reward for bearing a particular type of risk is set by the market. However, the extent to which an asset is exposed to that risk is determined by the characteristics of the asset itself. In other words, this factor exposure or factor loading, the b_i values, will be unique to each asset.

The systematic risk of an individual asset will be determined by the factor loadings and covariances of the factor returns. The systematic risk of a portfolio of assets will be determined by the covariances of the factor returns, the factor loadings and the weightings of each asset in the portfolio.

In order to operationalize the APT successfully it is necessary to have identified a manageable, i.e. relatively short, list of factors on which it is possible to measure the expected risk premium associated with each factor. This requires the ability to measure the unexpected changes in the factor. Consequently, some applications actually model the unexpected values of the factor by modelling the expected value and subtracting from the actual value. Some factors are considered to have so much noise in their time-series data that the change in the data is itself adequate as a measure of unexpected change.

It is also necessary to be able to measure the sensitivities of each security to each factor. In addition those sensitivities should be reasonably stable.

Roll and Ross (1980) used FA and found that only three and possibly four factors explained the return-generating process of US equities. However, Dhrymes, Friend and Gultekin (1984) noted that the number of factors may be dependent on the number of securities in each portfolio.

The results of tests of the UK market using FA have been inconclusive. Beenstock and Chan (1986) found 20 factors that explain security returns. Moreover, Diacogiannis (1986) found that the factors were not stable between time periods for the same portfolio. Both these studies also noted that the number of factors depended upon how many securities were included in each portfolio. A further study of the UK by Abeysekera and Mahajan (1987) concluded that they could not specify a unique number of factors across the seven portfolios in their study.

Exercises

1. Explain what you understand by the terms "factor analysis" and "principal components analysis". Differentiate between the two processes.

2. What are eigenvectors and eigenvalues? What is their function in factor analysis and in principal components analysis?

3. Using the techniques illustrated in this chapter and the following 2×2 variance–covariance matrix relating to variables X and Y, construct the eigenvectors and the eigenvalues.

$$\begin{bmatrix} 0.00020 & 0.00006 \\ 0.00006 & 0.00030 \end{bmatrix}$$

4. Normalize the eigenvectors calculated in question 3.

5. Construct the matrices QDQ^{-1} that relate to the matrix in question 3. Satisfy yourself that $C = QDQ^{-1}$.

6. Find the exposures expressed as linearly independent combinations of the variables X and Y for the variance–covariance matrix in question 3.

7. Explain how to use standardized variables in principal components analysis.

8. Using symbolic notation only, explain how to use principal components analysis to analyse the components of the variance of a four-year coupon paying bond paying annual coupons.

9. In the context of factor analysis, what do you understand by the term "communalities"?

10. Explain how factor analysis has been applied to test the arbitrage pricing theory.

Answers to selected questions

3.

$$\text{Eigenvector} \quad 1 = \begin{bmatrix} 1 \\ 2.135 \end{bmatrix}$$

$$\text{Eigenvector} \quad 1 = 0.00033$$

$$\text{Eigenvector} \quad 2 = \begin{bmatrix} 1 \\ -0.468 \end{bmatrix}$$

$$\text{Eigenvector} \quad 2 = 0.00017$$

4. Normalized eigenvectors are

$$\begin{bmatrix} 0.424 \\ 0.906 \end{bmatrix}$$

$$\begin{bmatrix} 0.906 \\ -0.424 \end{bmatrix}$$

5. QDQ^{-1} is

$$\begin{bmatrix} 0.424 & 0.906 \\ 0.906 & -0.424 \end{bmatrix} \begin{bmatrix} 0.00033 & 0 \\ 0 & 0.00017 \end{bmatrix} \begin{bmatrix} 0.424 & 0.906 \\ 0.906 & -0.424 \end{bmatrix}$$

6. $0.424X + 0.906Y$; $0.906X - 0.424Y$.

References and further reading

Abeysekera, S. and Mahajan, A. (1987) A test of the APT in pricing UK stocks. *Journal of Accounting and Finance*, **17**, 377–91.

Beenstock, M. and Chan, K. (1986) Testing the arbitrage pricing theory in the UK. *Oxford Bulletin of Economics and Statistics*, **50**, 27–39.

Dhrymes, P., Friend, I. and Gultekin, B. (1984) A critical reexamination of the empirical evidence on the arbitrage pricing theory. *Journal of Finance*, June, 323–46.

Diacogiannis, G. (1986) Arbitrage pricing model: a critical examination of its empirical applicability for the London Stock Exchange. *Journal of Business Finance and Accountancy*, Winter.

Kahn, R. H. (1989) Risk and return in the US bond market: a multi-factor approach. In F. J. Fabozzi (ed.) *Advances and Innovations in the Bond and Mortgage Markets*. Probus Publishing, Chicago.

Kahn, R.H., (1990) Estimating the US treasury term structure of interest rates. In F. J. Fabozzi (ed.) *The Handbook of US Treasury and Government Agency Securities: Instruments, Strategies and Analysis*. Probus Publishing, Chicago.

Kahn, R. N. and Gulrajani, D. (1993) Risk and return in the Canadian bond market. *Journal of Portfolio Management*, Spring, 86–93.

Karki, J. and Reyes, C. (1994) Model relationship. *Risk*, December.

Roll, R. and Ross, S. A. (1980) An empirical investigation of the arbitrage pricing theory. *Journal of Finance*, December.

Appendix: statistical tables

The standard normal distribution

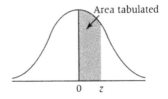

Area tabulated

Example

Pr $(0 \leqslant z \leqslant 1.96) = 0.4750$
Pr $(z \geqslant 1.96) = 0.5 - 0.4750 = 0.025$

z	0.00	0.01	0.02	0.03	0.04	0.05	0.06	0.07	0.08	0.09
0.0	0.0000	0.0040	0.0080	0.0120	0.0160	0.0199	0.0239	0.0279	0.0319	0.0359
0.1	0.0398	0.0438	0.0478	0.0517	0.0557	0.0596	0.0636	0.0675	0.0711	0.0753
0.2	0.0793	0.0832	0.0871	0.0910	0.0948	0.0987	0.1026	0.1064	0.1103	0.1141
0.3	0.1179	0.1217	0.1255	0.1293	0.1331	0.1368	0.1106	0.1113	0.1480	0.1517
0.4	0.1554	0.1591	0.1628	0.1664	0.1700	0.1736	0.1772	0.1808	0.1811	0.1879
0.5	0.1915	0.1950	0.1985	0.2019	0.2054	0.2088	0.2123	0.2157	0.2190	0.2224
0.6	0.2257	0.2291	0.2324	0.2357	0.2389	0.2422	0.2454	0.2486	0.2517	0.2549
0.7	0.2580	0.2611	0.2642	0.2673	0.2704	0.2734	0.2764	0.2794	0.2823	0.2852
0.8	0.2881	0.2910	0.2939	0.2967	0.2995	0.3023	0.3051	0.3078	0.3106	0.3133
0.9	0.3159	0.3186	0.3212	0.3238	0.3264	0.3289	0.3315	0.3310	0.3365	0.3389
1.0	0.3413	0.3438	0.3461	0.3485	0.3508	0.3531	0.3554	0.3577	0.3S99	0.3621
1.1	0.3643	0.3665	0.3686	0.3708	0.3729	0.3749	0.3770	0.3790	0.3810	0.3830
1.2	0.3849	0.3869	0.3888	0.3907	0.3925	0.3944	0.3962	0.3980	0.3997	0.4015
1.3	0.4032	0.4049	0.4066	0.4082	0.4099	0.4115	0.4131	0.4117	0.4162	0.4177
1.4	0.4192	0.4207	0.4222	0.4236	0.4251	0.1265	0.4279	0.4292	0.4306	0.4319
1.5	0.4332	0.4345	0.4357	0.4370	0.4382	0.4394	0.4406	0.4118	0.4429	0.4441
1.6	0.4452	0.4463	0.4474	0.4484	0.4495	0.4505	0.4515	0.4525	0.4535	0.4545
1.7	0.4554	0.4564	0.4573	0.4582	0.1591	0.4599	0.4608	0.4616	0.4625	0.4633

The standardized normal distribution (*continued*)

z	0.00	0.01	0.02	0.03	0.04	0.05	0.06	0.07	0.08	0.09
1.8	0.4641	0.4649	0.4656	0.4664	0.4671	0.1678	0.4686	0.4693	0.4699	0.4706
1.9	0.4713	0.4719	0.4726	0.4732	0.4738	0.4744	0.4750	0.1756	0.4761	0.4767
2.0	0.4772	0.4778	0.4783	0.47R8	0.1793	0.4798	0.1803	0.4808	0.4812	0.4817
2.1	0.4821	0.4826	0.4830	0.4834	0.4838	0.4842	0.4846	0.4850	0.1854	0.4857
2.2	0.4861	0.4864	0.4868	0.4871	0.4875	0.4878	0.4881	0.4884	0.4887	0.4890
2.3	0.4893	0.4896	0.4898	0.4901	0.4904	0.4906	0.4909	0.4911	0.4913	0.4916
2.4	0.4918	0.4920	0.4922	0.4925	0.4927	0.4929	0.4931	0.4932	0.4934	0.4936
2.5	0.4938	0.4940	0.4941	0.4943	0.4915	0.4946	0.4948	0.4919	0.4951	0.4952
2.6	0.4953	0.4955	0.4956	0.4957	0.4959	0.4960	0.4961	0.4962	0.4963	0.4964
2.7	0.4965	0.4966	0.4967	0.4968	0.4969	0.4970	0.4971	0.4972	0.4973	0.4971
2.8	0.4974	0.4975	0.4976	0.4977	0.4977	0.4978	0.4979	0.4979	0.4980	0.4981
2.9	0.4981	0.4982	0.4982	0.4983	0.4984	0.4984	0.4985	0.4985	0.4986	0.4981
3.0	0.4987	0.4987	0.1987	0.4988	0.4988	0.4989	0.4989	0.4989	0.4990	0.4990

Percentage points of the *t*-distribution

One-sided test

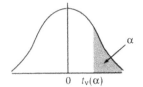

$Pr(T_v > t_v(\alpha)) = \alpha,$
for v degrees of freedom

Degrees of freedom	Area in upper tail				
	0.10	0. 05	0.025	0.01	0.005
	Level of confidence				
	90%	95%	97.5%	99%	99.5%
1	3.078	6.314	12.706	31.821	63.657
2	1.886	2.920	4.303	6.965	9.925
3	1.638	2.353	3.182	4.541	5.841
4	1.533	2.132	2.776	3.747	4.604
5	1.476	2.015	2.571	3.365	4.032
6	1.440	1.943	2.447	3.143	3.707
7	1.415	1.895	2.365	2.998	3.499
8	1.397	1.860	2.306	2.896	3.355
9	1.383	1.833	2.262	2.821	3.250
10	1.372	1.812	2.228	2.764	3.169
11	1.363	1.796	2.201	2.718	3.106
12	1.356	1.782	2.179	2.681	3.055
13	1.350	1.771	2.160	2.650	3.012
14	1.345	1.761	2.145	2.624	2.977
15	1.341	1.753	2.131	2.602	2.947
16	1.337	1.746	2.120	2.583	2.921
17	1.333	1.740	2.110	2.567	2.898
18	1.330	1.734	2.101	2.552	2.878
19	1.328	1.729	2.093	2.539	2.861
20	1.325	1.725	2.086	2.528	2.845
21	1.323	1.721	2.080	2.518	2.831
22	1.321	1.717	2.074	2.508	2.819
23	1.319	1.714	2.069	2.500	2.807
24	1.318	1.711	2.064	2.492	2.797
25	1.316	1.708	2.060	2.485	2.787
26	1.315	1.706	2.056	2.479	2.779
27	1.314	1.703	2.052	2.473	2.771
28	1.313	1.701	2.048	2.467	2.763
29	1.311	1.699	2.045	2.462	2.756
30	1.310	1.697	2.042	2.457	2.750
40	1.303	1.684	2.021	2.423	2.704
60	1.296	1.671	2.000	2.390	2.660
120	1.289	1.658	1.980	2.358	2.617
∞	1.282	1.645	1.960	2.326	2.576

Percentage points χ^2 distribution

The values tabulated are $\chi_v^2(\alpha)$, where $\Pr(\chi_v^2 > \chi_v^2(\alpha)) = \alpha$, for v, degrees of freedom.

0.995	0.990	0.975	0.950	0.900	0.750	0.500	α / v
392704.10^{-10}	157088.10^{-9}	982069.10^{-9}	393214.10^{-8}	0.0157908	0.1015308	0.454936	1
0.0100251	0.0201007	0.0506356	0.102587	0.210721	0.575364	1.38629	2
0.0717218	0.114832	0.215795	0.351846	0.584374	1.212534	2.36597	3
0.206989	0.297109	0.484419	0.710723	1.063623	1.92256	3.35669	4
0.411742	0.554298	0.831212	1.145476	1.61031	2.67460	4.35146	5
0.675727	0.872090	1.23734	1.63538	2.20413	3.45460	5.34812	6
0.989256	1.239043	1.68987	2.16735	2.83311	4.25485	6.34581	7
1.34441	1.64650	2.17973	2.73264	3.48954	5.07064	7.34412	8
1.73493	2.08790	2.70039	3.32511	4.16816	5.89883	8.34283	9
2.15586	2.55821	3.24697	3.94030	4.86518	6.73720	9.34182	10
2.60322	3.05348	3.81575	4.57481	5.57778	7.58414	10.3410	11
3.07382	3.57057	4.40379	5.22603	6.30380	8.43842	11.3403	12
3.56503	4.10692	5.00875	5.89186	7.04150	9.29907	12.3398	13
4.07467	4.66043	5.62873	6.57063	7.78953	10.1653	13.3393	14
4.60092	5.22935	6.26214	7.26094	8.54676	11.0365	14.3389	15
5.14221	5.81221	6.90766	7.96165	9.31224	11.9122	15.3385	16
5.69722	6.40776	7.56419	8.67176	10.0852	12.7919	16.3382	17
6.26480	7.01491	8.23075	9.39046	10.8649	13.6753	17.3379	18
6.84397	7.63273	8.90652	10.1170	11.6509	14.5620	18.3377	19
7.43384	8.26040	9.59078	10.8508	12.4426	15.4518	19.3374	20
8.03365	8.89720	10.28293	11.5913	13.2396	16.3444	20.3372	21
8.64272	9.54249	10.9823	12.3380	14.0415	17.2396	21.3370	22
9.26043	10.19567	11.6886	13.0905	14.8480	18.1373	22.3369	23
9.88623	10.8564	12.4012	13.8484	15.6587	19.0373	23.3367	24
10.5197	11.5240	13.1197	14.6114	16.4734	19.9393	24.3366	25
11.1602	12.1981	13.8439	15.3792	17.2919	20.8434	25.3365	26
11.8076	12.8785	14.5734	16.1514	18.1139	21.7494	26.3363	27
12.4613	13.5647	15.3079	16.9279	18.9392	22.6572	27.3362	28
13.1211	14.2565	16.0471	17.7084	19.7677	23.5666	28.3361	29
13.7867	14.9535	16.7908	18.4927	20.5992	24.4776	29.3360	30
20.7065	22.1643	24.4330	26.5093	29.0505	33.6603	39.3353	40
27.9907	29.7067	32.3574	34.7643	37.6886	42.9421	49.3349	50
35.5345	37.4849	40.4817	43.1880	46.4589	52.2938	59.3347	60
43.2752	45.4417	48.7576	51.7393	55.3289	61.6983	69.3345	70
51.1719	53.5401	57.1532	60.3915	64.2778	71.1445	79.3343	80
59.1963	61.7541	65.6466	69.1260	73.2911	80.6247	89.3342	90
67.3276	70.0649	74.2219	77.9295	82.3581	90.1332	99.3341	100

Percentage points χ^2 distribution (*Continued*)

For $v > 30$ take $\chi_v^2(\alpha) = v\left[1 - \dfrac{2}{9v} + u_\alpha\sqrt{\dfrac{2}{9v}}\right]^3$ where u_α is such that $\Pr(U > u_\alpha) = \alpha$, and $U \sim N(0,1)$.

α \ v	0.250	0.100	0.050	0.025	0.010	0.005	0.001
1	1.32330	2.70554	3.84146	5.02389	6.63490	7.87944	10.828
2	2.77259	4.60517	5.99146	7.37776	9.21034	10.5966	13.816
3	4.10834	6.25139	7.81473	9.34840	11.3449	12.8382	16.266
4	5.38527	7.77944	9.48773	11.1433	13.2767	14.8603	18.467
5	6.62568	9.23636	11.0705	12.8325	15.0863	16.7496	20.515
6	7.84080	10.6446	12.5916	14.4494	16.8119	18.5476	22.458
7	9.03715	12.0170	14.0671	16.0128	18.4753	20.2777	24.322
8	10.2189	13.3616	15.5073	17.5345	20.0902	21.9550	26.125
9	11.3888	14.6837	16.9190	19.0228	21.6660	23.5894	27.877
10	12.5489	15.9872	18.3070	20.4832	23.2093	25.1882	29.588
11	13.7007	17.2750	19.6751	21.9200	24.7250	26.7568	31.264
12	14.8454	18.5493	21.0261	23.3367	26.2170	28.2995	32.909
13	15.9839	19.8119	22.3620	24.7356	27.6882	29.8195	34.528
14	17.1169	21.0641	23.6848	26.1189	29.1412	31.3194	36.123
15	18.2451	22.3071	24.9958	27.4884	30.5779	32.8013	37.697
16	19.3689	23.5418	26.2962	28.8454	31.9999	34.2672	39.252
17	20.4887	24.7690	27.5871	30.1910	33.4087	35.7185	40.790
18	21.6049	25.9894	28.8693	31.5264	34.8053	37.1565	42.312
19	22.7178	27.2036	30.1435	32.8523	36.1909	38.5823	43.820
20	23.8277	28.4120	31.4104	34.1696	37.5662	39.9968	45.315
21	24.9348	29.6151	32.6706	35.4789	38.9322	41.4011	46.797
22	26.0393	30.8133	33.9244	36.7807	40.2894	42.7957	48.268
23	27.1413	32.0069	35.1725	38.0756	41.6384	44.1813	49.728
24	28.2412	33.1962	36.4150	39.3641	42.9798	45.5585	51.179
25	29.3389	34.3816	37.6525	40.6465	44.3141	46.9279	52.618
26	30.4346	35.5632	38.8851	41.9232	45.6417	48.2899	54.052
27	31.5284	36.7412	40.1133	43.1945	46.9629	49.6449	55.476
28	32.6205	37.9159	41.3371	44.4608	48.2782	50.9934	56.892
29	33.7109	39.0875	42.5570	45.7223	49.5879	52.3356	58.301
30	34.7997	40.2560	43.7730	46.9792	50.8922	53.6720	59.703
40	45.6160	51.8051	55.7585	59.3417	63.6907	66.7660	73.402
50	56.3336	63.1671	67.5048	71.4202	76.1539	79.4900	86.661
60	66.9815	74.3970	79.0819	83.2977	88.3794	91.9517	99.607
70	77.5767	85.5270	90.5312	95.0232	100.425	104.215	112.317
80	88.1303	96.5782	101.879	106.629	112.329	116.321	124.839
90	98.6499	107.565	113.145	118.136	124.116	128.299	137.208
100	109.141	118.498	124.342	129.561	135.807	140.169	149.449

F distribution

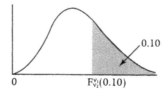

0 $F_{v_2}^{v_1}(0.10)$

Right tail of the distribution for $P = 0.05$ (lightface type), 0.01 (boldface type).

V_1 = degrees of freedom for numerator

V_2		1	2	3	4	5	6	7	8	9	10	11	12
	1	161	200	216	225	230	234	237	239	241	242	243	244
		4052	**4999**	**5403**	**5625**	**5764**	**5859**	**5928**	**5981**	**6022**	**6056**	**6082**	**6106**
	2	18 51	19.00	19.16	19.25	19.30	19.33	19.36	19 37	19 38	19.39	19 40	19.41
		98.49	**99.01**	**99.17**	**99.25**	**99.30**	**99.33**	**99.34**	**99.36**	**99.38**	**99.40**	**99.41**	**99.42**
	3	10 13	9.55	9.28	9.12	9.01	8.94	8 88	8.84	8.81	8.78	8.76	8.74
		34.12	**30.81**	**29.46**	**28.71**	**28.24**	**27.91**	**27.67**	**27.49**	**27.34**	**27.23**	**27.13**	**27.05**
	4	7 71	6 94	6.59	6.39	6.26	6.16	6.09	6.04	6.00	5.96	5.93	5 91
		21.20	**18.00**	**16.69**	**15.98**	**15.52**	**15.21**	**14.98**	**14.80**	**14.66**	**14.54**	**14.45**	**14.37**
	5	6.61	5.79	5 41	5.19	5.05	4.95	4 88	4.82	4.78	4.74	4.70	4.68
		16.26	**13.27**	**12.06**	**11.39**	**10.97**	**10.67**	**10.45**	**10.27**	**10.15**	**10.05**	**9.96**	**9.89**
	6	5.99	5.14	4.76	4.53	4.39	4.28	4.21	4.15	4.10	4.06	4.03	4.00
		13.74	**10.92**	**9.78**	**9.15**	**8.75**	**8.47**	**8.26**	**8.10**	**7.98**	**7.87**	**7.79**	**7.72**
	7	5.59	4.74	4.35	4.12	3.97	~ 87	3.79	3.73	3.68	3.63	3.60	3.57
		12.25	**9.55**	**8.45**	**7.85**	**7.46**	**7.19**	**7.00**	**6.84**	**6.71**	**6.62**	**6.54**	**6.47**
	8	5 32	4.46	4 07	3.84	3.69	3.58	3 50	3.44	3.39	3.34	3.31	3.28
		11.26	**8.65**	**7.59**	**7.01**	**6.63**	**6.37**	**6.19**	**6.03**	**5.91**	**5.82**	**5.74**	**5.67**
	9	5 12	4.26	3.86	3.63	3 48	3.37	3.29	3.23	3.18	3.13	3.10	3.07
		10.56	**8.02**	**6.99**	**6.42**	**6.06**	**5.80**	**5.62**	**5.47**	**5.35**	**5.26**	**5.18**	**5.11**
	10	4 96	4.10	3 71	3 48	3 33	3.22	3 14	3.07	3.02	2 97	2.94	2.91
		10.04	**7.56**	**6.55**	**5.99**	**5.64**	**5.39**	**5.21**	**5.06**	**4.95**	**4.85**	**4.78**	**4.71**
	11	4.84	3.98	3.59	3.36	3.20	3 09	3 01	2.95	2 90	2.86	2.82	2 79
		9.65	**7.20**	**6.22**	**5.67**	**5.32**	**5.07**	**4.88**	**4.74**	**4.63**	**4.54**	**4.46**	**4.40**
	12	4 75	3.88	3 49	3.26	3.11	3.00	2 92	2.85	2.80	2 76	2 72	2.69
		9.33	**6.93**	**5.95**	**5.41**	**5.06**	**4.82**	**4.65**	**4.50**	**4.39**	**4.30**	**4.22**	**4.16**
	13	4 67	3.80	3 41	3.18	3 02	2.92	2.84	2.77	2.72	2 67	2.63	2.60
		9.07	**6.70**	**5.74**	**5.20**	**4.86**	**4.62**	**4.44**	**4.30**	**4.19**	**4.10**	**4.02**	**3.96**
	14	4 60	3.74	3.34	3.11	2.96	2 85	2.77	2.70	2.65	2.60	2.56	2.53
		8.86	**6.51**	**5.56**	**5.03**	**4.69**	**4.46**	**4.28**	**4.14**	**4.03**	**3.94**	**3.86**	**3.80**
	15	4.54	3 68	3.29	3.06	2 90	2.79	2.70	2.64	2 59	2.55	2.51	2 48
		8.68	**6.36**	**5.42**	**4.89**	**4.56**	**4.32**	**4.14**	**4.00**	**3.89**	**3.80**	**3.73**	**3.67**
	16	4.49	3 63	3 24	3.01	2.85	2.74	2.66	2.59	2 54	2.49	2.45	2.42
		8.53	**6.23**	**5.29**	**4.77**	**4.44**	**4.20**	**4.03**	**3.89**	**3.78**	**3.69**	**3.61**	**3.55**
	17	4 45	3.59	3 20	2 96	2 81	2.70	2 62	2.55	2.50	2.45	2 41	2 38
		8.40	**6.11**	**5.18**	**4.67**	**4.34**	**4.10**	**3.93**	**3.79**	**3.68**	**3.59**	**3.52**	**3.45**
	18	4 41	3.55	3 16	2.93	2.77	2 66	2.58	2.51	2 46	2.41	2.37	2.34
		8.28	**6.01**	**5.09**	**4.58**	**4.25**	**4.01**	**3.85**	**3.71**	**3.60**	**3.51**	**3.44**	**3.37**
	19	4 38	3.52	3.13	2.90	2 74	2.63	2 55	2.48	2.43	2.38	2 34	2.31
		8.18	**5.93**	**5.01**	**4.50**	**4.17**	**3.94**	**3.77**	**3.63**	**3.52**	**3.43**	**3.36**	**3.30**
	20	4 35	3.49	3.10	2.87	2.71	2.60	2.52	2.45	2 40	2.35	2.31	2 28
		8.10	**5.85**	**4.94**	**4.43**	**4.10**	**3.87**	**3.71**	**3.56**	**3.45**	**3.37**	**3.30**	**3.23**
	21	4 32	3.47	3.07	2 84	2 68	2 57	2.49	2.42	2 37	2.32	2.28	2.25
		8.02	**5.78**	**4.87**	**4.37**	**4.04**	**3.81**	**3.65**	**3.51**	**3.40**	**3.31**	**3.24**	**3.17**
	22	4 30	3.44	3 05	2.82	2.66	2 55	2.47	2 40	2.35	2.30	2.26	2.23
		7.94	**5.72**	**4.82**	**4.31**	**3.99**	**3.76**	**3.59**	**3.45**	**3.35**	**3.26**	**3.18**	**3.12**
	23	4 28	3 42	3.03	2 80	2.64	2 53	2 45	2.38	2 32	2.28	2.24	2 20
		7.88	**5.66**	**4.76**	**4.26**	**3.94**	**3.71**	**3.54**	**3.41**	**3.30**	**3.21**	**3.14**	**3.07**
	24	4 26	3.40	3 01	2 78	2 62	2.51	2 43	2 36	2.30	2 26	2.22	2 18
		7.82	**5.61**	**4.72**	**4.22**	**3.90**	**3.67**	**3.50**	**3.36**	**3.25**	**3.17**	**3.09**	**3.03**
	25	4 24	3 38	2.99	2.76	2 60	2 49	2.41	2.34	2 28	2.24	2 20	2 16
		7.77	**5.57**	**4.68**	**4.18**	**3.86**	**3.63**	**3.46**	**3.32**	**3.21**	**3.13**	**3.05**	**2.99**
	26	4 22	3 37	2 98	2 74	2.59	2 47	2 39	2.32	2 27	2.22	2.18	2 15
		7.72	**5.53**	**4.64**	**4.14**	**3.82**	**3.59**	**3.42**	**3.29**	**3.17**	**3.09**	**3.02**	**2.96**

V_2 = degrees of freedom for denominator

F distribution (*continued*)

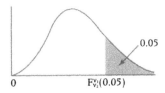

$F_{v_2}^{v_1}(0.05)$

				v_1 = degrees of freedom for numerator								
14	16	20	24	30	40	50	75	100	200	500	∞	
245	246	248	249	250	251	252	253	253	254	254	254	1
6142	6169	6208	6234	6258	6286	6302	6323	6334	6352	6361	6366	
19.42	19.43	19.44	19.45	19.46	19.47	19.47	19.48	19.49	19.49	19 50	19 50	2
99.43	99.44	99.45	99.46	99.47	99.48	99.48	99.49	99.49	99.49	99.50	99.50	
8.71	8.69	8.66	8.64	8.62	8.60	8.58	8.57	8.56	8.54	8 54	8.53	3
26.92	26.83	26.69	26.60	26.50	26.41	26.35	26.27	26.23	26.18	26.14	26.12	
5.87	5.84	5.80	5.77	5.74	5.71	5.70	5.68	5.66	5.65	5.64	5.63	4
14.24	14.15	14.02	13.93	13.83	13.74	13.69	13.61	13.57	13.52	13.48	13.46	
4.64	4 60	4.56	4.53	4.50	4.46	4.44	4.42	4.40	4.38	4.37	4.36	5
9.77	9.68	9.55	9.47	9.38	9.29	9.24	9.17	9.13	9.07	9.04	9.02	
3.96	3.92	3.87	3.84	3.81	3.77	3.75	3.72	3.71	3.69	3.68	3.67	6
7.60	7.52	7.39	7.31	7.23	7.14	7.09	7.02	6.99	6.94	6.90	6.88	
3.52	3.49	3.44	3.41	3.38	3.34	3.32	3.29	3.28	3.25	3.24	3.23	7
6.35	6.27	6.15	6.07	5.98	5.90	5.85	5.78	5.75	5.70	5.67	5.65	
3.23	3.20	3.15	3.12	3.08	3.05	3.03	3.00	2.98	2.96	2.94	2.93	8
5.56	5.48	5.36	5.28	5.20	5.11	5.06	5.00	4.96	4.91	4.88	4.86	
3.02	2.98	2.93	2.90	2.86	2.82	2.80	2.77	2.76	2.73	2.72	2.71	9
5.00	4.92	4.80	4.73	4.64	4.56	4.51	4.45	4.41	4.36	4.33	4.31	
2 86	2.82	2.77	2.74	2.70	2.67	2.64	2.61	2.59	2.56	2.55	2.54	10
4.60	4.52	4.41	4.33	4.25	4.17	4.12	4.05	4.01	3.96	3.93	3.91	
2.74	2.70	2.65	2.61	2.57	2.53	2.50	2.47	2.45	2.42	2.41	2.40	11
4.29	4.21	4.10	4.02	3.94	3.86	3.80	3.74	3.70	3.66	3.62	3.60	
2.64	2.60	2.54	2.50	2.46	2.42	2.40	2.36	2.35	2.32	2.31	2.30	12
4.05	3.98	3.86	3.78	3.70	3.61	3.56	3.49	3.46	3.41	3.38	3.36	
2.55	2.51	2.46	2.42	2.38	2.34	2.32	2.28	2.26	2.24	2.22	2 21	13
3.85	3.78	3.67	3.59	3.51	3.42	3.37	3.30	3.27	3.21	3.18	3.16	
2.48	2.44	2.39	2.35	2.31	2.27	2.24	2.21	2.19	2.16	2.14	2.13	14
3.70	3.62	3.51	3.43	3.34	3.26	3.21	3.14	3.11	3.06	3.02	3.00	
2.43	2.39	2.33	2.29	2.25	2.21	2.18	2.15	2.12	2 10	2.08	2.07	15
3.56	3.48	3.36	3.29	3.20	3.12	3.07	3.00	2.97	2.92	2.89	2.87	
2.37	2.33	2.28	2.24	2.20	2.16	2.13	2.09	2.07	2.04	2 02	2.01	16
3.45	3.37	3.25	3.18	3.10	3.01	2.96	2.89	2.86	2.80	2.77	2.75	
2.33	2.29	2.23	2.19	2.15	2.11	2.08	2.04	2.02	1.99	1.97	1.96	17
3.35	3.27	3.16	3.08	3.00	2.92	2.86	2.79	2.76	2.70	2.67	2.65	
2.29	2.25	2.19	2.15	2.11	2.07	2.04	2.00	1.98	1 95	1.93	1.92	18
3.27	3.19	3.07	3.00	2.91	2.83	2.78	2.71	2.68	2.62	2.59	2.57	
2 26	2.21	2.15	2.11	2.07	2.02	2.00	1.96	1.94	1.91	1.90	1.88	19
3.19	3.12	3.00	2.92	2.84	2.76	2.70	2.63	2.60	2.54	2.51	2.49	
2.23	2.18	2.12	2.08	2.04	1.99	1.96	1.92	1.90	1.87	1.85	1.84	20
3.13	3.05	2.94	2.86	2.77	2.69	2.63	2.56	2.53	2.47	2.44	2.42	
2.20	2.15	2.09	2.05	2.00	1.96	1.93	1.89	1.87	1.84	1.82	1.81	21
3.07	2.99	2.88	2.80	2.72	2.63	2.58	2.51	2.47	2.42	2.38	2.36	
2 18	2.13	2 07	2.03	1 98	1 93	1.91	1.87	1.84	1.81	1.80	1.78	22
3.02	2.94	2.83	2.75	2.67	2.58	2.53	2.46	2.42	2.37	2.33	2.31	
2.14	2.10	2.04	2.00	1.96	1.91	1.88	1.84	1.82	1.79	1.77	1.76	23
2.97	2.89	2.78	2.70	2.62	2.53	2.48	2.41	2.37	2.32	2.28	2.26	
2.13	2.09	2.02	1.98	1.94	1.89	1.86	1.82	1.80	1.76	1.74	1.73	24
2.93	2.85	2.74	2.66	2.58	2.49	2.44	2.36	2.33	2.27	2.23	2.21	
2.11	2.06	2.00	1.96	1.92	1.87	1.84	1.80	1.77	1.74	1.72	1.71	25
2.89	2.81	2.70	2.62	2.54	2.45	2.40	2.32	2.29	2.23	2.19	2.17	
2.10	2.05	1.99	1.95	1.90	1.85	1.82	1.78	1.76	1.72	1.70	1.69	26
2.86	2.77	2.66	2.58	2.50	2.41	2.36	2.28	2.25	2.19	2.15	2.13	

V_2 = degrees of freedom for denominator

F distribution (*continued*)

	V_1 = degrees of freedom for numerator											
m_2	1	2	3	4	5	6	7	8	9	10	11	12
27	4 21	3 35	2 96	2 73	2 57	2.46	2.37	2 30	2.25	2 20	2.16	2 13
	7.68	5.49	4.60	4.11	3.79	3.56	3.39	3.26	3.14	3.06	2.98	2.93
28	4.20	3 34	2.95	2.71	2.56	2.44	2.36	2.29	2.24	2.19	2.15	2 12
	7.64	5.45	4.57	4.07	3.76	3.53	3.36	3.23	3.11	3.03	2.95	2.90
29	4 18	3.33	2.93	2.70	2.54	2.43	2.35	2.28	2.22	2.18	2 14	2.10
	7.60	5.42	4.54	4.04	3.73	3.50	3.33	3.20	3.08	3.00	2.92	2.87
30	4.17	3.32	2.92	2.69	2.53	2 42	2.34	2.27	2.21	2.16	2.12	2.09
	7.56	5.39	4.51	4.02	3.70	3.47	3.30	3.17	3.06	2.98	2.90	2.84
32	4.15	3.30	2.90	2.67	2.51	2.40	2.32	2.25	2.19	2.14	2 10	2.07
	7.50	5.34	4.46	3.97	3.66	3.42	3.25	3.12	3.01	2.94	2.86	2.80
34	4.13	3.28	2.88	2.65	2.49	2.38	2.30	2.23	2.17	2.12	2.08	2.05
	7.44	5.29	4.42	3.93	3.61	3.38	3.21	3.08	2.97	2.89	2.82	2.76
36	4.11	3.26	2 86	2 63	2 48	2 36	2.28	2.21	2.15	2.10	2.06	2.03
	7.39	5.25	4.38	3.89	3.58	3.35	3.18	3.04	2.94	2.86	2.78	2.72
38	4.10	3.25	2.85	2.62	2.46	2.35	2.26	2.19	2.14	2.09	2 05	2.02
	7.35	5.21	4.34	3.86	3.54	3.32	3.15	3.02	2.91	2.82	2.75	2.69
40	4.08	3.23	2.84	2.61	2.45	2.34	2.25	2.18	2.12	2.07	2 04	2.00
	7.31	5.18	4.31	3.83	3.51	3.29	3.12	2.99	2.88	2.80	2.73	2.66
42	4.07	3.22	2.83	2.59	2.44	2.32	2.24	2.17	2.11	2.06	2.02	1.99
	7.27	5.15	4.29	3.80	3.49	3.26	3.10	2.96	2.86	2.M	2.70	2.64
44	4 06	3 21	2.82	2.58	2.43	2.31	2.23	2.16	2.10	2.05	2.01	1.98
	7.24	5.12	4.26	3.78	3.46	3.24	3.07	2.94	2.84	2.75	2.68	2.62
46	4.05	3.20	2 81	2.57	2.42	2.30	2.22	2.14	2.09	2.04	2.00	1.97
	7.21	5.10	4.24	3.76	3.44	3.22	3.05	2.92	2.82	2.73	2.66	2.60
48	4.04	3.19	2.80	2.56	2.41	2.30	2.21	2.14	2.08	2.03	1.99	1.96
	7.19	5.08	4.22	3.74	3.42	3.20	3.04	2.90	2.80	2.71	2.64	2.58
50	4.03	3.18	2.79	2.56	2.40	2.29	2.20	2.13	2.07	2.02	1.98	1.95
	7.17	5.06	4.20	3.72	3.41	3.18	3.02	2.88	2.78	2.70	2.62	2.56
55	4.02	3.17	2.78	2.54	2.38	2.27	2.18	2.11	2.05	2.00	1.97	1.93
	7.12	5.01	4.16	3.68	3.37	3.15	2.98	2.85	2.75	2.66	2.59	2.53
60	4.00	3.15	2.76	2.52	2.37	2.25	2.17	2.10	2.04	1.99	1.95	1.92
	7.08	4.98	4.13	3.65	3.34	3.12	2.95	2.82	2.72	2.63	2.56	2.50
65	3.99	3.14	2.75	2.51	2.36	2.24	2.15	2.08	2.02	1.98	1.94	1.90
	7.04	4.95	4.10	3.62	3.31	3.09	2.93	2.79	2.70	2.61	2.54	2.47
70	3.98	3.13	2.74	2.50	2.35	2.23	2.14	2.07	2.01	1.97	1.93	1.89
	7.01	4.92	4.08	3.60	3.29	3.07	2.91	2.77	2.67	2.59	2.51	2.45
80	3.96	3.11	2.72	2.48	2.33	2.21	2.12	2.05	1.99	1.95	1.91	1.88
	6.96	4.88	4.04	3.56	3.25	3.04	2.87	2.74	2.64	2.55	2.48	2.41
100	3 94	3.09	2.70	2.46	2.30	2.19	2.10	2.03	1.97	1.92	1.88	1.85
	6.90	4.82	3.98	3.51	3.20	2.99	2.82	2.69	2.59	2.51	2.43	2.36
125	3.92	3.07	2.68	2.44	2.29	2.17	2.08	2.01	1.95	1.90	1.86	1.83
	6.84	4.78	3.94	3.47	3.17	2.95	2.79	2.65	2.56	2.47	2.40	2.33
150	3.91	3.06	2.67	2.43	2.27	2.16	2.07	2.00	1.94	1.89	1.85	1.82
	6.81	4.75	3.91	3.44	3.14	2.92	2.76	2.62	2.53	2.44	2.37	2.30
200	3.89	3.04	2.65	2.41	2 26	2.14	2.05	1.98	1.92	1.87	1.83	1.80
	6.76	4.71	3.88	3.41	3.11	2.90	2.73	2.60	2.50	2.41	2.34	2.28
400	3 86	3 02	2.62	2.39	2.23	2.12	2.03	1.96	1.90	1.85	1.81	1 78
	6.70	4.66	3.83	3.36	3.06	2.85	2.69	2.55	2.46	2.37	2.29	2.23
1000	3.85	3 00	2.61	2.38	2.22	2.10	2.02	1.95	1.89	1.84	1.80	1.76
	6.66	4.62	3.80	3.34	3.04	2.82	2.66	2.53	2.43	2.34	2.26	2.20
∞	3.84	2.99	260	2.37	2.21	2.09	201	1.94	1.88	1 83	1.79	1 75
	6.64	4.60	3.78	3.32	3.02	2.80	2.64	2.51	2.41	2.32	2.24	2.18

V_2 = degrees of freedom for denominator

F distribution (continued)

14	16	20	24	30	40	50	75	100	200	500	∞	m^2
2.08	203	1.97	1.93	1.88	1.84	1.80	1.76	1.74	1.71	1.68	1.67	27
2.83	**2.74**	**2.63**	**2.55**	**2.47**	**2.38**	**2.33**	**2.25**	**2.21**	**2.16**	**2.12**	**2.10**	
2.06	2 02	1.96	1.91	1.87	1.81	1.78	1.75	1.72	1.69	1.67	1.65	28
2.80	**2.71**	**2.60**	**2.52**	**2.44**	**2.35**	**2.30**	**2.22**	**2.18**	**2.13**	**2.09**	**2.06**	
2.05	2.00	1.94	1.90	1.85	1.80	1.77	1.73	1.71	1.68	1.65	1.64	29
2.77	**2.68**	**2.57**	**2.49**	**2.41**	**2.32**	**2.27**	**2.19**	**2.15**	**2.10**	**2.06**	**2.03**	
2.04	1.99	1.93	1.89	1.84	1.79	1.76	1.72	1.69	1 66	1.64	1.62	30
2.74	**2.66**	**2.55**	**2.47**	**2.38**	**2.29**	**2.24**	**2.16**	**2.13**	**2.07**	**2.03**	**2.01**	
2.02	1.97	1.91	1.86	1.82	1.76	1.74	1.69	1.67	1.64	1.61	1.59	32
2.70	**2.62**	**2.51**	**2.42**	**2.34**	**2.25**	**2.20**	**2.12**	**2.08**	**2.02**	**1.98**	**1.96**	
2.00	1.95	1.89	1.84	1.80	1.74	1.71	1.67	1.64	1.61	1.59	1.57	34
2.66	**2.58**	**2.47**	**2.38**	**2.30**	**2.21**	**2.15**	**2.08**	**2.04**	**1.98**	**1.94**	**1.91**	
1.98	1.93	1.87	1.82	1.78	1.72	1.69	1.65	1.62	1.59	1.56	1.55	36
2.62	**2.54**	**2.43**	**2.35**	**2.26**	**2.17**	**2.12**	**2.04**	**2.00**	**1.94**	**1.90**	**1.87**	
1.96	1.92	1.85	1.80	1.76	1.71	1.67	1.63	1.60	1.57	1.54	1.53	38
2.59	**2.51**	**2.40**	**2.32**	**2.22**	**2.14**	**2.08**	**2.00**	**1.97**	**1.90**	**1.86**	**1.84**	
1.95	1.90	1.84	1.79	1.74	1.69	1.66	1.61	1.59	1.55	1.53	1.51	40
2.56	**2.49**	**2.37**	**2.29**	**2.20**	**2.11**	**2.05**	**1.97**	**1.94**	**1.88**	**1.84**	**1.81**	
1.94	1.89	1.82	1.78	1.73	1.68	1.64	1.60	1.57	1.54	1.51	1.49	42
2.54	**2.46**	**2.35**	**2.26**	**2.17**	**2.08**	**2.02**	**1.94**	**1.91**	**1.85**	**1.80**	**1.78**	
1.92	1.88	1.81	1.76	1.72	1.66	1.63	1.58	1.56	1.52	1.50	1.48	44
2.52	**2.44**	**2.32**	**2.24**	**2.15**	**2.06**	**2.00**	**1.92**	**1.88**	**1.82**	**1.78**	**1.75**	
1.91	1.87	1.80	1.75	1 71	1.65	1.62	1.57	1.54	1.51	1.48	1.46	46
2.50	**2.42**	**2.30**	**2.22**	**2.13**	**2.04**	**1.98**	**1.90**	**1.86**	**1.80**	**1.76**	**1.72**	
1.90	1.86	1.79	1.74	1.70	1.64	1.61	1.56	1.53	1.50	1.47	1.45	48
2.48	**2.40**	**2.28**	**2.20**	**2.11**	**2.02**	**1.96**	**1.88**	**1.84**	**1.78**	**1.73**	**1.70**	
1.90	1.85	1.78	1.74	1.69	1.63	1.60	1.55	1.52	1.48	1.46	1.44	50
2.46	**2.39**	**2.26**	**2.18**	**2.10**	**2.00**	**1.94**	**1.86**	**1.82**	**1.76**	**1.71**	**1.68**	
1.88	1.83	1.76	1.72	1.67	1.61	1.58	1.52	1.50	1.46	1.43	1.41	55
2.43	**2.35**	**2.23**	**2.15**	**2.06**	**1.96**	**1.90**	**1.82**	**1.78**	**1.71**	**1.66**	**1.64**	
1.86	1.81	1.75	1.70	1.65	1.59	1.56	1.50	1.48	1.44	1.41	1 39	60
2.40	**2.32**	**2.20**	**2.12**	**2.03**	**1.93**	**1.87**	**1.79**	**1.74**	**1.68**	**1.63**	**1.60**	
1.85	1.80	1.73	1.68	1.63	1.57	1.54	1.49	1.46	1.42	1.39	1.37	65
2.37	**2.30**	**2.18**	**2.09**	**2.00**	**1.90**	**1.84**	**1.76**	**1.71**	**1.64**	**1.60**	**1.56**	
1.84	1.79	1.72	1 67	1.62	1.56	1.53	1.47	1.45	1.40	1.37	1.35	70
2.35	**2.28**	**2.15**	**2.07**	**1.98**	**1.88**	**1.82**	**1.74**	**1.69**	**1.62**	**1.56**	**1.53**	
1.82	1.77	1.70	1.65	1.60	1.54	1.51	1.45	1.42	1 38	1.35	1.32	80
2.32	**2.24**	**2.11**	**2.03**	**1.94**	**1.84**	**1.78**	**1.70**	**1.65**	**1.57**	**1.52**	**1.49**	
1.79	1.75	1.68	1.63	1.57	1.51	1.48	1.42	1.39	1.34	1.30	1.28	100
2.26	**2.19**	**2.06**	**1.98**	**1.89**	**1.79**	**1.73**	**1.64**	**1.59**	**1.51**	**1.46**	**1.43**	
1.77	1.72	1.65	1.60	1.55	1 49	1.45	1.39	1.36	1.31	1.27	1.25	125
2.23	**2.15**	**2.03**	**1.94**	**1.85**	**1.75**	**1.68**	**1.59**	**1.54**	**1.46**	**1.40**	**1.37**	
1.76	1.71	1.64	1.59	1.54	1.47	1.44	1.37	1.34	1.29	1.25	1.22	150
2.20	**2.12**	**2.00**	**1.91**	**1.83**	**1.72**	**1.66**	**1.56**	**1.51**	**1.43**	**1.37**	**1.33**	
1.74	1.69	1.62	1.57	1.52	1.45	1.42	1.35	1.32	1.26	1.22	1.19	200
2.17	**2.09**	**1.97**	**1.88**	**1.79**	**1.69**	**1.62**	**1.53**	**1.48**	**1.39**	**1.33**	**1.28**	
1.72	1.67	1 60	1.54	1.49	1.42	1.38	1.32	1.28	1.22	1.16	1.13	400
2.12	**2.04**	**1.92**	**1.84**	**1.74**	**1.64**	**1.57**	**1.47**	**1.42**	**1.32**	**1.24**	**1.19**	
1.70	1.65	1.58	1.53	1.47	1.41	1.36	1.30	1.26	1.19	1.13	1.08	1000
2.09	**2.01**	**1.89**	**1.81**	**1.71**	**1.61**	**1.54**	**1.44**	**1.38**	**1.28**	**1.19**	**1.11**	
1.69	1.64	1.57	1.52	1.46	1.40	1.35	1.28	1.24	1.17	1.11	1.00	∞
2.07	**1.99**	**1.87**	**1.79**	**1.69**	**1.59**	**1.52**	**1.41**	**1.36**	**1.25**	**1.15**	**1.00**	

V_1 = degrees of freedom for numerator

V_2 = degrees of freedom for denominator

Durbin–Watson statistic

To test H_0: no positive serial correlation,
if $d < d_L$, reject H_0;
if $d > d_U$, accept H_0;
if $d_L < d < d_U$, the test is inconclusive.
To test H_0: no negative serial correlation, use $d = U - d$.

Level of significance $\alpha = 0.05$

n	$P = 2$		$P = 3$		$P = 4$		$P = 5$		$P = 6$	
	d_L	d_U	d_L	d_U	d_L	d_U	d_L	d_U	d_L	d_U
15	1.08	1.36	0.95	1.54	0.82	1.75	0.69	1.97	0.56	2.21
16	1.10	1.37	0.98	1.54	0.86	1.73	0.74	1.93	0.62	2.15
17	1.13	1.38	1.02	1.54	0.90	1.71	0.78	1.90	0.67	2.10
18	1.16	1.39	1.05	1.53	0.93	1.69	0.82	1.87	0.71	2.06
19	1.18	1.40	1.08	1.53	0.97	1.68	0.86	1.85	0.75	2.02
20	1.20	1.41	1.10	1.54	1.00	1.68	0.90	1.83	0.79	1.99
21	1.22	1.42	1.13	1.54	1.03	1.67	0.93	1.81	0.83	1.96
22	1.24	1.43	1.15	1.54	1.05	1.66	0.96	1.80	0.86	1.94
23	1.26	1.44	1.17	1.54	1.08	1.66	0.99	1.79	0.90	1.92
24	1.27	1.45	1.19	1.55	1.10	1.66	1.01	1.78	0.93	1.90
25	1.29	1.45	1.21	1.55	1.12	1.66	1.04	1.77	0.95	1.89
26	1.30	1.46	1.22	1.55	1.14	1.65	1.06	1.76	0.98	1.88
27	1.32	1.47	1.24	1.56	1.16	1.65	1.08	1.76	1.01	1.86
28	1.33	1.48	1.26	1.56	1.18	1.65	1.10	1.75	1.03	1.85
29	1.34	1.48	1.27	1.56	1.20	1.65	1.12	1.74	1.05	1.84
30	1.35	1.49	1.28	1.57	1.21	1.65	1.14	1.74	1.07	1.83
31	1.36	1.50	1.30	1.57	1.23	1.65	1.16	1.74	1.09	1.83
32	1.37	1.50	1.31	1.57	1.24	1.65	1.18	1.73	1.11	1.82
33	1.38	1.51	1.32	1.58	1.26	1.65	1.19	1.73	1.13	1.81
34	1.39	1.51	1.33	1.58	1.27	1.65	1.21	1.73	1.15	1.81
35	1.40	1.52	1.34	1.58	1.28	1.65	1.22	1.73	1.16	1.80
36	1.41	1.52	1.35	1.59	1.29	1.65	1.24	1.73	1.18	1.80
37	1.42	1.53	1.36	1.59	1.31	1.66	1.25	1.72	1.19	1.80
38	1.43	1.54	1.37	1.59	1.32	1.66	1.26	1.72	1.21	1.79
39	1.43	1.54	1.38	1.60	1.33	1.66	1.27	1.72	1.22	1.79
40	1.44	1.54	1.39	1.60	1.34	1.66	1.29	1.72	1.23	1.79
45	1.48	1.57	1.43	1.62	1.38	1.67	1.34	1.72	1.29	1.78
50	1.50	1.59	1.46	1.63	1.42	1.67	1.38	1.72	1.34	1.77
55	1.53	1.60	1.49	1.64	1.45	1.68	1.41	1.72	1.38	1.77
60	1.55	1.62	1.51	1.65	1.48	1.69	1.44	1.73	1.41	1.77
65	1.57	1.63	1.54	1.66	1.50	1.70	1.47	1.73	1.44	1.77
70	1.58	1.64	1.55	1.67	1.52	1.70	1.49	1.74	1.46	1.77
75	1.60	1.64	1.57	1.68	1.54	1.71	1.51	1.74	1.49	1.77
80	1.61	1.66	1.59	1.69	1.56	1.72	1.53	1.74	1.51	1.77
85	1.62	1.67	1.60	1.70	1.57	1.72	1.55	1.75	1.52	1.77
90	1.63	1.68	1.61	1.70	1.59	1.73	1.57	1.75	1.54	1.78
95	1.64	1.69	1.62	1.71	1.60	1.73	1.58	1.75	1.56	1.78
100	1.65	1.69	1.63	1.72	1.61	1.74	1.59	1.76	1.57	1.78

Index

Milton Keynes UK
Ingram Content Group UK Ltd.
UKHW052101141024
449526UK00001B/4